Fundamentals of Arc Spraying

Vladislav Boronenkov · Yury Korobov

Fundamentals of Arc Spraying

Physical and Chemical Regularities

Springer

Vladislav Boronenkov

Yury Korobov
Ural Federal University
Ekaterinburg
Russia

This book originally published as Основы дуговой металлизации. Физико-химические закономерности published in Russian by Ural Federal University Press, Ekaterinburg, Russia, 2012.

ISBN 978-3-319-22305-6 ISBN 978-3-319-22306-3 (eBook)
DOI 10.1007/978-3-319-22306-3

Library of Congress Control Number: 2015945332

Springer Cham Heidelberg New York Dordrecht London

Printed on acid-free paper

Springer International Publishing AG Switzerland is part of Springer Science+Business Media (www.springer.com)

Preface

Among the technologies of efficient use of resources, the group of thermal spraying methods is intensively developed. Studies from the standpoint of gas dynamics, heat transfer, mechanics, metallurgy, and physical chemistry allowed reliably reflection of the processes of motion, the heating of the particles, and their contact with the substrate, studying the structure and properties of coatings applied to the flame, plasma, detonation spraying. Great contribution was made by V.A. Barvinok, M.A. Belotserkovsky, G.V. Bobrov, Y.S. Borisov, V.S. Klubnikin, V.V. Kudinov, S.V. Petrov, A.F. Puzryakov, and others. On this basis, the equipment and the technologies were developed to produce coatings of stable and high quality.

In this group of methods, the Arc-Sprayed coatings are 3–10 times cheaper than the coatings produced by other methods according to the relative costs. Currently, arc spraying takes more than 20 % of the market of thermal spraying coatings, which has been developed dynamically over the last decade. However, its development is constrained by poor quality of the obtained coatings. This is due to the method specificity, which is expressed in a combination of a powerful influence on the sprayed material of the carrier gas, similar to other methods of thermal spraying, and the arc, spraying similar to the arc welding processes. This combination leads to rapid oxidation and splashing of sprayed metal.

Current state of researches does not reflect physical and chemical processes in arc spraying accurately, fairly and adequately. It makes it impossible to improve purposefully the equipment and technologies for this method, to predict the chemical composition and properties of the coatings.

This paper deals with the development and systematization of modern representations of the arc-spaying method. The processes of forming the droplets from the sprayed metal by the influence of the arc, movement, heating along the spraying distance, the chemical interaction of spray material and a gas phase and oxygen supply into the droplets. The interrelation of the processes with the properties of the obtained coatings was distinguished. On the basis of these studies, the possible ways of improving the equipment and improving the properties of the coatings were analyzed; and the proposals for their implementation were developed.

This monograph is the result of co-work of experts who have successfully engaged in the study of high temperature metallurgical processes and in the practical implementation of arc spraying in various branches of industry.

The monograph is designed for scientific and engineering specialists who are engaged in the field of welding and thermal spraying. It can also be used by students, undergraduates, and graduate students who are trained in relevant specialties.

The authors deeply appreciate their colleagues at the Department of Welding Technology, from the Ural Federal University, especially Shalimov M.P., Shanchurov S.M., Shumyakov V.I., as well as employees from the sided organizations Gudnev N.Z., Mishin V.M. ("Uraltransmash", Ekaterinburg), Lukanin V.L. ("Novator", Ekaterinburg), Pryadko A.S., Cherepko A.E. (Joint Institute of Mechanical Engineering, Belarus), Baranovsky V.E. (Unique Coat Technologies, USA) for their valuable comments during the discussion of materials and participation in bringing the experiments.

Contents

About the Authors

Vladislav Boronenkov Doctor of Technical Sciences, Professor

Created methods for predicting the composition and properties of multicomponent metal and slag melts in the Metal Refining, Welding and Thermal Spraying. From 1983 to 2000 worked as Head of the Welding Technology Department in the Ural Federal University. Under his leadership, three doctoral and 19 master's theses were defended. He is the author of 230 publications, including four monographs. He was president of the Ural Welding Association, Deputy Chairman of the National Certification Committee for Welding, Vice-President of Russian Welding Society.

Yury Korobov Doctor of Technical Sciences, Professor

At Uraltransmash developed the technologies of Welding and Thermal Spraying for defense and civil products. Since 2012 has been working as Head of the Welding Technology Department in the Ural Federal University. Works on modeling of physics-chemical high temperature processes, studies the properties of the coatings. Author of 240 publications including the monograph. More than 20 Thermal Spraying bays were organized under his technical supervision. Director of the Ural Welding Institute, Honored Inventor of Russia.

Introduction

Discussion of materials presented in the book "Fundamentals of Arc Spraying. Physical and Chemical Regularities" began a long time ago. Vladislav Boronenkov made the main scientific contribution, of course. His fundamental knowledge of the theory of metallurgical processes combined with unique abilities in solving various technical problems. He clearly identified the major problem and was able to find the best way to solve it. He was a world-recognized scientist in the field of modeling of high-temperature metallurgical processes.

He was known and appreciated in Germany, Israel, the USA, Ukraine, and Japan. Vladislav Boronenkov was doing his research with a great enthusiasm. I felt it physically while working with him. Unfortunately, we were not able to finish the work together. Surely, the scientific value of this work would have been much higher. Thank my lucky stars that I have been given the opportunity to work with such a great scientist, and I hope that the work will be interesting for readers as well as for its authors.

Yury Korobov

Chapter 1
Conception of Arc Spraying Method

Thermal Spraying makes an essential contribution to the economy of the developed countries. Among the methods for applying coatings the Arc Spray process (AS) seems more preferable by criteria of thermal efficiency, cost of spraying materials, and ease of maintenance. The AS shows its similarity to other TS methods. Here high-velocity gas jet of a high mass flow rate is used for the formation and transportation of the sprayed material. AS is also close to Arc Welding by nature of application of the arc as a power source. Direct transfer of arc heat to atomized metal causes its positive features. By the specific energy consumption of sprayed metal, the AS is on the same level with Arc Welding and several times more efficient than Plasma Spraying. In addition to high heat, the benefits of arc spraying include:

- high feed rate of spraying;
- the possibility of obtaining the coatings of high thickness, up to a few millimeters;
- ease of implementation compared to other thermal spray processes (except flame spraying);
- compatibility with the typical welding engineering;
- low cost of equipment.

But specific features of AS also determine intense burnout of alloying elements from the metal and the saturation of the sprayed coating with atmospheric gases. It leads to lowering of the alloying elements concentration, the excess content of oxides in the coating. In addition, electromagnetic forces in arc zone differ from desired direction of particle's trajectory. It lowers the deposition efficiency of the sprayed material and reduces the density of the coating on the spray pattern periphery. All these factors lead to a decrease of the coating quality. Additionally, Arc Spraying disadvantages include: sprayed materials are limited with conductive wires, nevertheless using of cored wires widen a range of applicable materials; high feed rate causes a large amount of emitted aerosols, high level of noise, ultraviolet, and infrared radiation, so high safety requirements take place. All these factors limit the widespread use of easy in operation and inexpensive spraying method. Basing on AS system analysis, the ways of improving of coating performance are discussed in the monograph. All influencing factors were divided into the following groups of

V. Boronenkov and Y. Korobov, *Fundamentals of Arc Spraying*,
DOI 10.1007/978-3-319-22306-3_1

parameters: input, internal, external, and output ones. Output parameters describe the system as a whole; internal parameters show the elements of the system; input parameters define the range of the internal parameters; external parameters have no relation to the AS process but influence on it. The present work is devoted to the analysis of physical and chemical relationships of processes at AS, their mutual influence and usability.

1.1 Essence of the Method

Metal losses through wear and corrosion resistance of machine components and steelworks account for about 30 % of their mass. The promising direction to reduce losses is to improve the properties of surface contacting with the external environment. The results of the research and practical developments show that it can increase the life of products 2–3 times.

The group of methods of thermal spraying (TS) is developing intensively. The analysis results show that the use of thermal spraying makes an essential contribution to the economy of the developed countries. The technologies world market volume of TS in 2000 amounted to 1,600 million euros, today it has increased to 6,500 million euros, Table 1.1.

According to an estimate of DVS (Deutscher Verband für Schweißen und verwandte Verfahren—German Welding Society), EWF (European Federation of Welding) and GTS (Association of Thermal Spraying), the use of TS in the European Community brought profits of more than €900 million and provided 20 thousand workplaces [3].

The market volume of TS in Japan in 2009 amounted to $500 million, over the last 50 years it has been increased 50 times. TS is on the list of 20 key technologies, supported by government programs [4].

The rapid development of TS we also can see in the emerging economies. In China, the market volume of TS in 2002 amounted to 145 million dollars [5]; by the end of the first decade, it rose to $500 million. Since 1992, the TS market in India has been developing rapidly: the number of shops increased from 5 to 55 in 2008, the consumption of powders was $3 million applied to the most developed the process of plasma spraying [6].

The TS share in gross domestic product was: the USA—0.21 %, Japan—0.17 %, Germany—0.25 %, China—0.12 % [2].

Table 1.1 Technologies world market volume, billions of dollars [1, 2]

Region	Years		
	1990	2000	2013
Europe	300	400	2100
North America	600	800	2100
Asia, Australia, Middle East	300	400	2300
Total	1200	1600	6500

The inventor of a method of thermal spraying was Max Ulrich Shoop (Fig. 1.1), who in 1909 filed a patent for spraying technology using flame as a heat source. Over 35 years of work, he had proposed technical solutions for flame spraying and arc spraying equipment and technologies. In particular, he developed processes for interrupting the metal spraying without cooling the metal bath, for spinning on meltable material, for electrical metal spraying, for creating photographic images, for repairing castings, for soundproofing walls, floors and ceilings, for manufacturing metallic paints, and for a mass metallisation machine [7–9].

Industrial applications, however, were only possible after scientific research focused on fundamentals of this type of thermal spray process in the 1950s and 1960s [10–13]. Arc spraying is very economical and therefore, has prevailed in many applications in spite of newly developed high-energy techniques such as plasma, detonation gun, and high-velocity flame spraying. Fields of application are broad due to system variety and mobility, further developments in systems design, and extended range of appropriate spray materials [14–18].

In the USSR, the basic organizations dedicated to flame and arc spraying were Avtogenmash Institute, Remdetal [19]. So, in 1944 on tank repair plants the recovery technology of crankshaft journals and crankpins of tank engines basing on spraying was developed, these companies were actively engaged in the application of anti-corrosion zinc and aluminum coatings [20].

Due to the development of mechanical engineering, first rocket and space and aviation industry, the coatings, which could withstand increased thermal and mechanical loads, were needed. This led to the rapid development of plasma spraying in the 1950–1960s of the 20th century. The pioneer of this field in the United States was the company Thermal Dynamic Corp. (Lebanon, NH) [21]. In the USSR, such academic institutions as Metallurgical and Metal Science Institute named after A. Baykov, Moscow; Bauman Moscow State Technical University; Novosibirsk Heat and Mass Transfer Institute, Electric Welding Institute named after E. Paton; and others were actively engaged in the process [22–24].

Fig. 1.1 Max Ulrich Shoop, inventor of thermal spraying

The basic principles of thermal spraying are defined in DIN EN 657 [25]. In general, the sprayed material is fed into a heat source within or outside the apparatus in the form of powder, wire, or rod. Here, it is heated to the plastic or molten state. A gas jet atomizes the heated material and accelerated it toward the prepared substrate where it is deposited as a coating. During coating deposition the substrate usually remains unmelted.

Graphically, it can be represented by the following scheme, Fig. 1.2. Under the influence of an external energy source the sprayed material is atomized as discrete particles, which are accelerated under the influence of gas jets. The particle's condition, softened or melted, will depend on the heating intensity by the heat source. Under impacting on the substrate the single droplet is strongly deformed, hardens, and takes the form of a thin disk. A set of such particles reaching the surface at a given point in time forms a layer. The coating is formed by sequential layers overlay.

This process is accompanied by a high cooling rate in the range of 10^4–10^6 degrees per second, due to the heat transfer into base substrate. Due to this feature small thermal deformation and, in many cases, the absences of structural changes in the detail are typical for TS processes. In addition, there are small restrictions on the composition of the sprayed materials. Entering the subsequent particles leads to the growth of coating thickness and the formation of layer structure (Fig. 1.2).

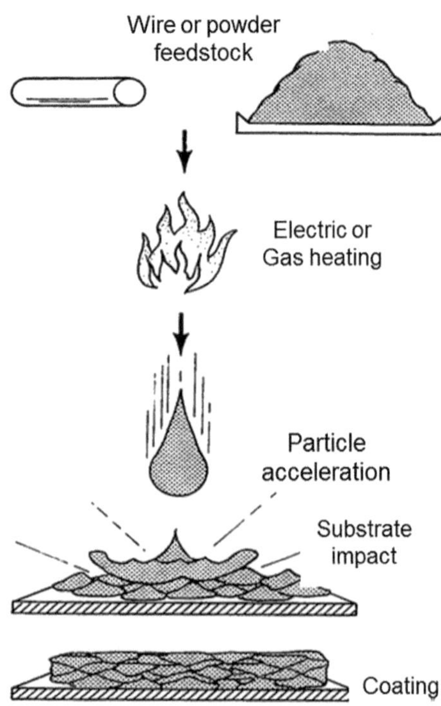

Fig. 1.2 Scheme of thermal spraying process

The TS process scheme determines some of its features:

- Practically no restrictions on the type of coating material, it could be metals, ceramics, plastics, cermet (metal ceramics);
- The substrate, which is coated, remains in a solid state; it is characterized by a low level of thermal stress and by lack of structural changes;
- It is possible to perform coatings in local areas of the surface;
- It is possible to run processes on-site;
- There are almost no restrictions of the base material;
- It is possible to perform coatings with high capacity in a wide range of thicknesses (from 20 μm to several millimeters).

These features provide TS attractiveness for improving the performance of products.

The spraying conditions and features of the sprayed material exert a dominant influence on the properties of sprayed coatings. They, in turn, determine the characteristics of the coating-base system.

The formation of the coating takes place as follows. The individual particles are deformed or splashed while hitting the surface of the substrate. Under such conditions, their rapid solidification occurs due to intensive heat transfer to substrate. Consequently, thermal spraying coatings are usually characterized by a lamellar structure having pores and inclusions of micro cracks, they differ in properties' anisotropy. Except lamellas formed from the initial sprayed material, the structure may include particles not fully molten or secondarily crystallized before hitting the surface, Also products of reactions between the particles and atmospheric gases take place (Fig. 1.3).

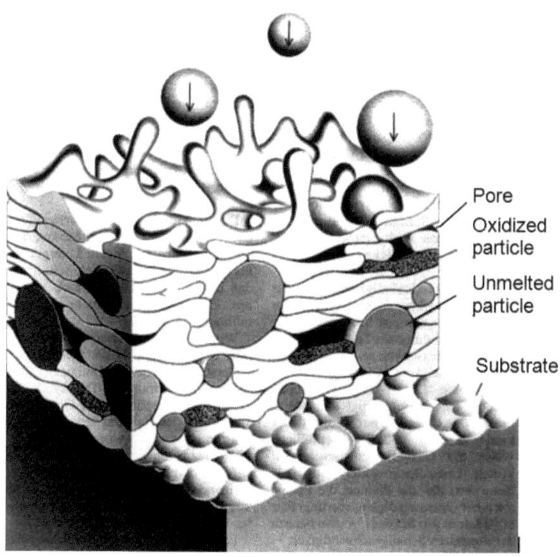

Fig. 1.3 Scheme of coating formation

Concentrated heat input into the sprayed material in conjunction with the interaction of the particles with the atmosphere when flying lead to chemical changes in the composition and the structure of the sprayed coatings, namely:

- Selective evaporation of alloying elements;
- Changes in the phase composition of the metal components, such as changes in the solid phases in the presence of oxygen;
- The formation of stable oxides and nitrides of active metals. These compounds on the one hand increase the surface hardness of the integral, which leads to increased durability. On the other hand, they will weaken the connection of the individual layers of the coating.

In comparison with cast metal of similar composition, these processes are faster and more complete due to the large specific surface of the particles.

Oxides resulting from the surface oxidation of metal atomizing particles are also involved in the formation of the laminar structure of the coating, providing both positive and negative effects on its properties. They are characterized by high hardness what leads to improved durability of the coating. However, it reduces the ductility and mechanical tensile strength. Furthermore, in many cases, the melting point of oxides is considerably higher in comparison with the initial metal, which increases the heat resistance of coatings. Their shares in the volume of coating determine resulting influence of oxides.

To control the oxidation degree of the particles inert or shielding gases are used and deoxidizing agents are added to the initial sprayed material.

The surface pretreatment largely determines the adhesive strength of the coating to the substrate. According to DIN En 13507 [26], it includes the steps of pre-cleaning, blasting, post-cleaning. At the stage of pre-cleaning, all impurities and oil are removed from the surface. For abrasive blasting highly rigid abrasive particulate materials of angular form, such as corundum or hardened steel grit are used. Because of the blasting, the surface activation takes place, it means increasing free surface energy, vacancy concentration, dislocation density, and the occurrence of stacking faults due to plastic deformation in near-surface zone. Furthermore, the surface roughness increases, which contributes to mechanical engagement of the particles upon contact with the surface [27]. When subsequent cleaning, the residues of the abrasive materials are removed from the surface, for example, by ultrasonic machining in alcohol [28].

The ratio of temperatures and velocities of the sprayed particles (Fig. 1.4) can classify TS methods.

The ratio of these parameters determines the possible application fields of TS specific methods to improve the surface properties.

Among the methods for applying thermal coatings the Arc Spray process (AS) seems more preferable due to the thermal efficiency, the cost of spraying materials, ease of maintenance. According to the technical—economic evaluation of Y.A. Kharlamov [39] on the relative value AS coatings are 3–10 times cheaper than TS coatings made with other methods while ensuring their high strength (Fig. 1.5).

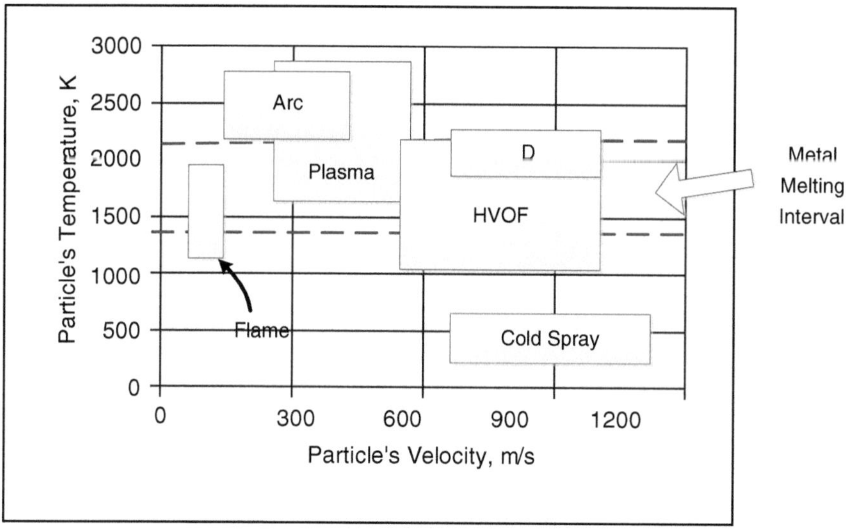

Fig. 1.4 The ratio of temperatures and velocities of particles for various thermal spraying methods [24, 29–38]. *Flame* flame spraying, *Plasma* plasma spraying in air and vacuum (dynamic), *HVOF* high veliocity oxygen-fuel spraying, *D* detonation spraying, *Arc* arc spraying, *Cold spray* cold gas-dynamic spraying

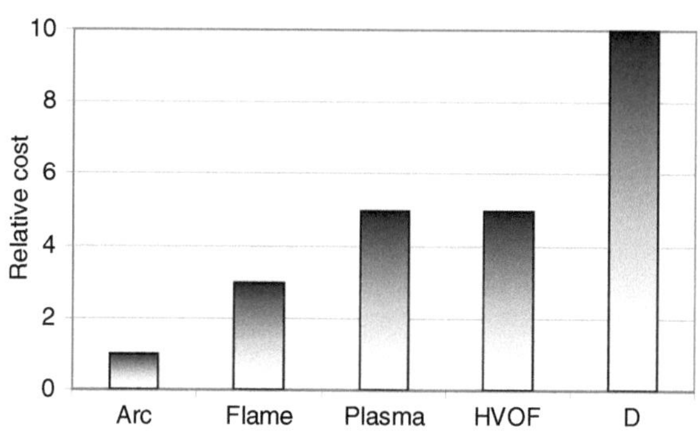

Fig. 1.5 Relative cost of coatings performed by different thermal spraying: *Flame* flame spraying, *D* detonation spraying, *Plasma* plasma spraying in air and vacuum (dynamic), *Cold spray* cold gas-dynamic spraying, *Arc* arc spraying

Based on the scheme shown in Fig. 1.2. TS methods can be classified according to the source of energy for heating and acceleration of particles, Fig. 1.6.

Let us compare arc spraying and plasma spraying, using identical energy source of heating—electric arc energy generated in the discharge between the electrodes through the gas gap.

At Arc Spraying (Fig. 1.7), the coating is formed from the droplets of the liquid metal in the stream of moving carrier gas. Heating and melting of the sprayed

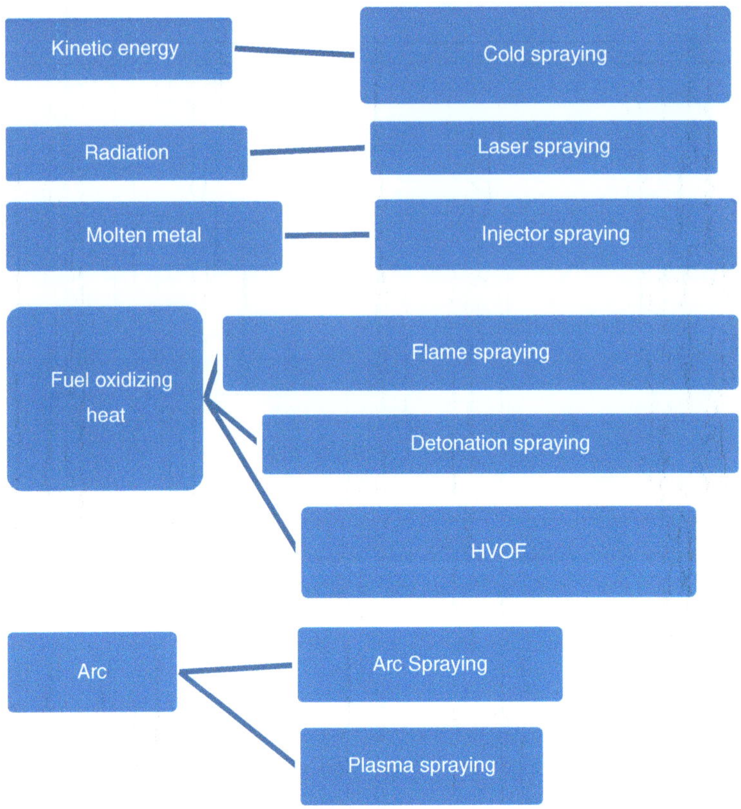

Fig. 1.6 Classification of TS methods by energy source effects on spray material

Fig. 1.7 Scheme of AS process

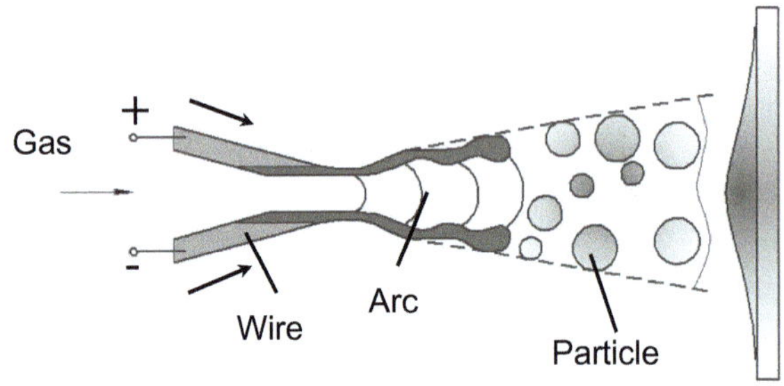

material (SM) is due to the electric arc heat which is burning between the consumable wires—electrodes, from which SM is formed. The molten metal is blown away from the end faces of the electrodes, atomized under the influence of gas-dynamic and electromagnetic forces, and in the form of droplets moves in the direction of resulting force.

The AS scheme shows its similarity to other TS methods. Here, high-velocity gas jet of a high mass flow rate is used for the formation and transportation of the SM. AS is also close to Arc welding and metal melting processes by nature of application of an electric arc as a power source.

Therefore, these processes have the equal level of high heating efficiency, which is a part of the heat energy used directly for metal heating and melting.

For arc welding, its value is 60.5 % of the thermal energy supplied to the electrode, according to the calculations of the heat balance at the end face of the electrode at arc welding and according to the experimental data of A.A. Erokhin [40]. Effective heating efficiency for AS is 61.3 % (average data for Zn, Sv-08A—EN 10016, Al) [41].

Similar values of effective heating efficiency for metal melting at arc spraying and arc welding are determined by a similar scheme of heat transfer from the power source to the metal. The energy generated by the power source during AS consumed for electrodes melting through the stages of heat release in the arc and heating of electrode by Joule heat (Fig. 1.8).

We can see that sprayed metal is heated and melted directly by the heat of the arc burning between the electrodes, which form the sprayed metal. This reduces the amount of heat loss during the heating and melting of the metal in comparison with other TS processes where generation and consumption of heat for melting of sprayed metal are divided into time and space.

The effective heating efficiency is several times lower at plasma spraying. Its value is 2–27 %, depending on the type of SM (wire or powder) and the design of the nozzle unit [24]. Such a large difference is caused by differences in schemes of heat transfer from the arc to the SM.

While PS the arc heat is transferred to sprayed metal indirectly through plasma as an intermediary. It leads to thermal losses for heating of forcibly cooled walls of the plasma torch channel, and SM heat transfer. Moreover, the heat losses in electrodes at current run are significant (Fig. 1.9).

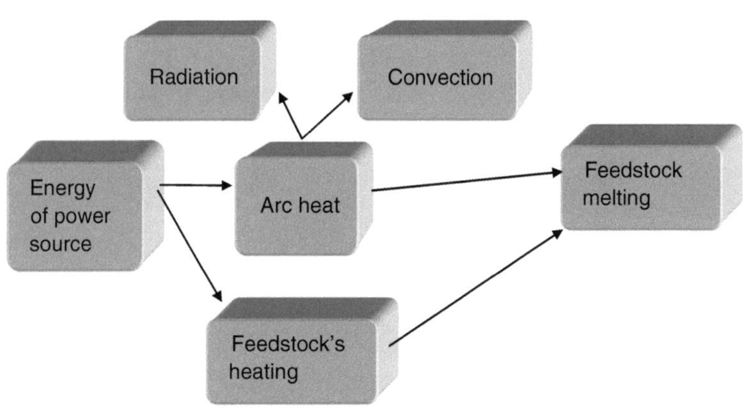

Fig. 1.8 Distribution of input energy at arc spraying

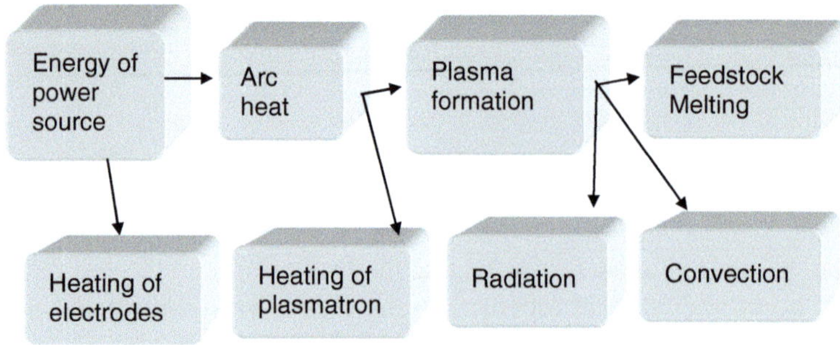

Fig. 1.9 Distribution of input energy at plasma spraying

Concerning the heat losses plasma arc spraying of a conductive wire is most effective among other PS processes. Here, the ballast energy discharge is heating the cathode (Fig. 1.10a). In the most used scheme of PS with powder feedstock (Fig. 1.10b), the heating of both electrodes leads to heat losses.

As for AS and PS, the items of heat balance of feedstock material are similar: melting, evaporation, overheating, and heat transfer. However, the absolute values of heat losses is higher at PS due to higher temperatures in the process.

Due to the difference in effective heat efficiency the specific energy consumption per 1 kg of SM (N_{sp}) in AS is 3–10 times less than in PS (Table 1.2). For comparison the characteristics of commercially available Russian spraying gun EM-17 and basic types of Atmospheric Plasma Spray Guns are used: single cathode-single anode, single cathode-cascade anode, triple cathode-single anode.

Thus, the analysis of available data shows that due to the using of power source energy the AS is on the same level with arc welding and several times more efficient

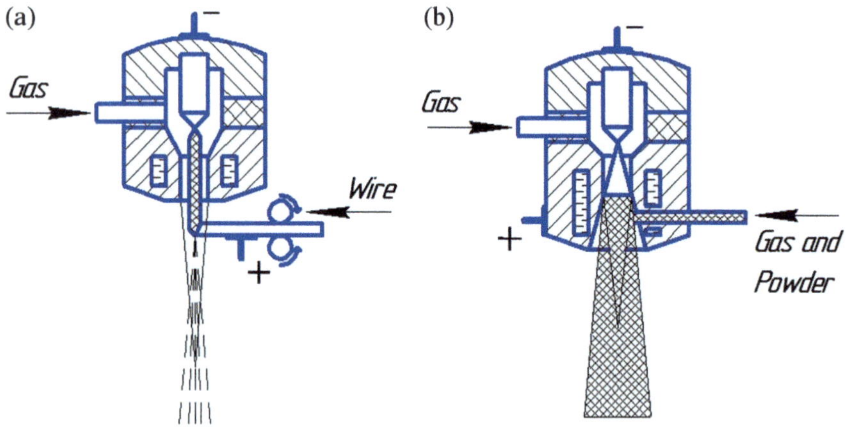

Fig. 1.10 Scheme of plasma spraying, **a** using conductive wire, **b** using powder [24]

Table 1.2 Comparison of the efficiency of electricity consumption for AS and PS processes

	Electricity consumption (kW h)	Feed rate (kg/h)	N_{sp} (kW/kg)
Arc spraying, typical parameters [42]	6	10	0.6
Plasma, wire spraying [43]	12	6	2
Plasma Guns, mean values [44]	15–40	3–6	6
Traditional single cathode F4MBXL plasma gun [45]	40	4.8	8.3
Cascaded single cathode SinplexPro plasma gun [45]	60	9	6.7
Three-cathode air plasma gun Triplex-Pro 200 [46]	60	9	6.7

than PS. The latter is caused by the presence of ballast chains in the heat transfer paths to the feedstock at PS.

In addition to high heat, the benefits of arc spraying include:

- high feed rate of spraying;
- the possibility of obtaining the with the coatings of high thickness up to a few millimeters;
- ease of implementation compared to other thermal spray processes (except flame spraying);
- compatibility with the requirements for a typical welding engineering;
- low cost of equipment.

The process scheme (Fig. 1.7) determines intense burnout of alloying elements from the metal and the saturation of the sprayed layer with atmospheric gases. It leads to changes in chemical composition lowering of the alloying elements concentration, the excess content of oxides in the coating. In addition, electromagnetic forces in arc zone differ from desired direction of particle's trajectory. It lowers the deposition efficiency of the sprayed material and reduces the density of the coating on the spray pattern periphery. All these factors lead to a decrease of the coating quality.

For example, commercially available AS gun EM-14 (Russia) is marked by wide jet spray pattern whereby the deposition efficiency of the material does not exceed 0.5–0.6, and the density of the coating on the periphery of the spray pattern is reduced.

Additionally, AS disadvantages include:

- sprayed materials are limited with conductive wires;
- high feed rate causes a large amount of emitted aerosols, high level of noise, ultraviolet and infrared radiation, so high safety requirements take place.

These factors lead to a decrease in the quality of the coating. They limits the widespread use of easy to operation and inexpensive spraying method.

The compressed air is the most widespread as atomizing gas. Its production is cheaper in comparison with other gases. Also, neutral and reducing gas mixtures may be used. C.S. Yakimavichus in 1955 suggested the use of an inert gas for atomizing metal at arc spraying [47]. The company Air Products and Chemical, Inc., U.S. proposed the using of argon, nitrogen, carbon dioxide, sulfur hexafluoride, or their mixture for spraying of reactive metals [48]. The results show, in particular, that the use of nitrogen, pure, or with additives as methane, hydrogen, helium, can increase the corrosion resistance of Zn–Al sprayed coatings [49].

There are schemes when compressed air is fed in a mixture with different fuels. This allows to use the heat (*Q*) generated during the combustion process and to reduce the oxidation potential of the gas phase. In general, the reaction is as follows:

$$C_XH_Y + \text{oxidizer} \rightarrow CO_2 + H_2O + Q$$

Such hydrocarbons as acetylene (C_2H_2), propane (C_3H_8), kerosene, natural gas, MAP gas are used as fuel. Oxygen and compressed air (approximately 21 % oxygen + 79 % nitrogen) are used as oxidizer. When using compressed air as the oxidizer the combustion temperature is 30–40 % lower due to heat consumption for the heating of nitrogen [50, 51]. Unlike the flame spraying methods, in arc spraying the material is melted due to the arc heat. Therefore, a high temperature of atomized gas is not an essential condition. Furthermore, for the same flow rate of oxygen in both oxidizers the mass flow rate of compressed air is five times higher, So the force impulse mv (m—mass, v—velocity) impact on sprayed material is increased. It makes possible to atomize a larger mass of sprayed material, i.e., to increase the spraying feed rate.

In one variation gaseous or liquid hydrocarbons are added to the air jet after the arc, where their ignition and burning take place. Kerosene is used as liquid agent [52], as gaseous are used combustible gases (acetylene, methane, propane, etc.) [53].

In another variation, the products of combustion of hydrocarbon fuel are applied as the gas for transporting and atomizing [54].

According to this scheme (Fig. 1.11), the air-fuel mixture which is burned in the combustion chamber 5 flows through the nozzle 2 into the space of chamber 1, and further through the spray nozzle 6 to atmosphere. The gas jet flowing out of the nozzle 6 atomizes the liquid metal of melting electrodes 4, on which the voltage is applied through current leads 3.

The wires of characteristic diameter of 1.6–3.17 mm are used as a feedstock at AS. In comparison with the powders, they are of lower cost and the feed process is simplified. The feed process is of great importance for AS because of the high feed rate. Requirement to use only electrically conductive wires is caused by the scheme of the process. It limits a range of the feeding materials. Solid and flux cored wires are used for spraying. Typical solid wires are successfully used in the following areas [55]:

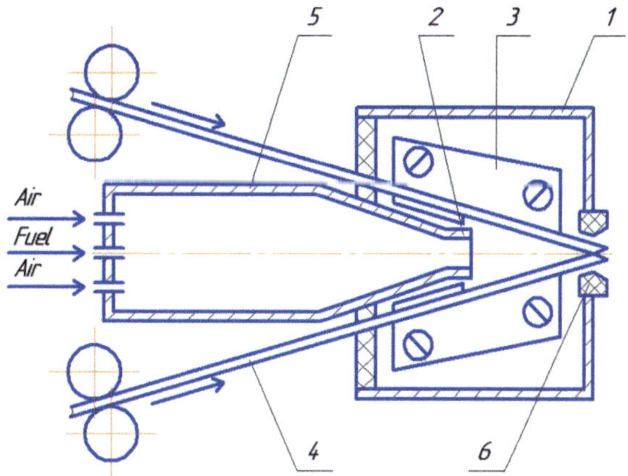

Fig. 1.11 The scheme of AS unit with combustion chamber. Explanations in the text

Al, AlMg	for restoration of Al parts, silencers (mass production), heat exchangers, offshore technology: pipelines and oil rigs, boilers, slab markings, and also food processing industry (butcher's machines);
Cu	resistors, capacitors, conductor paths on plastic and ceramic substrates, continuous-casting molds;
CuAl	roller-bearing seats, fittings;
Mo	adhesive layers on steel, wear-protection coatings;
NiCr, NiAl, NiCrMo	radiation shielding (X-rays), adhesive layers for plasma spray coatings, oxidation barrier layer below plasma-sprayed ceramic coatings;
NiCuMn	valve housings, fitting parts, pump shafts;
Sn-alloys	friction bearings, capacitors;
Steel	wear resistant and repair coatings: plunger, transmission cases, aluminum parts (automobile valves);
Zn	bridges and steel constructions, boilers, gates, pipelines, casting molds, car body joints.

Flux cored wire is a cylindrical shell filled with powder filler (other terms: charge, core), Fig. 1.12. The amount of material in the core is estimated by fill factor, representing the ratio of the core weight to the total weight of the wire. Fill factor lies in the range 15–40 % depending on the diameter of the wire, its purpose and the density of the core material.

Wires of tubular structure are the most common, Fig. 1.12a–c. The inserting the shell part into the core, Fig. 1.12d–g, provides its more uniform melting and the more effective metal protection from interaction with atmospheric gases [56].

Retaining technological properties of solid wire, cored wire allows varying in the flexible way the content of alloying elements when complex chemical composition of the coating is necessary.

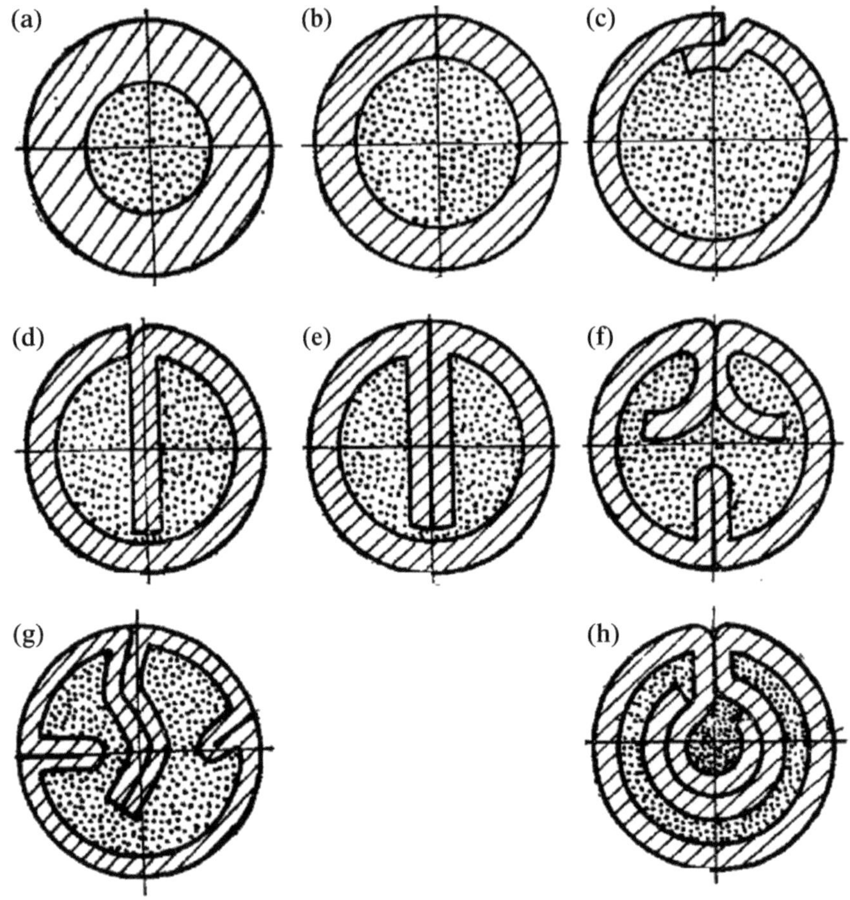

Fig. 1.12 Design of core wires

Several groups of components are enclosed into the core composition:

- alloying elements and deoxidizing elements: silicon, manganese, titanium, etc., used in the form of alloys of these elements jointed with iron (so-called ferro-alloys) or in the form of one-component core;
- ionizing or stabilizing components containing elements with a low ionization potential (potassium, sodium, etc.) to improve the arc stability;
- processing additives to facilitate the wire manufacturing process;
- iron powder to increase the productivity and to improve the arc stability.

Cored wires extend the range of materials that can be used at AS. Examples of spray material combinations (sheath/core) for cored wire spraying are the following: metal/metal, alloys (Ni/NiCrB), intermetallic compounds (Al/Ti), metal/carbide (Co/WC), metal/boride (Al/B_4C), metal/oxide (Al/Al_2O_3), metal/oxide/carbide

($Al/Al_2O_3/Cr_3C_2$), and metal/short fibers (Al/C short fibers) [57, 58]. They include materials of high hardness for wear-resistant applications (oxides, nitrides, borides, carbides, carbonitrides); heat resistant flux cored wires based on nickel and iron for protection of objects exposed to the combined effect of aggressive environments, high temperatures and erosion [59–61].

Difficulties in performing of high-quality coatings, related with the above-mentioned features, caused a variety of AS-units designs that realize the base scheme. They can be classified according to the following criteria:

- mutual positioning of wires and spray nozzles;
- the kind of atomizing gas.

The applied schemes of mutual positioning of wires and spray nozzles are divided into open and closed [62]. In a closed scheme, the electrodes are disposed inside the nozzle and atomizing gas is fed to the nozzle from the closed cavity, where current-carrying elements are also installed (Fig. 1.13a). At open scheme, one or more gas nozzles are disposed between the wires electrodes (Fig. 1.13b).

Nozzle orifices can be profiled for supersonic gas flow. Figure 1.14 shows a variant of an open scheme, where gas flows out of nozzles arranged coaxially with wires [63].

Furthermore, there are some research works, which use a combination of close and open systems in various embodiments. For example, the central nozzle, made in a closed scheme, may be supplemented by the covering cap, which provides an additional squeeze of the spray jet, Fig. 1.15 [64].

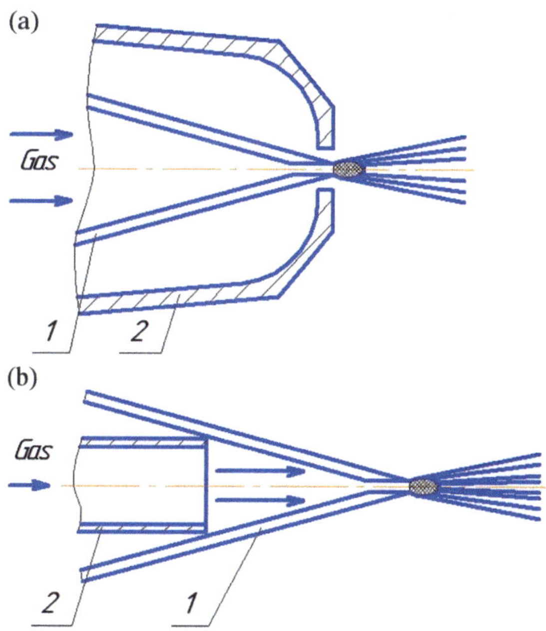

Fig. 1.13 Schemes of mutual positioning of the nozzle unit and the current-carrying wires. *1* electrode; *2* nozzle. **a** A closed scheme; **b** open scheme

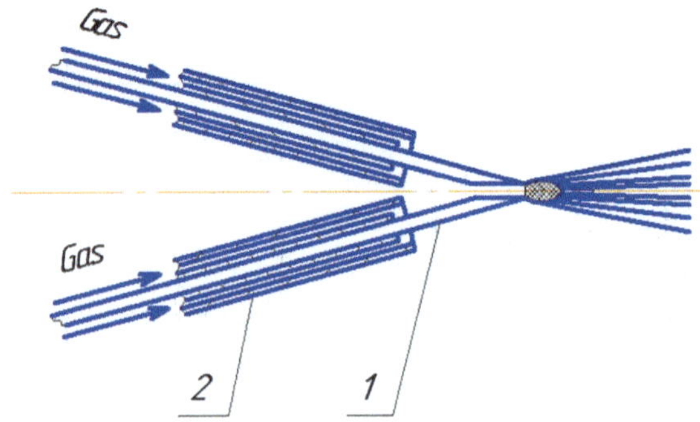

Fig. 1.14 Open AS scheme with gas feeding via nozzles coaxial with wires. *1* wire, *2* nozzle

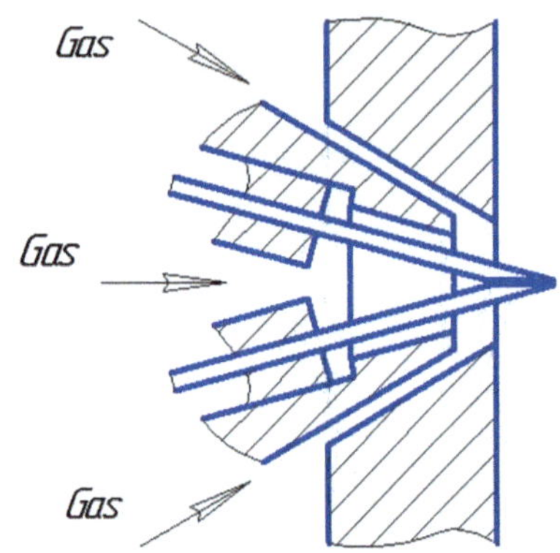

Fig. 1.15 Example of the combined spray nozzle

The variety of options that implement the basic scheme of the method highlights the complexity of the physical processes at AS. Also, a lack of understanding of the nature of their occurrence and the relationship takes place.

1.2 Research Plan of Arc Spraying

The improving of coating performance is possible based on AS system analysis.

AS process can be described as a model system having a set of parameters related to some kind of relation

$$\bar{Y} = F(\bar{X}, \bar{Q}, \bar{N}) \tag{1.1}$$

where $\bar{Y}$ are output parameters that describe the system as a whole; $\bar{X}$ are internal parameters describing the elements of the system; $\bar{Q}$ are input parameters defining the range of the internal parameters; $\bar{N}$—are external parameters that have no relation to the AS process but have an influence on it.

The ultimate goal of the AS, as well as other material processing is to perform products of desired properties. Kudinov conducted an analysis of spatial–temporal conditions of thermal coatings formation [13]. He concluded that their properties are determined by the processes of deformation and heat transfer in contact of sprayed material particles with the substrate, as well as physical and chemical interaction of droplets with the gas atmosphere at moving to the substrate. Thus, the knowledge of the laws of changing the basic parameters of the AS process—velocity, temperature in the two-phase flow, the intensity of gas input in metal drops—allows to predict the properties of the coatings. If we take the operating characteristics of coatings as output parameters, the model system can be represented as a diagram (Fig. 1.16).

Output parameters (Fig. 1.17), can be classified by purpose of coatings. The main of them are wear resistance, corrosion resistance, thermal conductivity, and heat resistance. To provide products survival it is often necessary to provide the combination of characteristics relating to different groups of output parameters.

Internal parameters as coating properties can be represented in the form of interconnected subsystems defining geometrical parameters, thermal and structural characteristics, strength, residual stress in the coating (Fig. 1.18).

External parameters include a set of subsystems that determine the composition of the surrounding atmosphere, the conditions of production, part characteristics (Fig. 1.19). The atmosphere can be characterized by its chemical composition, temperature, humidity. Parts to be coated differ in thermophysical properties, operating conditions, and geometry. Possible stability of technological parameters, production output can be attributed to the operating conditions.

Internal, external and output parameters are similar to other processes; they can be studied by similar methods. The set of input parameters is also similar—these are

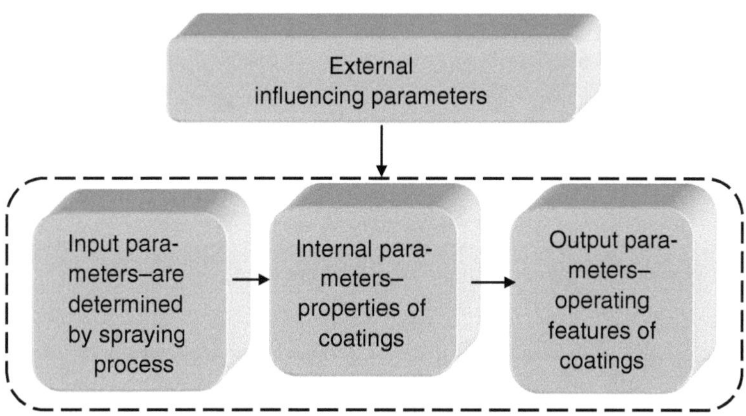

Fig. 1.16 AS process model

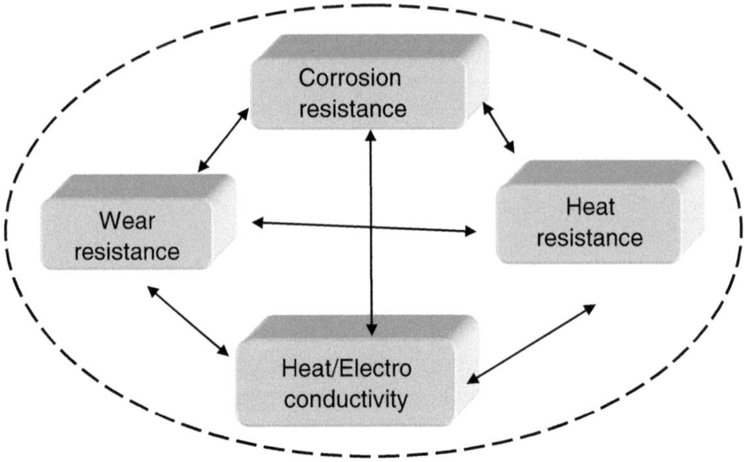

Fig. 1.17 Output parameters of AS process

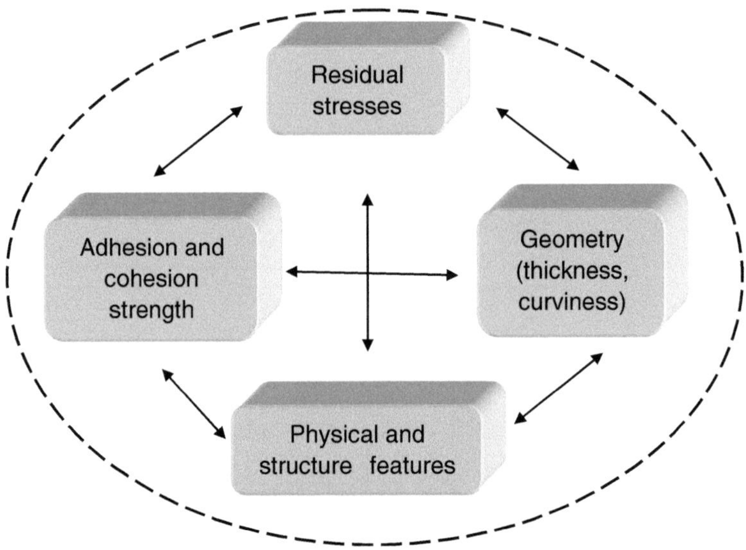

Fig. 1.18 System of coatings properties—internal parameters

the characteristics of the carrier gas, sprayed material, energy supply options, which are subject to further divide due to the criterion of impact nature (Fig. 1.20).

However, the input parameters will differ from other TS processes by the level of values, method and place of impact on the formation of the coating.

The present work is devoted to the analysis of physical and chemical relationships of processes at AS, their interrelationships and usability.

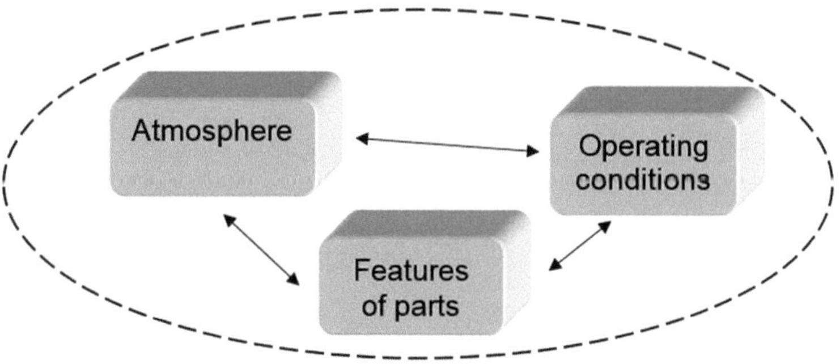

Fig. 1.19 External parameters which impact on the AS process

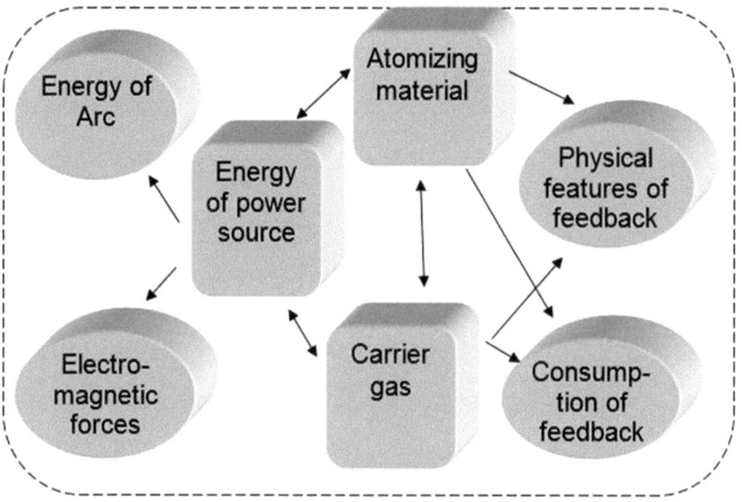

Fig. 1.20 Classification of input parameters by influence on the atomizing material

References

1. http://www.cetim.fr
2. Dorfman MR, Sharma A (2013) Challenges and strategies for growth of thermal spray markets: the six-pillar plan. J Therm Spray Techn 22(5):559–563
3. Middeldorf K (2009) Thermal spraying technology—significant economic factor. Therm Spray Bull 2(1):6–10
4. Nakahira A (2009) Current status and future prospect of thermal spray coating application and coating service market of job shops in Japan. In: Expanding thermal spray performance to new markets and applications. Proceedings of ITSC-2009, Las-Vegas, USA
5. Xiaoou H, Yufen L (2004) The current situation and future of thermal spraying industry in China. In: Thermal spray solutions. Advances in technology and application proceedings of ITSC-2004, Osaka, Japan

6. Sundararajan G, Mahajan YR, Joshi SV (2009) Thermal spraying in India: status and prospects. In: Expanding thermal spray performance to new markets and applications. Proceedings of ITSC-2009, Las-Vegas, USA
7. Abert C (2010) Thermal spraying—the inventor's life. Therm Spray Bull 2
8. Kretzschmar E (1966) Spraying of metals, ceramics and plastics. Mechanical Engineering, Moscow
9. Hartman S (2012) The History of Thermal spraying—from Max Ulrich Schoop to the Present Time. In: HVOF spraying. Proceeding of 9 colloquium, Erding, Germany
10. Steffens H-D (1963) Haftung und schichtaufbau beim lichtbogen–und flammspritzen. Dissertation, University of Hannover
11. Busse KH (1989) Das Verhalten von Spritzteilchen beim atmosphärischen Lichtbogenspritzen. Dissertation, University of Dortmund
12. Troitsky AF (1960) Basics of metallization. Gosizdat, Tashkent
13. Kudinov VV (1977) Plasma coatings. Nauka, Moscow
14. Hock VF, Benary R, Ganertz R, Herman H (1997) Automated Thermal Spray Technology for Rehabilitation and Maintenance of Civil Works Infrastructure. In: Thermal spray 1997. Proceedings of united forum for scientific and technological advances, Materials Park, OH, USA
15. Marantz DR (1991) State of the arc spray technology. In: Bernecki (ed) Research and applications. Proceedings of thermal spray 1990, Materials Park, OH, USA
16. Grant LJ (1989) How recent advances in arc spraying broaden the range of applications. In: Proceedings of international thermal spray conference, UK
17. Steffens HD, Wilden J, Duda T (1994) Thermal spraying. High Temp Chem Processes 3:653–664
18. Sampson ER (1998) Cored wire application using electric arc spray. In: Coddet C (ed) Meeting the challenges of the 21st century. Proceedings of thermal spray 1998, Materials Park, OH, USA
19. Katz NV, Antoshin EV, Vadivasov DG (1966) Metallization by spaying. Mechanical Engineering, Moscow
20. Nedzelsky MD (1963) Application of arc spraying in industry. Irkytsk Publishing House, Irkytsk
21. Fauchais P, Vardelle A, Dussoubs B (2001) Quo vadis thermal spraying? J Therm Spray Techn 10(1):44–66
22. Bobrov GV, Ilyn AA (2005) Deposition of inorganic coatings. Theory, technology, equipment. Intermet Engineering, Moscow
23. Borisov YS (1987) Thermal sprayed coatings from powder materials. Naykova Dymka, Kiev
24. Kudinov VV, Pekshev IY, Belaschenko VE et al (1990) Spraying coatings by Plasma. Nauka, Moscow
25. DIN EN 657: 2003-04. Thermal Spraying—Terminology, Classification
26. DIN EN 13507: Thermisches spritzen—vorbehandlung von oberflächen metallischer werkstücke und bauteile für das thermische spritzen, 2001-07. Thermal spraying—pre-treatment of surfaces of metallic parts and components for thermal spraying
27. Drozak J (1992) Teilcheneigenschaften und haftung beim thermischen spritzen von metall und keramik. Dissertation, University of Dortmund
28. DVS Merkblatt/Instructions 2307-1: arbeitsschutz beim entfetten und strahlen von oberflächen zum thermischen spritzen. DVS, Düsseldorf, 1999-01
29. Bobzin K Warda T, Bruhl M (2009) Significance of plasma spraying within thermal spray technologies. In: Proceedings of 8 HVOF colloquium, Erding, Germany
30. Alkhimov AP, Klinkov SV, Kosarev VF et al (1998) Gasdynamic spraying. Status and prospects. In: Films and coatings'98. Proceedings of 5 international science-technic conference, S-Peterburg, pp 20–25
31. Astakhov EA (2005) Scientific and technological basis of the properties control of detonation coatings. Dissertation, University of Kiev

32. Nestler MC, Erning U (1998) Characteristics and advanced industrial applications using the "Diamond Jet Hybrid"—the third generation of HVOF systems. In: Films and coatings'98. Proceedings of 5 international science-technique conference, S-Peterburg
33. Breitsameter MJ, Prosperini M (1997) JP-5000, the HVOF of the 21st Century. In: HVOF spraying. Proceedings of 4 colloquium, Erding, Germany
34. Thorpe RJ et al (2003) The new JP-8000 Model 8200 advanced HP/HVOF® Gun. In: HVOF spraying. Proceedings of 6 colloquium, Erding, Germany
35. Verstak A, Baranovski V (2003) Activated combustion HVAF coatings for protection against wear and high temperature corrosion. In: Proceedings of thermal spray 2003: advancing the science and applying the technology, Osaka
36. UniqueCoat Technologies (2005) AC-HVAF process information: bulletin. Richmond, VA
37. Kreye H (1997) A comparison of HVOF systems—behavior of materials and coating properties. In: HVOF spraying. Proceedings of 4 colloquium, Erding, Germany
38. Beczkowiak J, Keller H, Schwier G (1997) Carbide materials for HVOF applications powder and coating properties. In: HVOF spraying. Proceedings of 4 Colloquium, Erding, Germany
39. Kharlamov YA (2000) Thermal spraying of coatings and ecological production, operation and repair of machines. Tyazheloye mashinostroyeniye 2:10–13
40. Erokhin AA (1973) Fundamentals of fusion welding. Mechanical Engineering, Moscow
41. Vakhalin VA, Kudinov VV, Belashchenko VE (1985) Study of efficiency of electrode's heating and the deposition efficiency of materials at arc spraying. Phys Chem Mater Process 5:65–69
42. BAMZ (1987) Arc spraying gun EM-17: specification. Barnaul
43. Rusev GM et al (2000) Technology of restoration and strengthening of cylindrical parts and flat surfaces by plasma spraying. Autom Weld 12:61–62
44. Uschenko KA Borisov YS Pereverznev YN (1998) Microplasma spraying of coatings. In: Films and coatings'98. Proceedings of 5 international science-technique conference, S-Peterburg
45. Colmenares-Angulo Sabouni JO Ibanez C (2013) The definition of power in plasma spraying: kilowatts or powder melting capacity? In: Innovative coating solutions for global economy. Proceedings of ITSC-2013, Busan, Korea
46. Bobzin K et al (2007) TriplexPro 200—potential and advanced applications. In: Global coating solutions. Proceedings of ITSC 2007, Beijing, China
47. Yakimavichus CS (1955) A method of deposition of molten metal by arc spraying. USSR certificate of authorship 107023
48. Kaiser JJ et al (1991) Electric Arc Spraying of Reactive Metals. EP 0445353
49. Kroemmer W (2013) High performance arc sprayed coatings for active corrosion protection in offshore applications. In: Innovative coating solutions for global economy. Proceedings of ITSC-2013, Busan, Korea
50. Nekrasov YI (1974) Gases—substitutes of acytelene. Mechanical Engineering, Moscow
51. Sokolov II (1978) Gas welding and cutting. High School, Moscow
52. Lyalyakin VP, Sablukov AS, Litovchenko NN (2000) Investigation of the processes of dynamic metal atomization at arc spraying. Welding Production 1:16–24
53. Matthäus HD, Busse K (1986) Verfahren und Lichtbogenspritzdüse zum Beschichten von Werkstückoberflächen durch Schmelzen von Drähten in einem elektrischen Lichtbogen. DE Patent 3533966
54. Dorozhkin NN, Baranovski VE, Yelistratov AP, Pryadko AS (1987) Device for arc spraying. USSR Certificate of Authorship 1291216
55. Bach Fr-W, Laarman A, Wenz T (2006) Modern surface technology. WILEY-VCH Verlag GmbH & Co. KGaA, Weinheim
56. Pokhodnyua IK, Syptel AM, Shlepakov VN (1972) Welding by cored wire. Naykova Dymka, Kiev
57. Steffens H-D, Babiak Z, Kaczmarek R (1990) Thermisches spritzen—verfahren, anwendungen, tendenzen: DVS report. Thermische Spritzkonferenz TS'90, Essen

58. Borisov Y, Voropaj N, Netesa I, Korzhyk V, Kozjakov I (1995) Composite flux cored wires for thermal spraying. In: Berndt CC, Sampath S (eds) Thermal spray 1995: science & technology, Materials Park, OH, USA
59. Dallaire S, Levert H, Legoux J-G (2001) Erosion resistance of arc-sprayed coatings to Iron Ore at 25 and 315 °C. J Therm Spray Techn 10(2):337–350
60. Tillmann W, Klusemann B, Nebel J, Svendsen B (2011) Analysis of the mechanical properties of an arc-sprayed WC-FeCSiMn coating: nanoindentation and simulation. J Therm Spray Techn 20(1–2):328–350
61. Wilden J, Drescher V, Djahanbakhsh M (2010) Wire arc sprayed iron-based anti-corrosion coatings for waste incinerator plants. In: Thermal spray: global solutions for future application. Proceedings of ITSC-2010, Singapore
62. Ageev VA, Belashchenko VE, Feldman IE, Chernoivanov AV (1989) Analysis of control methods of the atomizing particle's parameters at arc spraying. Weld Prod 12:30–32
63. Pokhmursky VI, Student MM, PIkh VS, Tikhan MA (1991) Head of arc spraying gun. USSR Certificate of Authorship 1657230
64. Harrington KW, Stevens JL, Thorpe ML (1984) Metal spray device. US Patent 470071

Chapter 2
Input Parameters of the Coating Process

2.1 Stages of the Arc Spraying Process

To study the input parameters, it is rational to divide the process of AS into several stages (Fig. 2.1). They differ in the effect caused by input parameters on properties of atomized material, which is resulted in final coating properties.

The stages have the following peculiarities. Heating and melting of metal take place on the end face of electrode. It is occurred due to arc heat and joule heat that exuded due to the current run through the electrodes as well as via chemical reactions between the metal and the carrier gas components.

Supplementary liquid metal is fractured in the arc-heating zone into separate particles under the impact of gas-dynamic head of the carrier gas and the electromagnetic forces. The latter occurs because of the current run. The exchange of thermal, kinetic energy, and chemical reaction of the liquid metal drops with carrier gas and surrounding atmosphere takes place along the spraying distance. These processes determine the type of interaction "atomized metal–substrate" while formation of coating. Stages have overlapping zones due to continuous change of the parameters. However, for the sake of modeling it is preferential to describe the stages as discrete objects followed by the analysis of the accuracy of the taken assumption.

Accuracy and adequacy of the model of thermal, chemical gas-dynamic, and electro-dynamic ongoing processes are to be evaluated by correlating results of the model analysis with the data about metal coating properties. Effective development of equipment and technological processes is possible based on the modeling and coating properties studying.

V. Boronenkov and Y. Korobov, *Fundamentals of Arc Spraying*,
DOI 10.1007/978-3-319-22306-3_2

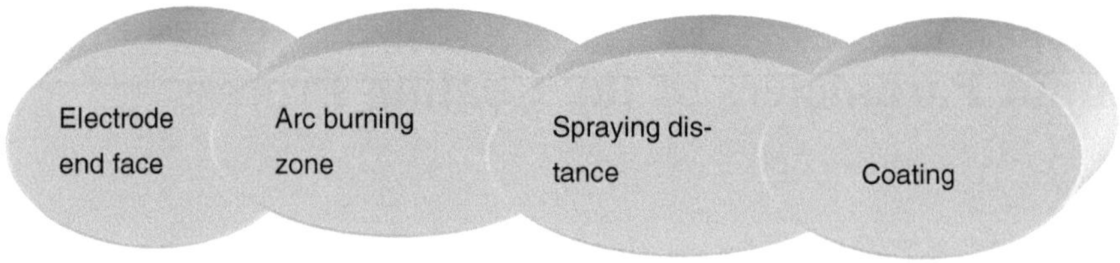

Fig. 2.1 Stages of coating process in AS

2.2 Temperatures, Velocities, Gas Atmosphere Components—Initial Values

At PS, the temperature of the arc in its axis reaches 50,000 K [1]. There is no such purposeful force compression of the arc at AS comparing to PS. Consequently, temperatures' level here is lower. The closest analog of this type of arc and gas burning is the process welding in atmosphere of air, where the arc temperature is 5000–6000 K [2–4].

At AS, the gas carrier temperature when it reaches the arc burning is varied from 300 K, in case of using compressed air, to the several thousand degrees, in case of using combustion products of various hydrocarbons as a carrier gas. Fuels' burning temperature depends on the type of oxidant and the chemical bond of characteristics (Table 2.1).

The propane is the most favorable for AS among hydrocarbon fuels. First, the use of propane is well compatible with wide spread gas-welding and cutting technological processes, and therefore, it does not require any significant starting costs. Natural gases with similar characteristics require a trunk pipeline, which quite often might be an obstacle for on-site applications. Second, propane is safer than acetylene which is also widely used gas in a welding processes. It is determined by the fact that propane inflammation velocity (1.31 m/c) is four times lower than acetylene inflammation velocity [5]. Therefore, the danger of the backfire decreases. It is significant due to the presence of melted particles as they can perform multidirectional high-velocity fly away. To reduce the costs and to secure the technological process in case of high flow rate of a carrier gas, it is highly recommendable to use a compressed air as an oxidant.

Table 2.1 Burning temperatures of the hydrocarbon fuels in the stoichiometric mixture [5, 6]

	Oxidant	Gas			Liquid		
		Methane CH_4	Acetylene C_2H_2	Propane C_3H_8	Ethyl Alcohol C_2H_6O	Patrol C_8H_{18}	Kerosene $C_{7,21}H_{13,5}$
T, K	Air	2148	2598	2198	2090	2243	2203
	Oxygen	2673	3423	2973	3053	2773	2673

Let us evaluate the maximum gas temperature for AS. The initial gas temperature T_{g0} that triggers its interaction with atomized metal will differ from the initial temperature of a carrier gas $T_{g0}^{исх}$:

$$T_{g0} = T_{g0}^{исх} - \Delta T_{к} + \Delta T_{H} \tag{2.1}$$

where $\Delta T_{к}$ corresponds to the gas cooling due to the heat irradiation toward the combustion chamber walls (by using combustion products of the "fuel-oxidant"; $\Delta T_{н}$ corresponds to the rising of the temperature due to gas heating while transiting through the electric arc.

Gas cooling calculation in the combustion chamber is to be done according to the scheme of recuperative parallel flow heat exchanger with the heat irradiation through monolayer cylindrical wall [7]. Air is used as a cooling heat carrier (coolant). Change of the temperature is:

$$\Delta T_{k} = \frac{\left(T_{гор} - T_{в}\right)\cdot\left(1-\exp\left(-\left(1+\frac{G_1\cdot c_{p1}}{G_2\cdot c_{p2}}\right)\cdot\left(\frac{k\cdot F}{G_1\cdot c_{p1}}\right)\right)\right)}{\left(1+\frac{G_1\cdot c_{p1}}{G_2\cdot c_{p2}}\right)}, \tag{2.2}$$

where $k = 1\Big/\left(\frac{1}{\alpha_1\cdot d_1} + \frac{1}{2\lambda}\cdot\ln\frac{d_2}{d_1} + \frac{1}{\alpha_2\cdot d_2}\right)$; $\alpha = Nu\cdot\lambda_g/d$; $Re = W_g\cdot d/v$; $Nu_1 = 0.18\cdot Re_1^{0.8}$; $Nu_2 = 0.245\cdot Re_2^{0.6}$; $T_{гор}$ is fuel burning temperature, K; Tb $T_{в}$ is initial temperature of cooling air, K; $G = \rho_g\cdot W_g\cdot f$ is mass flow rate of a heat carrier, kg/s; ρ_g is gas density, kg/m^3, W_g is gas velocity m/s, c_p is gas thermal capacity, J/(kg K); F is surface area of heat emission, m^2; α is gas heat emission factor (coefficient), W/(m^2 K); v—kinematic viscosity factor, m^2/c; Nu—Nusselt number; Re—Reynolds number; combustion chamber operation parameters: f is outlet duct, m^2; k is coefficient of heat transfer through the wall W/(m^2 K); δ is wall thickness, m; λ is wall thermal conductivity, W/(m^2 K); d is diameter, m, indexes "1", "2" refer to the combustion products and cooling air, consequently.

Calculations (Table 2.2) were done for a case, when compressed air and combustion products of propane–air mixture are used as a carrier gas.

Combustion chamber is constructed according to the arc spraying gun [8], W_g is adopted due to the equality relations of carrier gas and outlet of combustion outlet

Table 2.2 Calculating results of combustion products cooling from the combustion chamber walls (line 1) and cooling air heating (line 2)

	$T_0^{исх}$	F	f	ρ_g	W_g	λ_g
1	2100	5.2×10^{-3}	3.8×10^{-5}	0.17	100	0.149
2	300		6.3×10^{-6}	1.18	300	0.026
	δ	λ	ν	c_p	α	ΔT_k
1	0.001	22	1.5×10^{-5}	1400	85600	224
2			4.8×10^{-4}	1000	372	92

chamber and nozzle; Nu_1 is considered as for combustion flow in a tube when turbulence occurs; Nu_2 is considered as for that transfer of external tube flow.

Thermal-physics characteristics of gases (ρ_g, λ_g, v, c_p) are considered in accordance with [9], material used for combustion chamber construction—as for [10].

To evaluate the heating of a gas while passing through the arc, let us assume the heating of the gas is a result of the heat emitted by arc column minus radiation loss. In this case:

$$\Delta T_H = (q_\partial - q_u)/\alpha_T \text{ is gas heating from the arc;} \tag{2.3}$$

$$q_\partial = I \cdot E/(\pi \cdot d) \text{ Is heat emittedin the arc column;} \tag{2.4}$$

$$q_u = \beta \cdot \sigma \cdot T_\partial^4 \text{ is radiation heat loss of the arc column;} \tag{2.5}$$

$$\alpha_T = Nu \cdot \lambda_\partial / d, \tag{2.6}$$

$$Nu = 2 + 0.459 \cdot Re^{0.55} \cdot Pr^{0.333} \quad [11], \tag{2.7}$$

$$Re = v \cdot \rho \cdot d/\eta \tag{2.8}$$

In Eq. (2.4) electric field intensity $E = 2.5$ V/mm is considered according to [2, 11]. This corresponds to the arc length $l_\partial = 4$ mm and to a voltage drop in the arc column $U_c = 10$ V. According to Eq. (2.5), we assume the arc column to be opaque and completely black, with respect to the assumption made by Leskov [2]. In the Eq. (2.8), we consider the electrode's diameter size equals the diameter of the arc column.

According to the results after applying Eq. (2.1) with the data from Tables 2.2, 2.3 the temperature of carrier gas will be approximately 450 K. If using combustion products of the propane–air mixture the temperature will be 2200 K.

For the gases used in the arc spraying the most thermodynamically expected reactions are the reactions of thermal dissociation of the molecules on atoms and radicals: H_2, O_2, H_2O

$$H_2 \leftrightarrow 2H \tag{2.9}$$

$$O_2 \leftrightarrow 2O \tag{2.10}$$

$$H_2O \leftrightarrow OH + 0.5H_2 \tag{2.11}$$

$$H_2O \leftrightarrow 2H + O \tag{2.12}$$

The temperature of plasma jet formed during gas heating by plasma arc is 3500–16,000 K depending on the type of plasma forming gas [15–17] in the outlet nozzle. Calculations show that temperatures are responsible for greater share of the ionization products and dissociation gases (Fig. 2.2.) According to the thermo dynamical calculations for these gases (Table 2.4.), the presence of the sizable

Table 2.3 Gas heating while passing through the electric arc

Parameters	Gas type	
	Air	Combustion products
Arc temperature T_∂, K [2]	6000	
Arc current I, A (Typical mode)	200	
Electric field intensity, E, V/mm	2.5	
Arc column diameter, d, mm	2	
Arc column thermal conductivity, λ_∂, W/(m K) [12]	0.5	1
Emissivity factor, β	1	
Stefan-Boltzmann constant, σ_0, W/(m^2 K^4) [2]	5.67×10^{-8}	
Gas velocity, W_g, m/s [13]	300	900
Gas density, ρ_g, kg/m^3 [9]	1.29	0.17
Dynamic viscosity, η, Pa s [9, 13]	1.7×10^{-5}	5×10^{-5}
Prandtl number, Pr, [14]	0.7	
Reynolds number, Re	41404	3750
Nusselt number, Nu	146	40
Heat transfer coefficient, α_m, W/(m^2 K)	3.8×10^4	2.6×10^4
Radiation heat loss of the arc column, q_∂, W/m^2	8×10^7	
q_u, W/m^2	7.3×10^7	
Gas heating from the arc, ΔT_u, K	160	239

Table 2.4 The equilibrium constants of dissociation reactions of gases (k_i) [18]

T, K	lg K_i			
	Reaction (2.9)	Reaction (2.10)	Reaction (2.11)	Reaction (2.12)
1000	−17.2924	−19.6128	−10.545	−17.213
2000	−5.5816	−6.3553	−3.400	−5.719
3000	−1.6974	−1.8984	−1.002	−1.822
4000	0.4005	0.3393	–	0.149
5000	1.6120	1.6846	–	–
10000	4.0393	4.400	–	–

amount of the atomic particles and radicals shall be considered when the temperature of the atmosphere gas is >2500 K.

Comparing the calculation results applying Eq. (2.1) and data on gas dissociation (Fig. 2.2, Table 2.3) showed that the content of atoms and radicals is negligible and therefore the molecule dissociation can be ignored. This simplifies the calculation of the atmosphere content at AS compared to PS.

Initial gas flow velocity is determined by nozzle geometry, pressure, and gas temperature. In modern arc spraying guns, transonic velocity of gas is provided. Gas temperature has an indirect influence on the dynamic pressure of the jet $\left(\rho_g \cdot W_g^2\right)/2g$. Change in temperature causes the change in gas density and sound

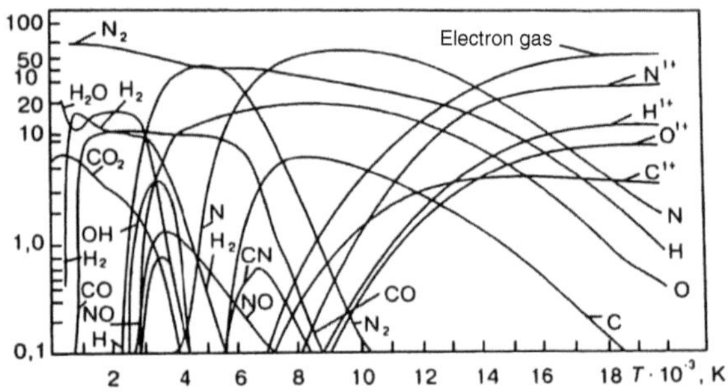

Fig. 2.2 Function of equilibrium plasma composition of combustion products of neutral mixture "methane-air" on temperature [17]

velocity rate. Velocity of sound ($a_{зв}$) determines Mach number, which influences the gas velocity that by turn depends on gas flow temperature:

$$a_{зв} = \sqrt{k \cdot g \cdot R \cdot T_g}, \tag{2.13}$$

where ρ_g is gas density, kg/m^3; W_g is gas velocity, m/s; T_g is gas temperature, K; k is adiabatic index (depends on temperature and gas composition); g is standard acceleration of gravity, m/s^2; R is universal gas constant.

For instance, for nozzle providing gas flow rate (0.8–0.9 M), in case of cold air $W_g \sim$ 300 m/s, propane–air mix combustion in case of $W_g \sim$ 900 m/s. The density of cold and hot air is 0.17 and 1.29 kg/m^3 consequently [9].

2.3 Metal Heating and Melting on the End Face of Electrode

Metal heating and melting on the end face of electrode for AS are similar to arc welding. They are provided by the heat from chemical reactions, electric arc, Joule's heat. Heat retrieves when the electric current passes through metal wires. Metal temperature (steel) on the end face of the electrode when welding is 2000–2700 K, although the most characteristic temperature is closer to 2500 K [4]. In AS researches, it is viewed similar to welding intervals—in between the boiling and melting points of the material used for electrodes [19, 20]. At AS the most reliable experimental metal temperature values on the end face of electrode are 2323–2573 K. The results were obtained via laser spectroscopy [21]. As shown by W. Tillmann et al., [103], and A.P. Newbery et al., [104], particle temperature is 2700 K near arc burning zone at high voltage (34 V) and 2500 K at 28 V. However, higher voltage leads to intensive burning of atomized metal [34], so more preferable is voltage 28 V. Although if the gap between the electrodes is small (nearly 0.1–1.0 mm), the results of the measurement refer to the metal drops escaped from the electrodes according to Royanov

[22]. Influenced by high-velocity airflow their temperature is lower than those at the surface of the end face. For our calculations, let us assume the temperature of the steel end face of the electrode is 2500 K.

There are some differences in the gas flow blow-off of the electrodes. For arc welding, a lamellar gas flow is more preferable. Transition to turbulent mode lowers the arc stability. In this case, the gas flow velocity which is flowing around liquid metals is 10–30 m/s and gas consumption does not exceed 1.2 m^3/h [23]. During AS process, the gas flow of high kinetic energy and high turbulence is forcedly formed. Duct gas flow rate is 250–1500 m/s and its consumption is between 90 and 150 m^3/h. That is, by 1–2 orders higher as compared to arc welding [24].

The direction of gas flow influence on electric arc also changes compared to welding. When welding gas flow goes along the electrode axe, whereas during AS, the latter forms an angle 15–300 toward the electrode axe. All this leads to the reinforced convective heat output of electrode extension (the length of the wire from the current-carrying contact tip to the fused end). This heat output should be taken into account when calculating the heat balance in contrast to welding [10, 25].

Such calculation seems interesting if using enlarged electrode extension, when wire notably heats from the passing current. Researches of arc welding process with enlarged electrode extension showed that it leads to the increasing of the metal fusing efficiency, rising of the average particles size, and decrease of spattering [4, 23]. Therefore, AS is also arc-type process, we can assume that the consequences of the enlarged electrode extension at AS will be similar to the shown above for welding. Total power density obtained by the electrode ($q_э$, W/m^3) can be retrieved from the heat power equation:

$$q_э = q_{мок} + q_m - q_{конв} \tag{2.14}$$

where $q_{мок}$ is heat generated in the electrode due to passing current, q_m is electric arc heat, $q_{конв}$ is convective heat losses.

Let us expand the equation:

$$c\gamma v_n \frac{dT}{dL} = j^2 r + \frac{c\gamma v_n (T_k - T)}{L - L_x} \times \exp\left(\frac{c\gamma v_n (L - L_x)}{\lambda}\right) - \left(\frac{4\alpha(T - T_g)}{d}\right) \tag{2.15}$$

where j is current density, A/m^2; r is specific electric resistance, ом м; γ is material density of solid electrode, kg/m^3; v_n is wire feed rate, m/s; L, L_x are electrode extension (general and actual value), m; T, $T_к$ are particle temperatures (actual value and on the electrode end face measurement), K; T_g is gas temperature, T_g is λ-conductivity coefficient of metal droplet, W/(m k); λ is surfacing heat transfer coefficient calculated according to Nusselt number, W/(m^2 K); d is diameter of the electrode, m.

Thermal heating of the electrode by the arc active spots due to rapid heat supply can be calculated according to the equation of heat spreading from the continuously functioning flat source in the rod moving with a constant velocity [10].

Temperature dependences of specific electrical resistivity (r), specific heat metal capacity (c) were approximated by the following equation type: $y = A_1 \cdot T^2 + A_2 \cdot T + A_3$ (Table 2.5).

Gas velocity at the electrode meeting point is 300 m/s. Calculating results are shown in Fig. 2.3. Qualitative ratio differs from the data, acquired in arc welding [23] in the following: the influence of heat of electrode extension from the arc heat starts from 3 mm distance, which is 5–8 times more compared to AS.

This is due to the physical characteristics of the AS, where most of the molten metal is blown away by gas flow from the electrode end-face.

Moreover, an intensive gas cooling of the electrode leads to a decrease in electrode extension heating by 20–40 K in comparison to welding. Wherein he heat transfer coefficient from the surface increases to α = (860–1500) W/(m^2 K), which is two orders more than in welding, where α = (0.5–5) W/(m^2 K) [27].

Table 2.5 Temperature dependences of the electrode parameters [23, 26]. AWS70S-6 solid wire is taken

Parameter	Temperature interval, K	A_1	A_2	A_3
c, J/(kg K)	300–1033	0.0021	−1.8221	866.28
	1033–1184	0.0429	−99.038	57866
	1184–1809	0.0004	−0.86	1118.3
$r \times 10^8$, Ω m	300–800	9.57×10^{-5}	2.41×10^{-2}	24.41
	>800	–	2.45×10^{-2}	89.98

AWS70S-6 wire is taken [23, 26]

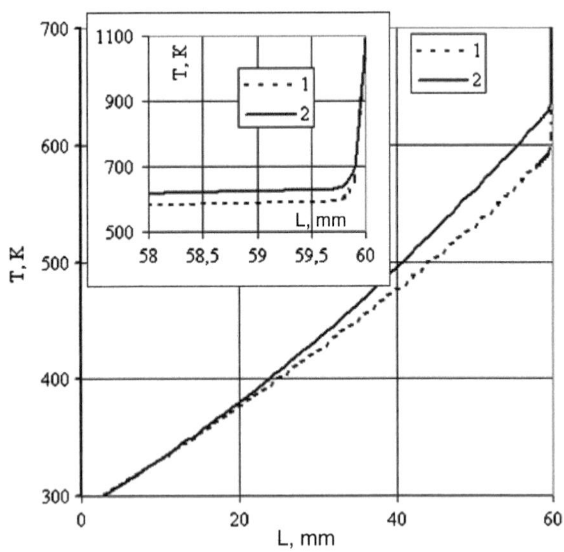

Fig. 2.3 The temperature distribution in the electrode extension at AS. *1* is accounted for heat transfer from the electrode; *2* is accounted for heat transfer from the surface of the electrode is not accounted

Currently, the use of core wires (CW) as an electrode material is expanding. However, they differ by the unevenness of thermal and physical properties of the cross-section, which drastically changes the nature of the thermal and physical–chemical processes at the end face of the electrode. Thus, for example, the study of the coating produced by CW spraying showed that the charge could be partially separated from the electrode end face in a non-fused state [28].

We introduced our model of heat distribution in CW from arc heating at AS. Obtained simulation data can be used independently for temperature prognoses of particles formed by melting of the wire, as well as a source of initial data for describing physical and chemical interactions. Analysis of ongoing processes at CW spraying will significantly reduce the volume of study, which is necessary to develop the new CW.

Heat distribution in the end face area of CW is considered as an axe-symmetrical task and described by the differential equation of thermal conductivity

$$\partial T/\partial t = (\lambda/c \cdot \lambda) \cdot (\partial^2 T/\partial x^2 + \partial^2 T/\partial y^2), \tag{2.16}$$

where T—temperature, K; t—time, s; λ—coefficient of thermal conductivity, W/(m K); c—specific thermal heat capacity, J/(kg K), γ—density, kg/m^3; x, y—coordinates of cross-section and generating line of CW, respectively, m.

The task was solved using explicitly difference method. End face zone was divided on the radius (r) by coordinate lines i = const with a step $\Delta r = r/10$, and to the depth (s) with coordinate lines j = const with a step $\Delta s = s/10$ (see Fig. 2.4).

Within an element, the temperature is considered to be constant and refers to its middle. The following equation was adopted for the obtained elements of the mesh:

$$\partial T\Big/\partial t \cong \Delta T\Big/\Delta t = \left(T_{i,j}^{k+1} - T_{i,j}^{k}\right)\Big/\Delta t \tag{2.17}$$

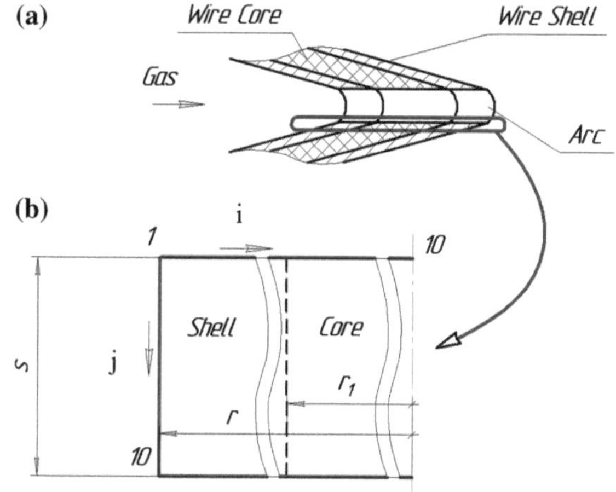

Fig. 2.4 End face electrode area model; *a* end face sketch at arc spraying; *b* scheme of heat propagation in electrode end face accepted

A differential, for instance on r, can be expressed as: $\partial T/\partial r \cong \Delta T/\Delta r$,

$$\partial^2 T/\partial r^2 = \frac{\partial}{\partial r}\cdot\left(\frac{\partial T}{\partial r}\right) \cong \frac{\partial}{\partial r}\cdot\left(\frac{\Delta T}{\Delta r}\right) \cong \frac{1}{\Delta r}\left(\left(\frac{\Delta T}{\Delta r}\right)_+ - \left(\frac{\Delta T}{\Delta r}\right)_-\right)$$
$$= \frac{1}{\Delta r}\cdot\frac{\left(T^k_{i+1,j} - T^k_{i,j}\right) - \left(T^k_{i,j} - T^k_{i-1,j}\right)}{\Delta r} = \left(T^k_{i+1,j} - 2T^k_{i,j} + T^k_{i-1,j}\right)/\Delta r^2 \tag{2.18}$$

Differential with respect to s is expressed similarly. Then for the wire section under consideration, difference equation which approximates differential Eq. (2.16) can be represented as follows:

$$T^{k+1}_{i,j} = T^k_{i,j} + (\lambda/c\cdot\gamma)\cdot\Delta t\cdot\left(1/\Delta r^2\cdot B_1 + 1/\Delta s^2\cdot B_2\right), \tag{2.19}$$

where $B_1 = T^k_{i+1,j} - 2T^k_{i,j} + T^k_{i-1,j}$; $B_2 = T^k_{i,j+1} - 2T^k_{i,j} + T^k_{i,j-1}$; κ, Δt are a step number and time step. The time interval is divided into 10^4 steps.

When solving (2.19) the equitation, a restriction arising from the convergence conditions for the selected method is imposed:

$$(\lambda/c\cdot\gamma)\cdot\left(1/\Delta r^2 + 1/\Delta s^2\right) < 1/2 \tag{2.20}$$

Since the conductivity of the charge is 1–2 orders lower than that of the shell, it is assumed that the active arc spots are located at the ends of the shells. The surface temperature of the shell at the end is assumed 2500 K. Overheating of the electrode end face to such temperature regarding the melting point (T_m) is typical feature to arc processes [4]. It is assumed that the surface of the charge receives the heat radiated from the arc column and heated up to 1300 K.

Heat distribution from the end face onto downward direction is affected by thermal exchange “shell-gas” and “shell-charge”. Heated by the arc, the shell gives off the heat to the streamline gas flow and the charge.

When calculating the heat “shell-gas”, the heat transfer coefficient of the surface is defined by the equation

$$\alpha_1 = Nu\cdot\lambda/2r \tag{2.21}$$

Nusselt number is calculated according the dependency used for the cross-flow tubes calculation [7]:

$$Nu = 0.245\cdot Re^{0.6} \tag{2.22}$$

According to the calculations, when using air and propane–air mixture as the carrier gas, the value of α_1 is in a range of 2000–4000 W/(m^2 k), depending on changes in transfer characteristics of the gas and its velocity. The heat transfer coefficient on the border “shell-charge” is adopted as α_2 = 1400 W/(m^2 K) [29].

The change in temperature on the surface of the electrode due to the heat loss while electrode blow-off of by the carrier gas was calculated as following:

$$\lambda_1 \cdot \frac{\left(T_{i,j}^k - T_{i-1,j}^k\right)}{\Delta r} = \alpha_1 \cdot \left(T_{i-1,j}^k - T_g\right) \tag{2.23}$$

The change in temperature at the border of the "shell-flux wire core" due to the heat loss calculated by the equation:

$$\lambda_1 \cdot \frac{\left(T_{i,j}^k - T_{i-1,j}^k\right)}{\Delta r} = \alpha_1 \cdot \left(T_{i-1,j}^k - T_g\right) \tag{2.24}$$

where "1", "2"—indices related to metal shell and core charge correspondently.

Thermo and physical properties of the charge are adopted in the composition range, typical for these CW components. Low carbon steel (0.08 % C) is used as the shell material. Necessary for calculating thermophysical characteristics as thermal conductivity (λ), density (ρ), heat capacity(c) are adopted according to [30, 31] and approximated by polynomials (Table 2.6).

In calculation, according to Eq. (2.16), the enthalpy is taken into account as follows. At each step, the cell temperature was compared with the melting temperature of the shell and the charge, respectively. On reaching the latter, the cell temperature remains constant for the following time. At all subsequent points of time, the heat accumulated in the cell was summed:

$$\sum_{k=1} q_{ij}^k = \sum_{k=1} c_{ij}^k \cdot \left(T_{ij}^k - T_{ij}^{k-1}\right) \tag{2.25}$$

This heat at each step was compared with the specific melting enthalpy of cell material ($h_{пл}$). Under the following condition: $\sum_{k=1} q_{ij}^k > h_{пл}$ for further calculation steps, the temperature rise of the cell was taken into account.

Table 2.6 Equations of thermal parameters of the approximation

Parameter	Temperature range, K	The polynomial coefficients $Y = A_1T^3 + A_2T^2 + A_3T + A_4$			
		A_1	A_2	A_3	A_4
λ, W/(m K)	300–2500	-2×10^{-8}	0.0001	0.1742	123.89
	2500–3500	0	-3×10^{-6}	0.0175	22.42
ρ, kg/m^3	300–3500	0	8×10^{-5}	0.463	7971.1
c, J/(kg K)	323–1073	0	0.0003	0.0969	409.12
	1073–1273	0	0	−1.045	1987.8
	1273–2473	0	0	0.0658	575.03

Calculation of the layer thickness at the end of the electrode, melted by the heat of the active arc spot, was carried out according to the solution of the task to melt the cold body of limited sizes under conditions of ongoing removal of resulting melt taking into account the enthalpy of fusion [32]. Balance equation of heat in this case can be written as:

$$q_1 \cdot F \cdot \mathrm{d}\tau = \mathrm{d}Q \tag{2.26}$$

The left-hand side of the equation ($q_1 \cdot F \cdot d\tau$) is the heat input on the electrode end face, which is going on due to the heat of the active arc spot (q_1 is specific heat flow from the active arc W/m^2; F is the electrode end face area, m^2). Right-hand side of the equation (dQ) is heat discharge for melting the metal layer at the end and for the heating of the solid part of the electrode shell.

B. Tillman and et al. cite that in arc spraying using air as the atomizing gas leads to unevenness of CW melting. On combined macro thin section of anode and cathode is evident that non-melted components are less presented on anode-wire then on cathode-wire (Fig. 2.5).

Note that according to the illustrated macro thin section, the shell thickness varies for anode and cathode. This will adjust the process of wires melting. The asymmetric melting behavior of solid wires and core wires was observed in many studies [48, 103]. The reason is that the anode and cathode are heated differently. The same effect is observed at arc welding [2–4]. But Erokhin also noted that for the case of the arc welding in the air, the thermal power released in anode and cathode spots is approximately the same [4]. In case of AS, the asymmetric melting behavior is lowered due to the heat exchange between the active arc spots, located at a distance of 0.1–1.0 mm [22]. For this model, the share of thermal power of the arc, for each active spot is assumed nearly the same.

After substitution and integration, according to [32], Eq. (2.26) turns out into the equation regarding to the parameter (Ψ), in which the unknown is the thickness of the liquid layer (s), melting by the active arc spot:

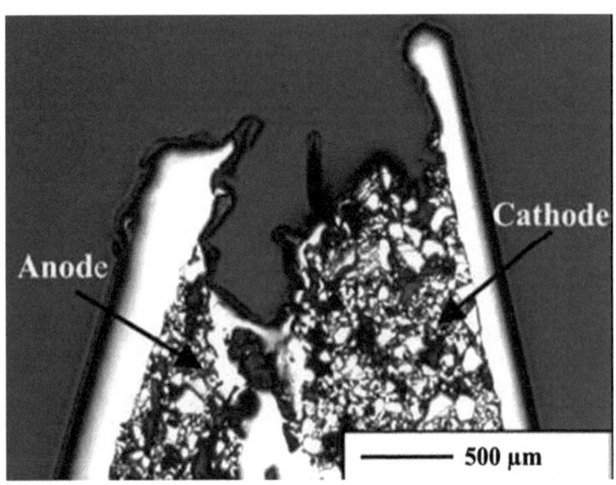

Fig. 2.5 Combined macro-section molten CW Ø1.6 mm, the composition of 28 % Cr, 1.4 % Si, 4.8 % C, <1 % B, Fe bal. [33]

$$2 \cdot a \cdot t/h^2 = \left(\Psi(N\varpi)/N^2\right), \tag{2.27}$$

where

$$\Psi(u) - 9 \cdot u \cdot \sqrt{(u+1/3)^2 + 8/9} \Big/ 16 + \sqrt{(u+1/3)^2 + 8/9} \Big/ 16$$
$$+ 2 \cdot \ln\left(6 \cdot \left(\sqrt{(u+1/3)^2 + 8/9} + u + 1/3\right)\right) \Big/ 12 - 3 \cdot u^2/16 - u/4;$$
$$N = c \cdot \Delta T/(2 \cdot h_{nn}); \quad \omega = (h - s)/h;$$
$$h = 2 \cdot \lambda \cdot (T_n - T_c)/q_1; \quad a = \lambda/(c \cdot \rho);$$

N—melting criteria, reflecting the effect of latent heat of melting; $\tau = 4 \times 10^{-4}$ s—a typical period of droplets removal from the end face of the electrode [20]; $T_s = 300$ K, the temperature of the unheated metal; $T_n = 2500$ K, the temperature of the surface melt at the end face of the electrode; a—thermal conductivity, m^2/s; $h_{n\pi}$ = 1,319,000 J/kg, a melting enthalpy of liquid iron [4], h—heating zone longitude from the surface end downward to the depth of the electrode, m.

Results of calculation have shown in Fig. 2.6, in the case of CW, the thickness of the tearing off molten metal interlayer increases several times in comparison to solid wire.

Furthermore, with decreasing of shell thickness (δ) the thickness of liquid layer rises. Calculations based on Eq. (2.27) are done with reference to solid wire and CW. The correctness of the results was checked for the solid wire at calculation assuming the amount of layer and drops were equal. The droplets data are obtained due to oscilloscoping breakdown of droplets [20, 34] (these data are indicated in Fig. 2.6 as a solid, test).

In calculations of heat transfer, inverse problem was solved with respect to thermophysical parameters of the charge, for the blow round gas, the geometric

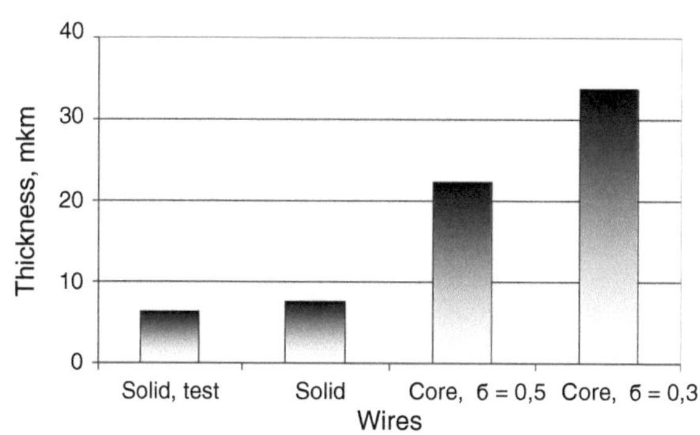

Fig. 2.6 The results of the calculation of the layer thickness of the liquid metal being broken away during AS for the solid wire and CW with different thickness of wire shell

Table 2.7 Parameters of corer wires and the modes used in calculations

Version number	r, mm	δ, mm	α, W/(m^2 K)		Parameters of the gas		c_{charge}, $\frac{J}{kg K}$	λ_{charge}, $\frac{W}{m K}$	s, μm
			Cover	Charge	T, K	Type			
1	1	0.5	2070	1400	300	1	1600	20	22
2	+	+	+	+	+	+	+	5	22
3	+	0.3	+	+	+	+	+	+	33
4	+	0.3	+	+	+	+	+	5	33
5	+	+	3820	+	2100	2	+	+	22

Notes In the column "Version", index 1 corresponds to the compressed air and the index 2—combustion products of propane-air mixture
Sign "+" in the cells corresponds to the variant 1

dimensions of the cross-section of CW. In the range of values of these parameters, the search of the area satisfying favorable conditions for drops formation of sprayed material was conducted. Table 2.7 shows the parameters of CW adopted for the calculation of the various options, differs from the basic option 1 by the thermal conductivity of the charge (variations 2, 4), the thickness of shell (variations 3, 4) and the conditions of heat transfer at the border "gas-wire" (variation 5).

Analysis of the results of calculations (Figs. 2.7, 2.8, 2.9 and 2.10) showed:

(1) CW shell throughout the thickness of the liquid interlayer is strongly overheated above the melting temperature (Fig. 2.7).
(2) The CW shell thickness along with the previously marked growth of the liquid layer height, results in decrease of its temperature (Fig. 2.7).
(3) The surface temperature of the shell is strongly depended on the heat-transfer conditions in system "gas-wire" (Fig. 2.8).
(4) The surface temperature of the shell is reduced in comparison with their inner-metal layers (Fig. 2.9) due to the heat exchange with cooler gas. It slows the droplets discharge.
(5) Some melting of the charge may be caused by the reduced heat-conductivity of the charge and the small thickness of the shell (Fig. 2.10).

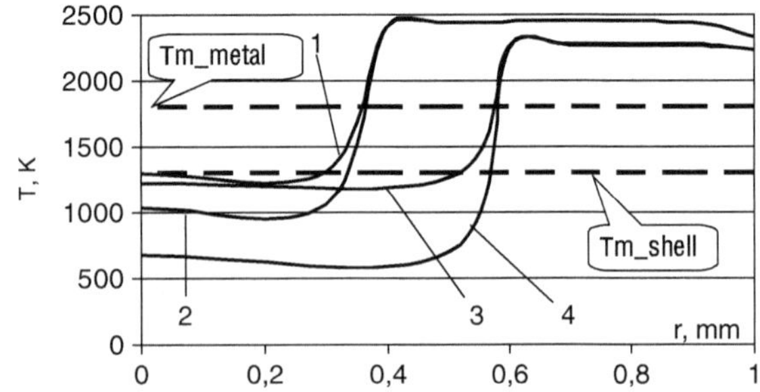

Fig. 2.7 The temperature distribution over the cross-section of core wire

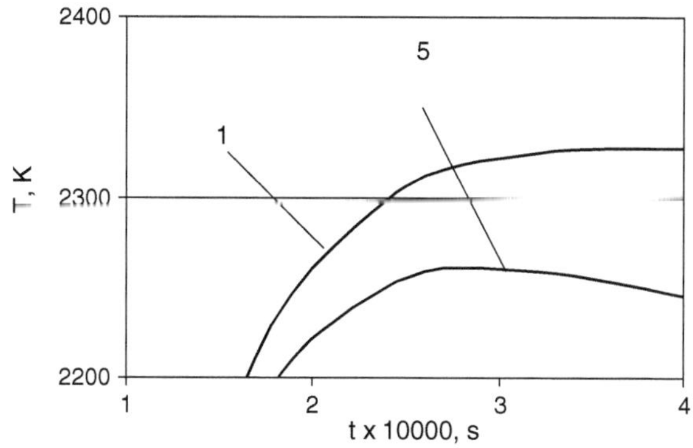

Fig. 2.8 Changes in the surface temperature of the shell according to the heat transfer conditions in the system "gas-wire"

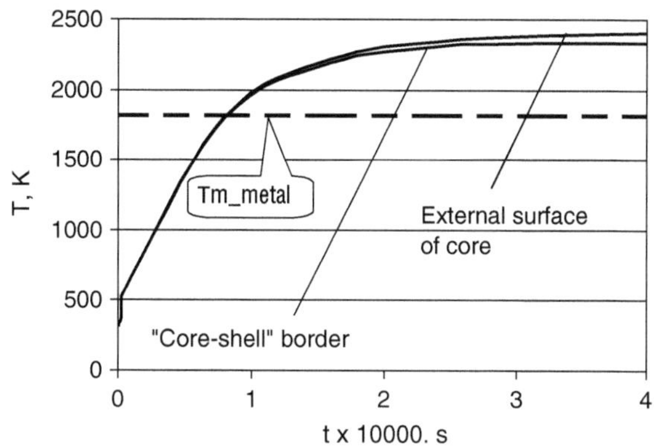

Fig. 2.9 Temperature changing in shell for the external and internal surfaces in variant 1

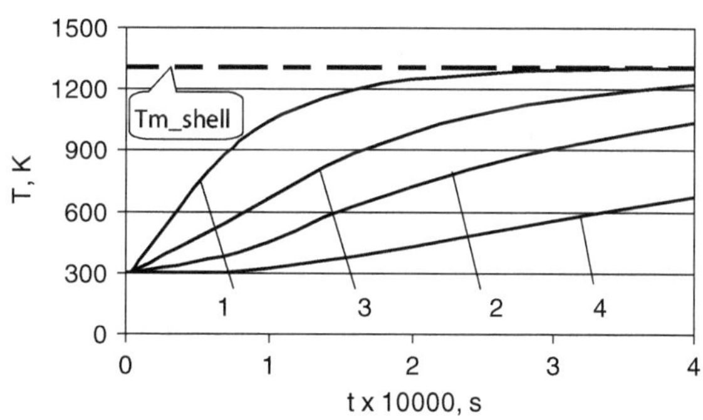

Fig. 2.10 Changing of the temperature in the center of the core for any combinations of thermal conductivity of the charge and the shell thickness

2.4 The Formation of Droplets of Liquid Metal

In the methods of powder thermal spraying, spherical particles flow gets into a stream of gas or plasma jet, where they are basically exposed to a drag force of the jet. This makes it easy to generate the spraying material jet. Other forces affected the droplet apart from aerodynamic force have been neglected in calculations. In addition, it is assumed that they have a spherical form [11, 16, 17, 35].

The processes of formation of droplets at AS significantly differ from those indicated. Aside from gas-dynamic pressure, droplets are strongly affected by the electromagnetic forces generated when electrical current is passing through the wire electrodes, the liquid metal layer on the ends of the electrodes and the electric arc. Studying of droplet formation features enables to specify the initial conditions for the droplet motion for the subsequent step, on spraying distance.

For AS, the evaluation of droplets maximum size d_{max}, broken away from the wire (diameter d_{np}) was done by Kudinov [15]. Condition that the surface tension and aerodynamic forces are equal was used.

$$d_{\max} = 1/\left|W_g - W_p\right| \cdot \sqrt{8 \cdot \sigma \cdot d_n/\left(C_d \cdot \rho_g\right)}, \qquad (2.28)$$

where $S = \pi \cdot d_{\max}^2/4$ is the cross-section area of the droplet, m^2; C_d is drag coefficient; ρ *is* density, kg/m^3; W is velocity, m/s; indices "p", "g" refer to the droplets and gas, respectively.

Calculation according to Eq. (2.28) using σ = 1.5–1.16 J/m^2 gives d_{max} = 600–900 μm.

The sizes of the discharged droplets can be obtained by welding arc oscillographic testing (Table 2.8). Based on the calculated and experimental data it can be stated that, in case of steel, the droplets that are breaking away from the end face of the electrodes corresponds to a midship section of a diameter of 550–900 μm.

A picture of the gas dynamic effect on the liquid surface is described by Levich [36]. Let us suppose that on the jet surface have occurred a small disturbance of any minimum size. The gas pressure above the roughness top will be reduced, whereas bottom pressure will be raised in comparison to the average. The reason for this is that the velocity of the gas passing over the roughness top is higher than at its bottom.

According to Bernoulli's equation, the pressure is lower where velocity is higher. Thus, the roughness formed on the liquid surface that is moving in relation to the gas tends to increase. This effect is stronger, if the relative velocity of movement is higher. The amplitude rise ultimately leads to the separation of the

Table 2.8 Characteristics of the droplets discharge for AS according to oscillographic testing, 2 mm wire

	Steel [20]	Steel [20]	Steel [34]	Al [20]	Al [20]	Zn [20]	Zn [20]
f, Hz	900	1750	1740	600	1750	500	850
$t \times 10^4$, s	11.1	5.71	5.74	16.7	5.71	20.0	11.8
$G \times 10^3$, kg/s	4.0	8.0	2.2	1.1	4.0	3.3	11.1
$m_{капли} \times 10^6$, kg	2.22	2.29	0.63	0.93	1.14	3.33	6.54
ρ, kg/m^3	7200	7200	7200	2500	2500	6500	6500
$d_{капли} \times 10^6$, m	838	846	551	891	956	993	1240

droplet from the surface. Separation occurs in violation of the balance between gas-dynamic head of the gas jet and the surface tension force.

With regard to the above-mentioned AS scheme of metal removal from the face assumes an elongated form of droplets in the form of a tongue. Deviation from the spherical shape of the droplet increases the possibility of its separation. Deformation processes and a subsequent separation were described by Levich for the case of the falling droplet [36]. The review was done considering that the dynamic head of the jet equals to the capillary pressure in the droplet. For the critical droplet radius, $r_{\kappa p}$, at which its disintegration occurs the equation is derived as:

$$r_{\kappa P} = \sigma \Big/ \left(\rho_g \cdot \left(W_p - W_g\right)^2\right) \cdot \sqrt[3]{6/C_d} \tag{2.29}$$

Further clarification of this equation was done taking into account dimensionless parameters like the Weber number (*We*), the Laplace (*Lp*) number, the Mach number (*M*), Bond (*Bo*) number and the, Reynolds (*Re*). Moreover, the degree of their influence on the destruction conditions is different [37]. Weber number is the most important among them.

$$\mathrm{We} = d \cdot \rho_g \cdot \left(\bar{W}_p - \bar{W}_g\right)^2 \Big/ \sigma \tag{2.30}$$

Knowing We, for taken hydrodynamic conditions, the estimation of critical droplet diameter can be made.

In work [38], based on a studied spark photo of it was concluded that the separation of the droplet occurs at a critical value of the Weber number $\mathrm{We}_{\kappa p} = 5{,}406$. During the separation, the droplet shape is closer to the disk, with the ratio of the semi axes 3.75.

Thinning liquid narrow disk is unstable and separated under the influence of small perturbations by disruption of wave crests of increasing amplitude. The midship section area of the deformed drop cross-section was estimated as the average between the sphere area $f_{ш}$ and ellipsoid area in a critical phase, then $f_{м} \approx f_{ш}$. In the critical stage, the droplet is closer to the ellipsoid with the semi axes ratio of 4–6.

During the experiments with water–glycerin droplets of diameter 3.5–5.3 mm in an air stream, the increased separation was studied [39]. To evaluate the nature of load change on the droplet of the aerodynamic forces, the *G* similarity number was introduced:

$$G = d\mathrm{We}/(d(\tau/\tau_\kappa)) = 1.66 \cdot \mathrm{We}^{0.5} \cdot \bar{B}o \tag{2.31}$$

where, τ is current time, s; $\tau_\kappa = 0.83 \cdot \left(\rho_p \cdot d_p^3 \big/ \sigma\right)^{0.5}$ is the period of natural droplet fluctuation, s; $\bar{B}o = \sqrt{\rho_g \cdot \rho_p} \cdot d_p^2 \cdot a/\sigma$ is modified Bond number; $a = \mathrm{d}u/\mathrm{d}\tau$ is droplet acceleration, m/s^2.

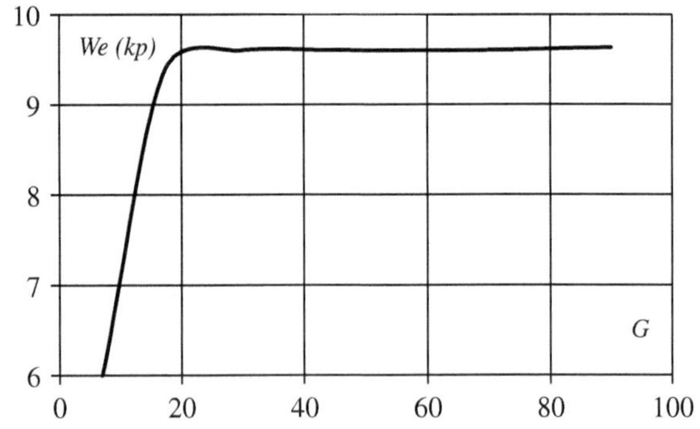

Fig. 2.11 Change of the critical Weber number ($We_{кр}$) depending on the parameter G

The result of the experimental data processing is presented in Fig. 2.11 and corresponds to the Laplace number

$$Lp = \sigma \cdot \rho_{ж} \cdot d / \eta^2 = 23200,$$

where $\rho_{ж}$ is liquid density, kg/m^3; d is diameter of droplets, m; η is coefficient of dynamic viscosity, kg/(m s).

In the paper, it was concluded that the low rate of rise of aerodynamic forces which means more time of exposure, ($G < 15$), the value of $We_{кр}$ can be nearly twice as different from the corresponding values for $G > 15$.

In the latter case, $We_{кр} \approx 10$ and the critical conditions for the separation are of self-similar with respect to G. At the point of reaching point of that value, the droplet is inevitably separated. It corresponds to the dumbbell shape of the droplets.

Research data [39] were used by Petrov et al. to evaluate the changes in the particle size of aluminum, tungsten, copper, nickel, and aluminum oxide at plasma spraying [40]. The main conclusion to be drawn from the above analysis is that at jet processes associated with the powder processing, the conditions of particles separation of different materials with a diameter of >50 μm are realized at a velocity of 1500 m/s and temperatures between 3000–6000 K of the plasma jet. Unfortunately, the paper does not present any data on the surface tension of the materials that were used in the calculations.

In the case of AS, to evaluate the critical size of particles of various materials $We_{кр} = 12$ was taken [41]. According to the calculations, the critical particle diameter in microns is: Al–80, Zn–72, Fe–156, Ni–180, Cu–118. These calculations truthfully reflect the quality correlation of critical droplet diameters of different materials. Thus, experience shows that metal coatings of Al and Zn are denser and smoother than those from Fe under the same coating conditions. However, the reasoning for taking values in the calculation of the surface tension $\sigma = 0.6$ J/m^2 is not provided.

Data that are more reliable are given in the works on the theory of metallurgical processes. According to Popel [42], when the temperature is slightly above the melting point, the changing of surface tension is described by the equation:

Table 2.9 Surface tension of liquid iron

T_{Me}, K	1809		2070		2500	2900	3148[a]
Source	[43]	[42]	[43]	[42]	Equation (2.33)		
σ, J/m^2	1.85	1.85	1.65	1.76	1.50	1.29	1.16

[a]$T_{кип}$ = 3148 K, according to [26]

$$\sigma(T) = \sigma_{nn} + (\mathrm{d}\sigma/\mathrm{d}T) \cdot (T - T_{nn}), \tag{2.32}$$

where $\sigma_{пл}$ is the surface tension at the melting point of the metal, J/m^2; $\mathrm{d}\sigma/\mathrm{d}T = -0.35 \times 10^{-3}\mathrm{J}/(\mathrm{m}^2\,\mathrm{K})$.

According to the generalization of the various studies experimental data made by Lynchevskiy, the surface tension of iron varies in the range of σ = 1.85–1.65 J/m^2 [43], with the temperature increase from 1809 to 2070 K, i.e., for the same area.

Approximate temperature dependence of the surface tension at higher temperatures between the melting and boiling points is expressed by equation [20]:

$$\sigma(T) = \sigma_{nn} - B(\rho_p/\mu)^{2/3} \cdot (T - T_n), \tag{2.33}$$

where ρ_p, μ is density, kg/m^3 and a molecular weight, kg/mol, of the liquid metal; T, $T_{пл}$ are current and melting temperatures, respectively, K;

$$B = 2.10^{-7}\,\mathrm{kg\,m^2}/(\mathrm{s^2\,K}).$$

The surface tension values for the intended temperature intervals for Fe droplets are summarized in Table 2.9 according to the cited studies. They are >2 times higher than those used in studies [41].

Dealing with arc processes, we can only speak about the approximate values of the metal surface tension. The temperature at the end faces of the electrodes is not the same due to the active arc spots wandering and various thermal contributions of cathode and anode. This leads to inhomogeneity of the surface tension at the end faces of the electrode. In addition, the presence of alloying elements and impurities in the metal influences the value of surface tension. In the case of liquid steel, for example, oxygen, nitrogen, sulfur, phosphorus, reduce the surface tension by 20–30 % [26].

We carried out the calculation of the critical diameter of the iron particle along the spraying distance for AS, Table 2.10. This analysis leads to the following conclusion: metal droplets of the above given diameter interval in the range of 550–900 μm (Table 2.8), can be separated near the nozzle cut. Herewith, the droplets with the diameter <165 μm are resistant to deformation and fracture. On the further movement along the distance, the droplets resistance increases. In the calculations for AS, it can be certainly assumed that the droplets have a sphere shape when the diameter is <300 μm.

However, the studied mechanism of droplets fragmentation under the gas-dynamic forces influence does not explain the appearance of droplets with a

Table 2.10 Calculating results of the critical diameter of droplets along the spraying distance (S) using Eq. (2.30)

S, mm	*σ*, J/m^2	*We*	*Wg*, m/s	*Wp*, m/s	ρ_g, kg/m^3	$d_{\kappa p}$, μm
0	1.29	12	900	7	0.118	165
5	1.29	12	582	43	0.137	332
10	1.5	12	426	91	0.131	1002
100	1.85	12	119	177	0.415	11011

The values of the surface tension are taken from Table 2.9. The data on the hydrodynamic parameters of two-phase flow for AS are according to [13]

diameter smaller than the critical, $d_{\kappa p}$ = 165 μm. Measurements of particle diameters during wire metallization were made by Vadivasov [44], Kretzschmar [34], Pohmurskiy et al. [45], Borisova et al. [46], Vahalin et al. [20], as well as by the authors by trapping the droplets from flying jet with different methods for the subsequent grain-size analysis. Experimental data suggest that the diameters of droplets during the metallization process are in the range of 1–600 μm.

Moreover, the share of droplets with diameter less than the critical value (for example, for iron) is 20–80 %, depending on the experimental conditions. Although, according to calculations, the surface tension forces should be enough to deter the droplets of this diameter interval from separating by the gas jet. The hypothesis about the secondary separation of droplets under the influence of the gas-dynamic forces [20] is not convincing enough, because it does not justify the possibility of reducing the surface tension for iron to $\sigma = 0.6$ J/m^2.

To resolve the contradictions in the results of the droplet sizes evaluation according to oscilloscoping and grain size analysis, the authors proposed another mechanism of droplets formation in the arc burning zone [47].

It is accepted that in the arc, the big diameter droplet separation from the electrode end face due to the gas-dynamic pressure of the gas occurs simultaneously with its fracture into smaller droplets under the action of electromagnetic forces.

In the works, where the melting and spraying process of the electrode material for the arc spraying is studied, the effect of electromagnetic forces on the droplet formation is not considered. Although, their presence is indicated. In addition, it is assumed that the gas-dynamic forces have a decisive effect on the character of the droplets formation [20, 48]. To assess the correctness of this conclusion, we performed a quantitative estimation of the main forces acting on the liquid metal in the combustion zone of the arc [98].

Under the influence of the gas-dynamic pressure, the metal is pushed away from the electrode end face and extends in a plane jet "tongue". It is being held on the end face due to the surface tension forces. The balance of these forces can be expressed by the equation:

$$0.5 \cdot C_d \cdot S \cdot \rho_g \cdot (W_g - W_p)^2 = \pi \cdot d_n \cdot \sigma, \qquad (2.34)$$

where C_d is drag coefficient, S is the area of the midship section, m^2; W_g, W_p are gas and particle velocities in m/s; d_n is the diameter of the separation necks, m.

According to the previously described scheme of Levich, the fracture of the liquid metal jet occurs in the necks formed in the tongue under the gas pressure fluctuations at the local perturbation spots [36]. The presence of passing current flow running through the tongue and electric arc add to the constricting mechanism with more powerful destroying mechanism of neck's breakage. It was described by Sheyhaliev applying to steel casting in arc furnace [49].

Joule heating release at the constrictions is more intensive, when compare with the other parts of the conductor due to the increased ohmic resistance. Further breakage of the jet progresses due to the joint processes of liquid metal boiling and shock expansion of the gas at the necks.

Furthermore, in the arc burning zone the electromagnetic forces arise from the interaction of running current and the induced induction flows. Interaction "liquid metal-electric arc" is considered in the number of research works with respect to arc welding [2–4, 50]. However, they do not allow an adequate description of what is happening in the case of arc spraying for several reasons.

(1) The difference of 1–2 orders in velocities and gas flow consumption leads to a decrease of 2–3 orders of the size and time of droplets fracture from the end faces of the electrodes. If for the arc welding the fracture period $= (10^{-1}–10^{-2})$ with a diameter of 1–3 mm [2], then for AS $= (10^{-3}–10^{-4})$ s (see Table 2.8) with a diameter of droplets 0.01–0.3 mm (see Table 2.10).
(2) The direction and a value of an impact of the gas stream on the electric arc are changing. When Arc Welding, the gas flow is directed along the axis of the electrode and the gas consumption rate does not exceed 1.2 m^3/h [23]. When the Arc Spraying direction of gas flow forms an angle (15–30)° to the electrode axis, and the gas flow rate is higher on average by 100 times. While welding, it is preferable to have a laminar flow of the gas stream. Transition to turbulent flow reduces the stability of the arc combustion. During AS, gas flow with a high kinetic energy of high turbulence is forcefully directed on the liquid metal in the arc burning zone.
(3) For AS, the cathode and the anode acts like electrodes simultaneously. The metal droplets do not reach the other electrode after breaking from the one. Whereas, the angled co-relative position of the electrodes causes specific interaction of the magnetic and electric fields of current flow.

These differences change the parameters of the processes at the end faces of the electrodes in comparison with the Arc Welding.

Let us consider the electromagnetic forces in the burning zone of the arc for the Arc Welding. The scheme of arc movement in the gap between the electrodes while the latter are being blown by the cross-flow (Fig. 2.12) was offered by Bron and Sushkov. It was confirmed by the high-speed shooting data. According to this scheme, arc column retains a cylindrical shape under the influence of the cross-flow of gas. Whereas the plasma flows are tied to the electrode end faces and are demolished under the joint action of the Ampere force and gas-dynamic pressure.

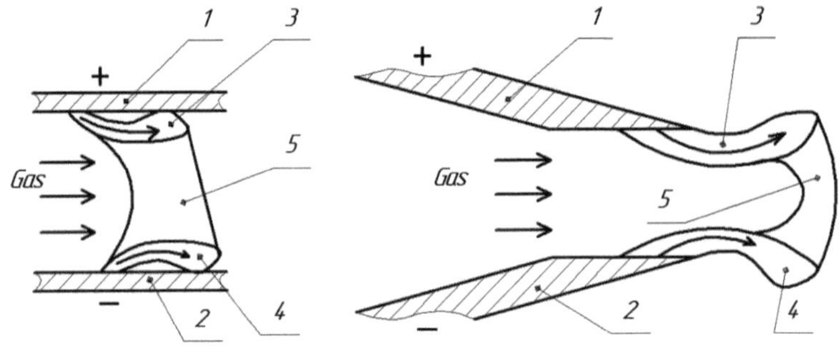

Fig. 2.12 The arc shape at the combined air and magnetic blast [51]. **a** In the gap between the electrodes: **b** In the extreme position at the end face of the electrodes (*1*, *2*), plasma flows (*3*, *4*), arc column (*5*)

For calculations, it is convenient to assume that the arc column has cylindrical shape and its axis is perpendicular to the end faces of the electrodes. However, in this case, the correctness of this assumption should be evaluated further due to the high velocity of the gas stream flowing around the end faces of the electrodes. Let us calculate the velocity of the arc and gas-plasma flows which is flowing around it.

Due to the pinch effect, i.e., the interaction proceeds in a current column with its own magnetic field, the compression of the arc column by plasma surrounding this column has occurred. Thus, there is a longitudinal pressure gradient, which in turn generates a flow of plasma from the spots of decreased cross-section. According to the decision of the Navier–Stokes equations for this case, the maximum velocity of the plasma in the flow can be determined from the equation [51]:

$$v_{\max} = \sqrt{\mu \cdot I \cdot j / \left(2 \cdot \pi \cdot \rho_g\right)}, \tag{2.35}$$

where $\mu = 4 \times 10^{-7}$ H/m is magnetic permeability of the medium corresponds to a vacuum; I is current, A; j is current density, A/m^2; $\rho_g = 0.129$ kg/m^3 is the density of the gas at $T = 2500$ K [30].

If we consider radius of the arc column to be equal to the radius of the electrode $r = 0.001$ m, then for a typical mode, $I = 200$ A, this corresponds to $j = 6.4 \times 10^7$ A/m^2. According to Eq. (2.35) for these parameters, the flow velocity of the plasma $v_{\max} = 140$ m/s. These calculation results are close to the experimental data obtained by Lenivkin and Dyatlov acquired with a help of drop movement record [3, 50].

The resulting value is in the interval between the flow velocity of the plasma in the arc column for the arc welding $v = 78$ m/c, where $j = 2.0 \times 10^7$ A/m^2 [2], and the flow rate of the plasma in the plasmatron $v = 393$ m/c, where $j = 5.0 \times 10^8$ A/m^2 [12]. From the structure of Eq. (2.35), we can see that an increase in the rate of the gas flow causes the compression of the plasma flows around the arc column, which leads to the increased stability of the arc in the inter electrode gap.

The velocity of the arc movement in the gap between the electrodes can be assessed by the frequency of droplets detachment. This follows from the physical representation about the mechanism of arcing between the end faces of the electrodes. First, the arc is bypassed in the most energetically favorable location—on the inner edges of the ends. Thereafter, the arc moves toward the front edges of the end faces. Here, the arc column bends and its length increases. While the potential difference across the cathode and anode is constant, it leads to the fracture of the arc. Then, the process begins again with the bypassing of the arc on the inside edges. When the diameter of the electrode is 2 mm, the angle of inclination toward the central axis is 15°. The breakdown period is 5.7×10^{-4s} and the velocity of the arc column is:

$$v_{nc} = 2.10^{-3}/\left(\sin 15^{\circ} \cdot 5.7 \times 10^{-4}\right) \approx 12\,\text{m/s} \tag{2.36}$$

The gas velocity in the gap between the electrodes is estimated from the liquid layer being teased from the end face of the electrode by the gas stream. From the Newton's equations for the case of thin liquid films flowing that are entrained in the gas stream, Levich derived the dependency of the viscous layer thickness (δ) from the gas velocity and the liquid metal properties [36]:

$$\delta = \left(\sigma \cdot v/\rho_p \cdot v_0^3\right)^{0.5}, \tag{2.37}$$

where v is the kinematic viscosity, m^2/s; v_0 is the gas velocity in the gap above the boundary layer is m/s. We take parameter values as for the liquid iron under the temperature of 2500 K: $\rho_p = 7265$ kg/m^3 [26], $\sigma = 1.5$ J/m^2 [20], $v = 7 \times 10^{-7}$ m^2/s [52].

Let us express the gas velocity (v_0) from the Eq. (2.37):

$$v_0 = \left(\sigma \cdot v/\rho_p \cdot \delta^2\right)^{1/3} \tag{2.38}$$

Thickness of the layer (δ) will be estimated based on the condition that the volume of separating layer and droplet according to oscilloscoping of droplets breakdown for Arc Welding. According to Table 2.8, the average value of breakdown period is 5.74×10^{-4} s. Then, at wire feed velocity 0.05 m/s the droplet mass is $m = 6.4 \times 10^{-7}$ kg If we assume that the thickness of the separated layer over the end face is uniform, then its value will be:

$$\delta = m/\left(\rho_p \cdot S_{пов}\right) = 6{,}4 \cdot 10^{-6} M, \tag{2.39}$$

where: $S_{пов} = 1.3 \times 10^{-5}$ m^2 is the surface area of the electrode end face, according to the scheme adopted for AS process (Fig. 1.3), where the gradient angle of the electrode to the axis of the gas jet is 15° and diameter is 2 mm.

The thickness of the layer at the end face of the electrode fusing by the heat of the near-electrode arc spot is $s = 7.6 \times 10^{-6}$ m. The ratio between this value and the thickness of the separated liquid layer calculated according to Eq. (2.39), $\delta = 6.4 \times 10^{-6}$ m is qualitatively corresponds with the physical picture of the process, when the liquid metal layer is carried away by the gas flow.

Table 2.11 The velocities in the gap

Velocities	Calculated equation	Value, m/s
Plasma flow around the arc column	(2.35)	140
Arc column transition along the end faces	(2.36)	12
The carrier gas flow	(2.38)	1.5

The calculating results of the velocities of the arc column movement, gas, and plasma flows in the gap between the end faces of the electrodes are shown in Table 2.11.

It is evident that the velocity of carrier gas flow is smaller comparing to the velocity of the plasma that compressing the arc column and to the velocity of column moving within the gap. Therefore, we can conclude that the arc column will likely have a cylindrical shape with a vertical axis, and the electromagnetic forces are the main reason for its motion.

We are to estimate the magnitude and direction of forces acting on a single droplet when it is separated from the electrode end face. Assuming that the mutual vapor pressure of the metal at both end faces is leveled, then the affecting forces are the following:

- aerodynamic drag;
- impact wave pressure of the at the explosive expansion of the metal vapor and the gas at the neck;
- Amper: F_1 (the interaction of the magnetic field of electrode extension and the current in the neck), F_2 (pinch–effect), F_3 (the interaction of the magnetic field of the arc column and the current in the neck).

The direction of the currents flows, the location of magnetic fields and Ampere forces are shown in Fig. 2.13. Let us perform the calculation of these forces.

The aerodynamic drag force can be calculated using Eq. (2.34):

$$F_{ac} = \pi C_d (v_g - v_p)^2 y_g d_k^2 / 8 \tag{2.40}$$

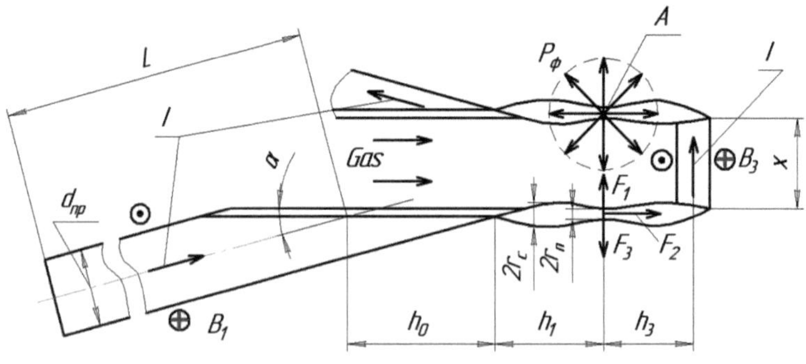

Fig. 2.13 Electromagnetic forces in the arc combustion zone

We will consider the following:

$C_d = 0.4$, as the self-similarity mode [53];
$(v_g - v_p) = 800$ m/s;
$\gamma_г = 0.168$ kg/m^3 (gas temperature is 2100 K);
d_κ − 10 μm, corresponds to an average droplet size of the smallest fraction interval according to the sieve analysis data obtained by authors.

The force value $F_ф$ from the impact wave on the surface of the neck of radius r_n and length Δh will be:

$$F_\phi = P_\phi \cdot 2\pi \cdot r_n \cdot \Delta h, \tag{2.41}$$

where $P_ф$ is the pressure at the front of the impact wave, MPa.

According to the mechanism of the "tongue" breakage which was suggested by us, the current running through the neck drastically enhances the heat release due to the increased ohm resistance. This all initiates the "tongue" breakage in the neck zones. Moreover, it stimulates the appearance of micro arcs at the breakage zones that leads to the explosive expansion of the vapors metal and gas. During the gas expansion, there is a sharp increase in the volume of the heated gas and spreading of the impact wave, followed by the energy release. According to [54], the pressure on the front of the impact wave can be expressed by the dependency:

$$P_ф = 2 \cdot \gamma_г^0 \cdot v_ф^2 / (\kappa + 1), \tag{2.42}$$

where $\gamma_Г^0$ is the initial gas density, kg/m^3; k is the adiabatic constant; $v_ф$ is the front velocity of the expanding gas, m/s.

$v_ф$ can be estimated from the equation of gas-dynamic pressure forces and capillary pressure in the bridge, which prevents its destruction [49]:

$$v_ф = \sqrt{(16 \cdot C_d \cdot \sigma)/(d_{пер} \cdot \gamma_г)} = 2857 \text{m/s} \tag{2.43}$$

Neck diameter d_{nep} is assumed to be the thickness of the separated layer (δ) to be calculated by the Eq. (2.39).

Aside from the gas-dynamic forces, the neck is exposed to electromagnetic forces caused by the current flow of in the tongue and in the electrodes. There are three main directions of current flow regarding to the axis of the gas flow: the section of extension of length (L) at an angle (α); along the layer of the liquid metal; along the arc column of length (x), Fig. 2.14. Magnetic induction B_1 of the electrode departure at the neck center (A) is calculated according to the equation of the magnetic field of linear conductor [55]:

$$B_1 = \mu \cdot I \cdot (\cos\varphi_2 - \cos\varphi_1)/(4 \cdot \pi_1), \tag{2.44}$$

where according to Fig. 2.14: r_1 is the shortest distance from point A (the neck center) to the electrode; φ_1, φ_2 are the angles between the vector of the current

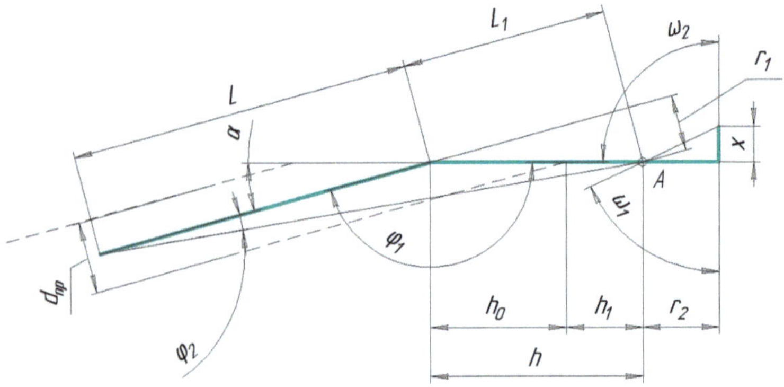

Fig. 2.14 The calculating scheme of electromagnetic forces in the arc burning zone. Equation (2.44) takes the form

density in the conductor and radus—vectors which are drawn to the point A from the beginning and the end of the conductor; μ—is the magnetic constant. Angles φ_1, φ_2 can be determined by the inclination angle of the electrode (α):

$$\cos\varphi_1 = \cos(180 - \alpha) = -\cos\alpha,\; r_1 = h \cdot \sin\alpha, \tag{2.45}$$

$$\cos\varphi_2 = (L + h \cdot \cos\alpha) \Big/ \sqrt{(h \cdot \sin\alpha)^2 + (L + h \cdot \cos\alpha)^2}, \tag{2.46}$$

where $h = d_3/(2 \cdot \sin\alpha) + h_1$,

$$B_1 = \frac{\mu \cdot I}{4 \cdot \pi \cdot h \cdot \sin\alpha} \cdot \left(\frac{L + h \cdot \cos\alpha}{\sqrt{(h \cdot \sin\alpha)^2 + (L + h \cdot \cos\alpha)^2}} + \cos\alpha \right) \tag{2.47}$$

The force F_1 that occurs from on the interaction of the magnetic field B_1 and the current flowing through the bridge will influence the neck of the length Δh:

$$F_1 = I \cdot B_1 \cdot \Delta h, \tag{2.48}$$

We are to take the length of the neck $\Delta h = 1 \times 10^{-6}$ m.

Due to the local narrowing of the conductor at the neck, the force F_2 is acting due to the pinch effect which caused by the interaction of current flow and the magnetic field (induction B_2):

$$F_2 = \mu \cdot I^2 \cdot \ln(r_c/r_n)/(4 \cdot \pi), \tag{2.49}$$

We are to consider: $2r_s = 7 \times 10^{-6}$ m, $2r_n = 6.5 \times 10^{-6}$ m are transverse dimensions of the “tongue” and the neck (Fig. 2.12).

Table 2.12 The estimated forces acting in the arc burning zone on the liquid metal

Forces	Equation	Estimated value, $\times 10^5$ N
Aerodynamic drag	(2.40)	16.9
The impact wave pressure	(2.41)	2.7
Ampere F1 (electrode extension magnetic field-the current in the neck)	(2.48)	5.2
Ampere F2 (pinch-effect)	(2.49)	9.4
Ampere F3 (the magnetic field of the arc column-current in the neck)	(2.51)	0.03

The numerical value of the magnetic induction B_3 at point (A) is similar to Eq. (2.44) and is going to be:

$$B_3 = \frac{\mu \cdot I}{4 \cdot \pi \cdot r_2} \cdot (\cos \omega_2 - \cos \omega_1) = \frac{\mu \cdot I}{4 \cdot \pi \cdot h_2} \cdot \left(1 - \frac{x}{h_2}\right), \tag{2.50}$$

where ω_1, ω_2 are the angles between the vector of the current density in the arc column and radius—vectors that are drawn to the point A from the beginning and the end of the column; $\omega_2 = 90°$ due to the assumption that the position with respect to the axis of the arc column of the gas stream is vertical.

The interaction of the magnetic field of the arc column with the current in the neck gives the force F_3:

$$F_3 = I \cdot B_3 \cdot \Delta h \tag{2.51}$$

The results of calculation of the forces separating droplets from the end faces of the electrodes are shown in Table 2.12.

These results are evaluative due to the lack of experimental data. Moreover, the calculation shows that the force of an electromagnetic nature, has the same magnitude with aerodynamic force. However, the direction of their actions differs from the desired motion for droplets. Thus, our assumption about the possibility of the droplets fracture in the burning zone of the arc under the influence of not only aerodynamic but also electro-dynamic forces is rather significantly accurate.

The accuracy of mechanism proposed by us of processes at the end face of the electrode at AS was confirmed later in experimental data obtained by Hussary and Heberlein [48] dealing with the high-speed shooting of the droplets formation. BP400 (Praxair) spraying gun was used during this experiment. Nitrogen was used as a carrying gas, the carbon wire Ø1.6 mm (0.37–0.44 C, wt%) was a feedstock, U = 30 V, I = 100 A.

On the anode (+) active spot heats a large portion of the electrode end face surface. This causes the formation of molten metal on the surface (Fig. 2.15). Carrier gas flow affects the layer, creating a wave which propels a metal toward the edge of the electrode end face, creating the anode metal strip. Anode spot tends to

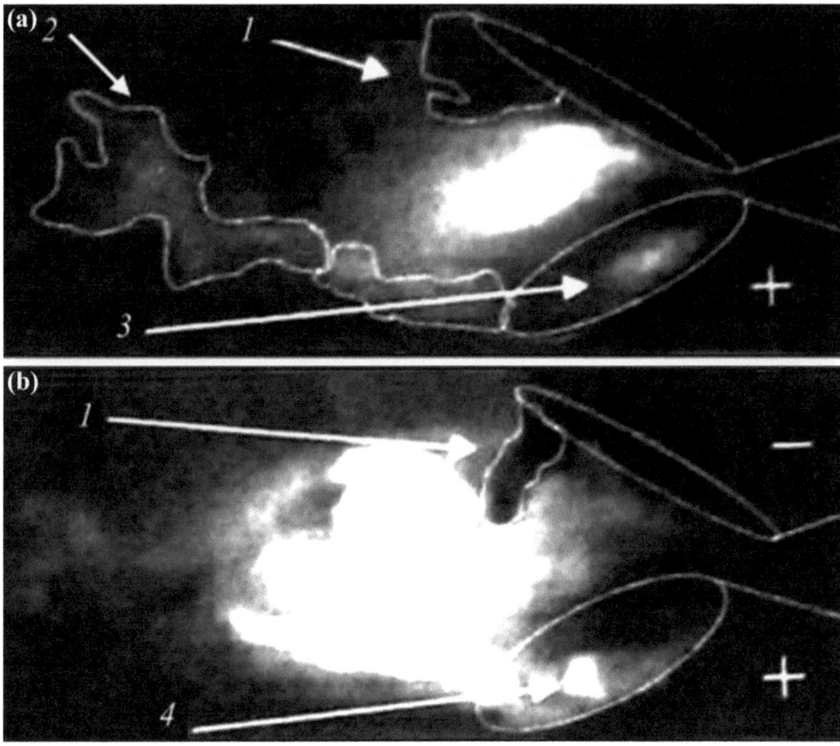

Fig. 2.15 Formation of the liquid metal droplets, f = 13,500 frames/s, P = 138 kPa. **a** t = 0; **b** t = 0.28 ms. *1* Tabs on the edge of the cathode spot and the cathode tongue; *2* anode tongue and its fracture; *3* wave on the anode surface; *4* wave motion on the anode surface

move in the direction with the gas flow, settling on the surface of the anode tongue (sheet according to Hussary and Heberlein). When the anode tongue breaks away, the arc is cut off and reignited in the shortest distance between the electrodes.

The cyclical behavior of the arc corresponds to the similar voltage fluctuations. Also, the arc fracture without abandoning the anode tongue was observed, but only under the influence of aerodynamic forces.

There are obviously the three mechanisms (Fig. 2.16) of metal separation occurred at the cathode (−).

(1) Ejection of the molten metal in the droplets form into the cathode stream. Due to the current change flowing through the cross-section of the arc (the arc compression at the cathode due to the pinch effect), the arc interaction with its own magnetic field leads to the ejecting step and separation of the molten metal in the arc column.
(2) The formation of small projections on the perimeter of the cathode end face and their separation before the formation of the tongue. Often, these projections are part of an extended ring (or jacket) of the molten metal on the surface

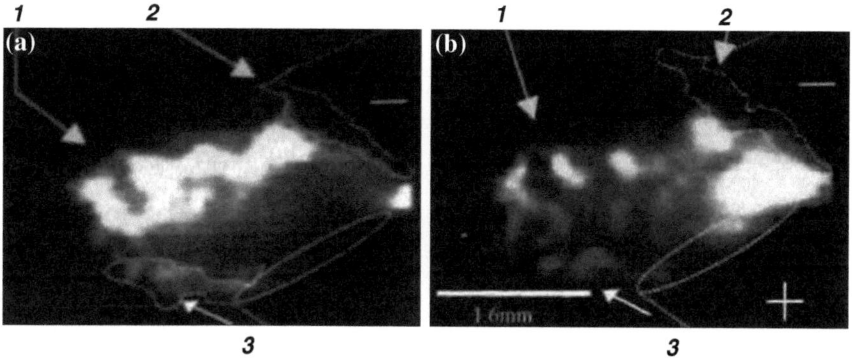

Fig. 2.16 Droplets separation from the cathode into the arc column, the two consecutive frames. f = 13,500 frames/s, P = 138 kPa. *1* Separated droplets; *2* cathode projections; *3* anode tongue

of the wire. The displacement of the binding cathode spot over its surface has occurred. The pressure gradient at the end face surface of the wire force the molten metal to shift toward end face perimeter creating the metal projections. The gradient occurs due to the impact of ion flux on the cathode spot.

(3) Small projections are grouped to form a cathode tongue on the front edge of the wire. It is similar to the anode, but usually is smaller in size.

The influence of atomizing gas pressure on the nature of the metal melting on the anode and the cathode was investigated in the range of P = 138–414 kPa, f = 27,000 frames/s. With increasing pressure all the effects were clearly observed (Fig. 2.17):

(1) According to the oscillation, the frequency of the voltage has increased. As the consequence, the velocity of the arc transferring has also increased. It was witnessed from the high-speed shooting data. There has been a reduction of time of the arc being in the stable and the rise in the number of resignations.

(2) The nature of the anode and cathode melting has changed. A bulbous metal formation has occurred on the anode near the end face of the electrode. They are increasing in accordance with a continuous displacement of the molten metal from the liquid layer at the end face of the electrode by the gas. Upon reaching a critical mass of bubbles, the molten metal is drawn into a long tongue and is separated (Fig. 2.17b, right).

However, at a certain point, when the forces of surface tension in the tongue exceed the aerodynamic resistance, the tongue pulled toward the end face of the electrode and the formation of bubbles begins again. This effect is most clearly observed at the intermediate pressure (345 kPa).

When the pressure increases to 414 kPa, the anode bubble is reduced in size, and the contraction of the anode tongue to the end face of the electrode eliminates (Fig. 2.17). At the cathode, the extrusions are increased in size by the end face perimeter forming a cathode tongue which is not formed at a low pressure (Fig. 2.17a).

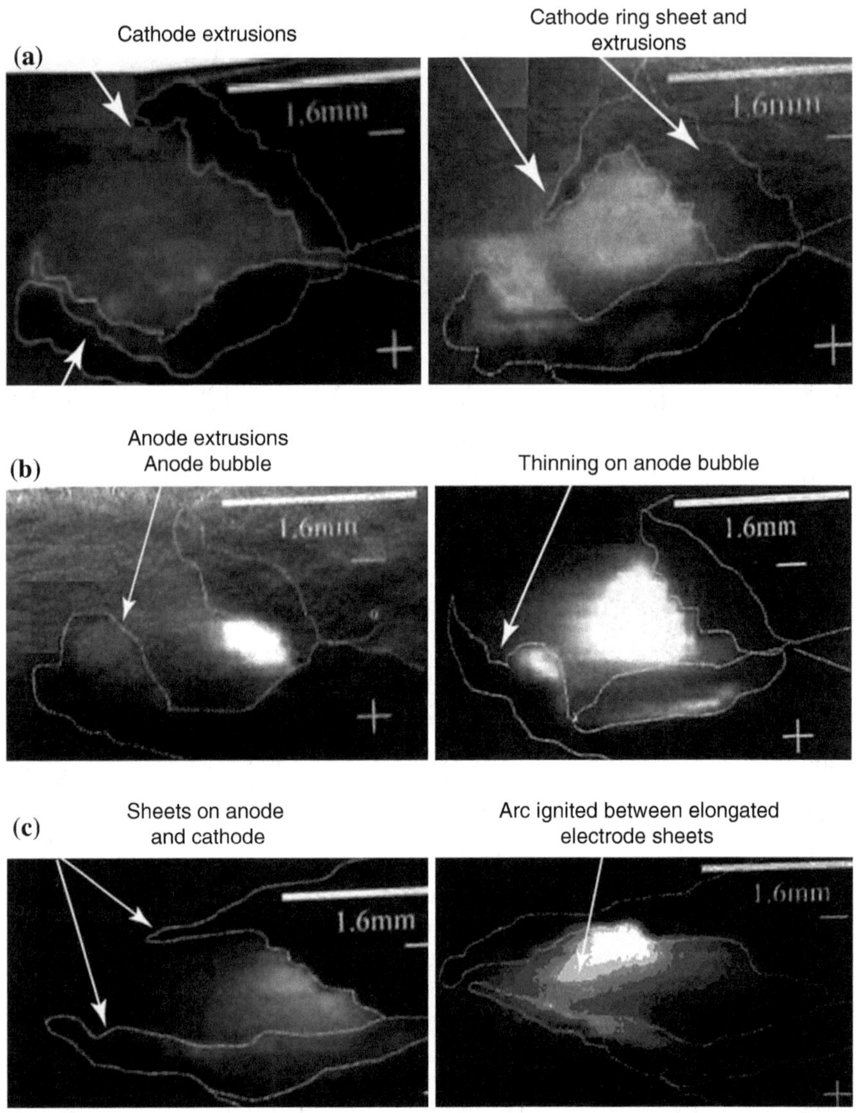

Fig. 2.17 The nature of the metal melting at AS **a** $P = 207$ kPa; **b** $P = 345$ kP **a** $P = 414$ kPa

It is noted that the shape and size of metal droplets that are separated from the anode and cathode can be the same, despite the differences in the mechanisms of their formation. This is due to the influence of high pressure of gas. Probably, an additional effect can be caused by heat transferring from the anode to the cathode, where the heat input is higher (Fig. 2.18).

Large whirling was marked on the anode and cathode tongues (Fig. 2.19). Although these pictures are not consistent, they demonstrate the process of tongue

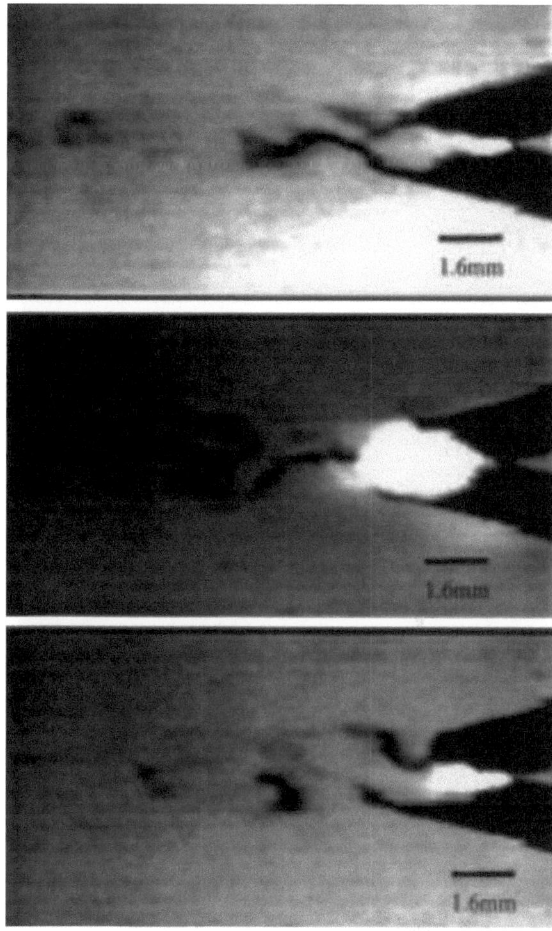

Fig. 2.18 The structure of the spraying metal. $f = 27{,}000$ frames/s, $P = 276$ kPa

formation. As soon as the tongues begin to exceed a certain critical length, the influence of aerodynamic forces gets stronger. This is reflected in their curves. These curves are increasing along the flow and changing its structure into a whirling form as on anode and on cathode.

At the critical moment, when the surface tension forces cannot keep the tongue of the curved shape, the formation of the droplets shower begins. Thus the breakage the tongues, the droplets may fly in directions other than axial (shown by arrows in Fig. 2.19).

From the calculated and the experimental data, it was derived the following:

(1) During the formation process, the metal droplets are subjected to the influence of the significant forces in directions other than axial;

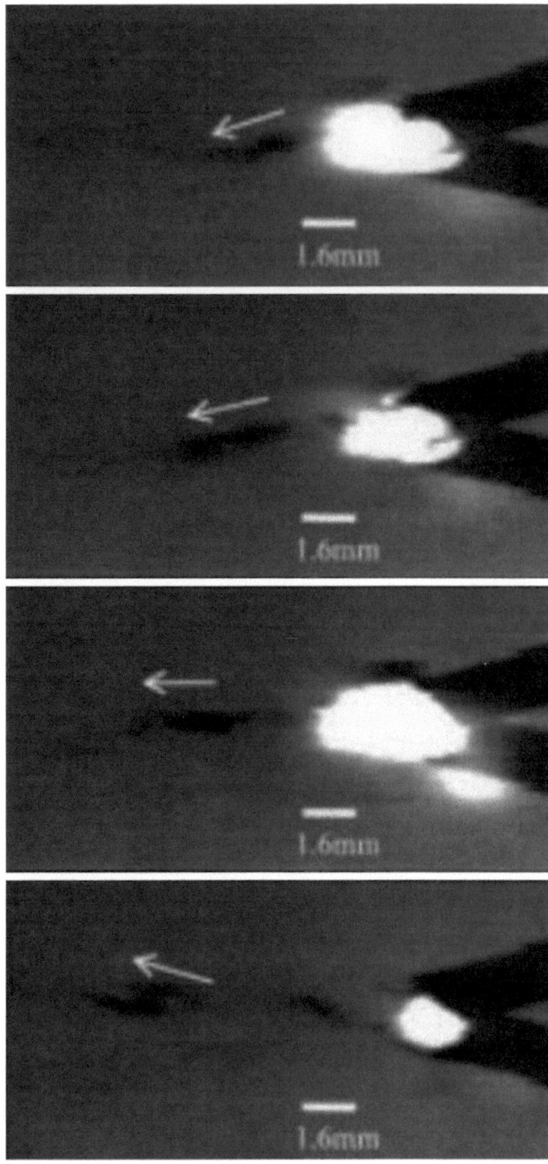

Fig. 2.19 Droplets formation cycle. $f = 27{,}000$ frames/s, $P = 276$ kPa

(2) The value and direction of the forces in the burning zone of the arc is sufficient to explain the possibility of small-diameter droplets;
(3) The metal droplets have different surface area of in the process of forming;
(4) Due to the additional heat release in the necks, the droplets temperature at this stage increases comparing to the end face temperature.

These features cause difficulties in obtaining high-quality coatings at AS:

(1) Combination of a large specific surface area of the droplets with their high temperature stimulates the undesirable input of the gases into the liquid metal.
(2) To design the nozzle subassembly we need to take special measures to prevent the deflection of droplet's way from the jet axis.

2.5 Description of the Geometry and the Velocity of Two-Phase Flow

In general, the motion of the particles in the two-phase flow can be considered as the motion of a viscous incompressible liquid, which is described by the Navier—Stokes equation:

$$\mathrm{d}\overline{W}/\mathrm{d}t = -\mathrm{gradp}/\rho + \mu \cdot \nabla^2 \cdot \overline{W}/\rho + \overline{f}, \quad (2.52)$$

where ρ is density; $\overline{W}$ is the velocity vector of the particles; p is pressure; μ is dynamic viscosity of the fluid; $\nabla^2 = \left(\partial/\partial x^2 + \partial/\partial y^2 + \partial/\partial z^2\right)$—Laplace operator; $\overline{f}$ is the resulting volume forces affecting an elementary volume of fluid.

With regard to the coating process, there are forces that caused an impact on moving particles [35]:

- aerodynamic resistance force;
- Basset force, owing to non-stationary processes and depending on the nature of the particle motion in the previous time span;
- Magnus force, occurred due to the rotation of the particles by gradient of velocity of an ambient around flow;
- the force caused by the pressure gradient in the flow;
- thermophoresis force caused by the temperature gradient in the flow;
- the gravity of the particle;
- the force, reflecting the acceleration of the gas layers adjusted to the particle;
- the force caused by the inertia of the gas volume displaced by the particle.

Besides the above mentioned, the motion of the particles must be also influenced by the electrostatic forces, the forces of light pressure, the forces caused by pulsations, and the diffusion of the gas components.

It was estimated that for the hydrogen plasma jet of with the temperature of 3000 K and the particle size of 50 μm, the density of 10^4 kg/m ^{3}, and the gas velocity of 1000 m/s, the aerodynamic resistance force value is approximately 5×10^{-2} N; Basset and Magnus forces are 0.3×10^{-2} H and 0.2×10^{-2} N, respectively, other forces are at $10^{-6} - 10^{-8}$ N [35]. At AS, the two-phase flow parameters are at the same level, which makes reasonable to calculate the aerodynamic resistance force only, as is done in modern studies of plasma spraying [16].

Other forces are noticeable only in the case of an external conditions change, such as, for instance, the impact against the substrate. The force caused by the pressure gradient when moving in a viscous medium is Magnus force, and so on.

A closed scheme was chosen for calculations (see Fig. 1.4a). Among nozzle designs used for AS, it is the most general and simple for calculations. It has a submerged gas jet, symmetric spraying axis. Other schemes of interaction between the gas and the particles can be reduced to the scheme under consideration. For the carrier gas composition, the variant in which the combustion products of propane–air mixture are supplied into the spray zone was used. All of the main components of the gas mixture are represented: nitrogen, hydrogen, oxygen, carbon, and their compositions. It can be reduced to other nozzle schemes by simplifying.

According to the AS scheme (see Fig. 1.3), the gas jet can be represented as a axisymmetric, flowing out into the motionless space. It is distinguished by low viscosity and high velocity. According to the Reynolds assessment for gas flowing out from the nozzle, at spraying distance is in the range of $Re = (5000–12{,}000)$. It is several times higher than the critical value, constituting $Re = (2100–2300)$ [56]. This suggests that the motion of the gas is in the turbulent regime.

Based on understanding the physical nature of gas flow, turbulent jets were described by Abramovich [53]. He proposed advanced mathematical apparatus describing the distribution of velocity, temperature, and concentration of the gases adjusted from the atmosphere in any point the jet. The following specific regulations and assumptions were made:

1. Thickening of the jet boundary layer, consisting of a involved particle of the environment and broken particles of the jet itself, leads, on the one hand, to a cross-section increase. On the other hand, it causes a gradual reduction of the potential core of the jet. It is the region lying between the inner rims of the boundary layer (Fig. 2.20). Part of the jet containing the potential core flow is the initial part with the length of X_n.
2. One of the properties of such jet is the constant static pressure over the flow region, whereby the potential velocity in the potential core remains constant.
3. Its common feature is the low value of the transverse velocity components in comparison with the longitudinal velocity. It is usually neglected in engineering calculations.

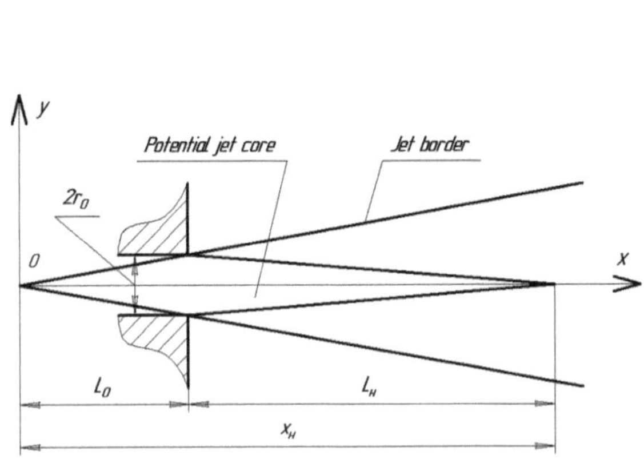

Fig. 2.20 Turbulent jet schemes

4. The simplified scheme of the jet is used frequently, assuming that there is no transition part between the initial and basic areas.
5. Jet spreading outside the initial section is expressed not only by its cross-section increasing but also by the velocity temperature changes along the axis. In the boundary layer, the parameters asymptotically approach the parameters of the surrounding space.
6. A distribution of excessive concentrations of involved particles and temperature of the jet is similar to the jet velocity kind of dependence.
7. During supersonic jet flow, its geometry is changing.
8. Axisymmetric turbulent jet behaves as if it emanates from a point source—pole (O), where it is convenient to place the beginning of coordinates (see Fig. 2.20). Pole is located inside the nozzle at a relative distance from its face:

$$a \cdot L_0/r_0 = a \cdot (x_H - L_H)/r_0 = 0.29, \tag{2.53}$$

where L_0 is the distance from the pole to the jet nozzle face $L_н$ is the distance from the nozzle face to the cross-section of the initial area end; r_0 is the radius of the nozzle, a is an empirical constant proportional to the length of the mean path of molecules before the collision. It is one of the features of turbulence jet. With uniform initial velocity field, the good agreement with experiment is reached for $a = 0.066$.

Within the framework of the theory of Prandtl, the model of involved particles influence on the turbulent structure of the jet is proposed. According to this model micro volume of the jet in its motion carries involved particles and is slowed by the accumulative force of aerodynamic resistance. In this case, the Navier–Stokes Eq. (2.52) takes the form [35]:

$$m \cdot \frac{\mathrm{d}W_p}{\mathrm{d}t} = 0.5 \cdot C_d \cdot \rho_g \cdot \left(W_p - W_g\right) \cdot \left|W_p - W_g\right| \cdot S \tag{2.54}$$

The correlation between gas velocities and particles is also determined by the equation of momentum:

$$\mathrm{d}W_g/\mathrm{d}x + \left(G_p/G_g\right) \cdot \left(\mathrm{d}W_p/\mathrm{d}x\right) = 0, \tag{2.55}$$

where m, S are weight and midship section area of the particle, respectively; C_d is aerodynamic resistance coefficient; ρ is density; G is mass flow; W is velocity; t is time; x is coordinate; indices "g", "p" refer to the gas and particle, respectively.

Its analytical solution is impossible to perform because of the complex dependencies on the movement coordinate of almost all the terms of the equation. We therefore use a numerical solution method for which we divide the transferring particles interval into sufficiently small segments Δx. considering constant values of the terms within its limits. Then, instead of (2.54), (2.55), we obtain:

$$W_p^{i+1} = W_p^i + 0.75 \cdot \frac{\rho_g^i}{\rho_p} \cdot \frac{C_d^i}{d_p} \cdot \frac{\left(W_g^i - W_p^i\right) \cdot \left|W_g^i - W_p^i\right|}{W_p^i} \cdot \Delta x, \quad (2.56)$$

$$W_g^{i+1} = W_g^i - \frac{G_p}{G_g} \cdot \left(W_p^{i+1} - W_p^i\right) \quad (2.57)$$

Scheme of the jet for calculations is shown in Fig. 2.21.

The confining nozzle of diameter ($2r_O$) is set in the middle section of the atomizing point of wires 1. The nozzle centers the gas stream that separating the drops of the molten metal from the end faces of the electrodes. The combustion products of propane–air mixture, which is generated in the closed space of the protective housing 3 are used as the carrier gas. Protective housing 3 with an installed nozzle 2 prevents the suction of ambient air into the jet prior to its ejection from the nozzle. An expansion of a gas jet at an angle of (2α) takes place at the ejection of the nozzle. The jet has a current width of (2δ) at a distance (x) from the pole of the jet. The length of the initial section (xH) consists of the distance from the jet pole to the nozzle face (L_O) and the distance from the nozzle face to the end point of the initial zone ($L_н$).

The initial parameters of the Eqs. (2.56) and (2.57) for $i = 1$, i.e., when $x = 0$ (at the nozzle face) and their change in the length and cross-section of the jet are selected as follows.

For particles ejected at an angle to the axis of the jet, according to the assumptions made, the particle velocity toward the x direction is calculated according to Eq. (2.56), whereas with the parameters related to the point with coordinate y_i, is defined by the equation:

$$y_j = x_i \cdot tg\alpha_j, \quad (2.58)$$

where α_j is the angle at which the particle flew out from the spray point (see Fig. 2.21).

This scheme assumes that there is no dynamic effect of the gas on the particles at y-axis. Therefore, the initial impulse that defined its movement at an angle to the x-axis was obtained at the point ($x = 0$, $y = 0$) due to the arc electro-dynamic impact and the expansion of the compressed gas at the exit of the confining nozzle. Half-angle of the jet spread α is divided into equal intervals. Thus, the position and micro volume characteristics the two-phase jet is determined by its coordinates x_i and y_j.

Geometric and velocity characteristics are connected by the following dependencies [53]:

$$\text{If } M_0 < 1.0: \quad \begin{array}{ll} x_H = 12.4 \cdot r_0, & \beta_W = 12.4, \\ \delta = 0.22 \cdot x & \text{at } x \geq x_H, \\ \delta = 0.27 \cdot x & \text{at } x < x_H \end{array} \quad (2.59)$$

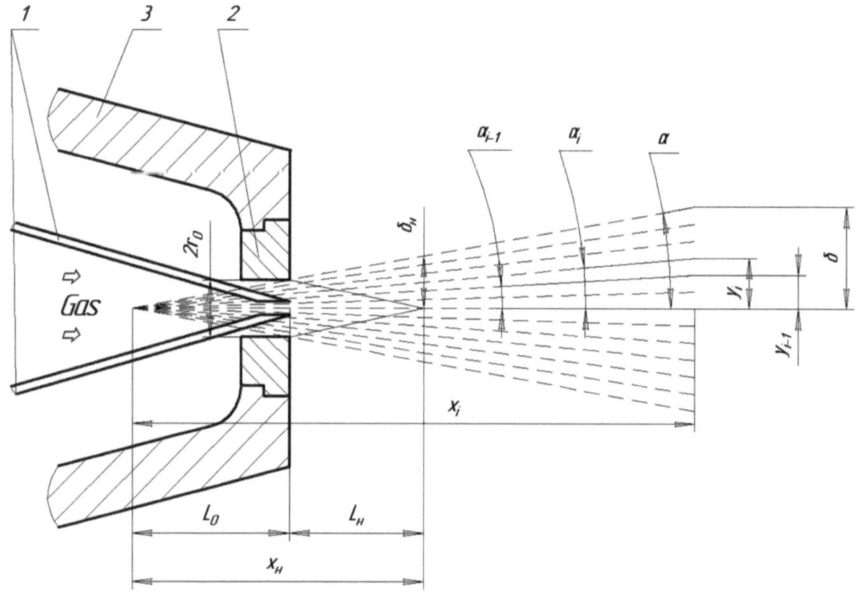

Fig. 2.21 Scheme of the jet for the calculations

$$\text{If } 1 \le M_0 \le 1.6: \quad \begin{array}{l} x_H = \left(6.4 + 2.2 \cdot M_0^2\right) \cdot r_0, \quad \beta_W = 14, \\ \delta = \left(0.165 + 0.045 \cdot M_0^2\right) \cdot x, \end{array} \tag{2.60}$$

where x, y are the coordinates of the point; δ is the radius of the jet for the present coordinate x; β_W is the coefficient of the velocity change along the axis of the jet; M_0 is the initial Mach number.

The velocity change of the gas jet along the axis (W_{gx}) and along the cross-section (W_{gy}) depends on the initial velocity of the gas (W_{g0}) and its distance from the pole of the jet:

$$W_g = W_{g0} \quad \text{if} \quad x < x_H \tag{2.61}$$

$$W_{gx} = \left(W_{g0} \cdot \beta_W \cdot r_0\right) / x \quad \text{if} \quad x \ge x_H \tag{2.62}$$

$$W_{gy} = W_{gx} \cdot \left[1 - (y/\delta)^{1.5}\right]^2 \tag{2.63}$$

The wires–electrodes positioned in the nozzle exit cause an additional turbulence of the gas jet. The high accuracy of the calculations can be reached by using the finite element analysis. Experiments show [57] sufficient for practice calculations accuracy can be achieved while using a coefficient of the flow disturbance ($k_{возъ}$), which reflects an influence of the wires (placed in the nozzle) on the width and

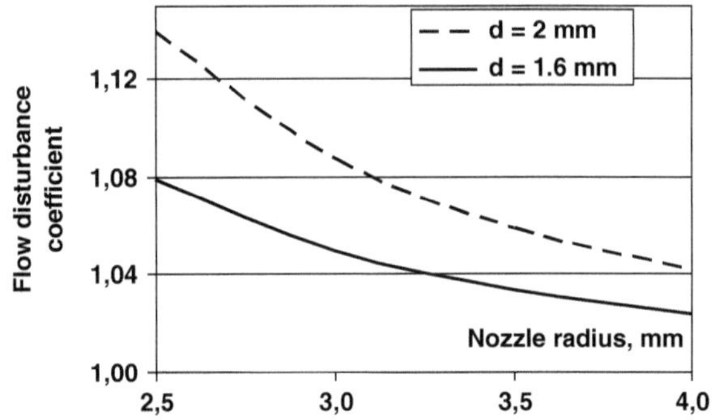

Fig. 2.22 The dependency of the flow disturbances ratio on the radius of the nozzle and diameter of the input wires

velocity of the jet: of the flow disturbance:

$$\kappa_{возм} = \left((0{,}4 \cdot d_{пр}/(2 \cdot r_0))^{2,56} + a\right)/a, \tag{2.64}$$

where $a = 0.066$ is empirical constant.

The radius of the jet is directly, and its velocity is inversely to $\kappa_{возм}$:

$$\delta \cdot \kappa_{возм}, \tag{2.65}$$

$$W'_{gx} = W_{gx} / \kappa_{возм} \tag{2.66}$$

Calculation according to the Eq. (2.65) was done for typical intervals (wire diameter d_{np} = 1. 6–2.0 mm, radius of the emanate nozzle r_0 = 2–4 mm). It shows that disturbance ratio is changing by 2–13 % at wire placing in the nozzle, Fig. 2.22). Proportional to it according to Eq. (2.66) is the spreading radius increase, whereas the velocity of the jet is decreased. It can be seen that the influence on the jet parameters caused by wires is reduced by increasing the wire nozzle diameter. For the limit case of the maximum wire diameter and the minimum radius of the nozzle, the velocity of the gas is reduced by 14 % when you input the wires into the nozzle. For the maximum radius of the nozzle wires input will not have a strong impact, as the deviation of parameters is ~3 %.

Aerodynamic resistance coefficient C_d in the formulas (2.54, 2.56) is determined by the Reynolds number:

$$Re = |W_g - W_p| \cdot d_p \cdot \rho_p / \eta_g, \tag{2.67}$$

where η_g is a coefficient of the dynamic viscosity of the gas, kg/(ms).

For a single non-deformable spherical particle, which is moving with a constant velocity in a stationary isothermal and incompressible fluid of infinite extension at Re = 0.05–1.0 the Stokes formula shall be used [35]:

$$S_d = 24/Re, \tag{2.68}$$

Consideration of various factors amends this formula as the follows:

$$C_d = 0.5 + 29.2\Big/\sqrt{\mathrm{Ar}} + 430/\mathrm{Ar} \quad [58], \tag{2.69}$$

If $2 < Re < 10^3$, the free flow regime is

$$C_d = 24 \cdot Re^{-1} + 3.6 \cdot Re^{-0.317}, 0.2 < Re < 4.0 \quad [35], \tag{2.70}$$

$$C_d = 24 \cdot Re^{-1} + 4 \cdot Re^{-0.333}, 4 < Re < 400 \quad [35], \tag{2.71}$$

$$C_d = 1.1,\ 400 < Re < 5500, f = 1.15 - 1.2 \quad [59], \tag{2.72}$$

$$C_d = 1.8,\ 300 < Re < 500, f = 1.4 - 1.5 \quad [59], \tag{2.73}$$

$$C_d = 37.8 \cdot Re^{-0.6},\ 30 < Re < 400 \quad [59], \tag{2.74}$$

where $f = (S/S_M)_{V=\mathrm{idem}}$ is a geometric shape factor, $\mathrm{Ar} = \rho_0 \cdot W_0^2/(\rho \cdot g \cdot d_0)$ is a criterion of Archimedes at the nozzle outlet.

It is accepted as it is stated above, that the gas jet is a submerged axisymmetric turbulent jet out flowing directly into immobile space as shown in Fig. 2.20.

In calculations of the geometric characteristics of the gas jet: the length of the initial section (x_n), the jet radius (δ) for the coordinates (x); the dependencies for the jet zone, consisting of two wires to determine the gas velocity and the jet radius, the flow disturbance coefficient is used as in the Eq. (2.65).

The spreading angle of the jet is determined experimentally based on the spray spot size of the spray jet at a working distance and from the photographs of the jet (Fig. 2.23). Although, on the photographs, it can be seen that the spreading angle of the jet varies in distance, the calculations has been assumed that: $2\alpha = 10°$, which corresponds to a straight boundary line of the jet and the absence of the transition area.

Initial particle velocity was equated to the wire feeding velocity.

Let us estimate the velocity of the elementary volume of metal movement in the tongue. According to Sect. 2.4, "tongue" is a portion of the liquid metal that was drawn off from the electrode face end, but has being retained at the expense of the gas-dynamic pressure balance and surface tension. Its thickness is $2r_s = 7 \times 10^{-6}$ m, where m is the mass of drawn off metal portion; $m = 6.4 \times 10^{-7}$ kg. Let us assume that transverse dimension of the "tongue" will be equal to the diameter of the wire: $b = 0.002$ m, the period of breakage of the liquid metal is $t = 5.74 \times 10^{-4}$ s (Table 2.8). In this case:

$$s = m/\left(\rho_p \times 2r_s \times b\right) = 6.3 \cdot 10^{-3}\mathrm{M}; \quad v = s/t = 11.7\,\mathrm{M/S},$$

where s is the length of the "tongue"; v is the velocity of movement of the elementary volume in the "tongue".

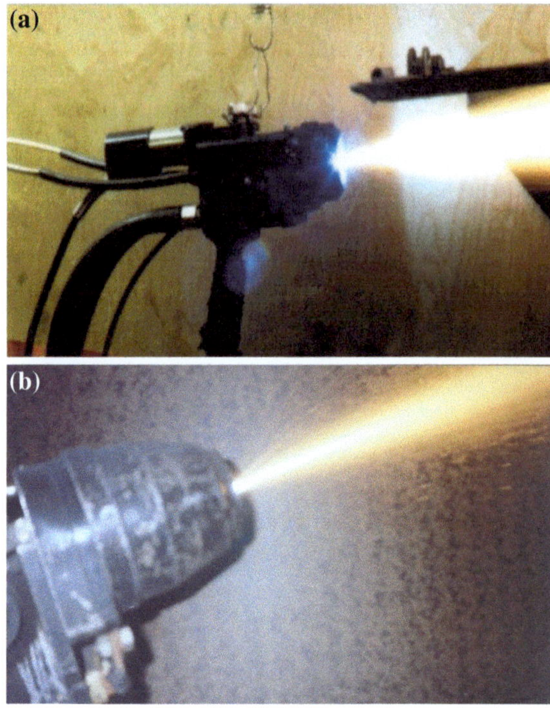

Fig. 2.23 AS guns that were adopted as the basic equipment for calculating the parameters of two-phase jet. Steel spraying [8]. **a** Gun with combustion chamber for pre-burning of propane-air mixer of carrier gas. **b** Gun where propane-air mixer of carrier gas is used without pre-burning

The estimated value of the "tongue" length is comparable with the speed cinema shooting data [20, 21]. In the "tongue" the specific surface area is gradually increased comparing with the electrode face end surface. Consequently, there is a further increase in temperature due to the additional heat release in necks.

Velocity changing of the gas in this sector occurs as in the momentum Eq. (2.55). Gas engages with the teared-off portion of the molten metal which mass is similar to 550 μm droplet in diameter. This droplet size is adopted according to the oscillocoping records data according to Table 2.8, [34].

How it was shown above, the portion of the metal in the form of "tongue" in gas velocity calculations was replaced by the scheme shown in Fig. 2.24. In this scheme, the "tongue" is substituted by the droplets of spherical shape. Its diameter is equal to the smallest droplet size $d_p = 10 \times 10^{-6}$ m. At the end of this sector, droplet size corresponds to its distribution according to sieve analysis, (Table 2.15.)

Assume that density, diameter, and overall consumption of particles remain unchanged in distance. At the same time, adopted density is the same as for the liquid iron $\rho_p = 7265$ kg/m^3 [27]. The initial velocity of the gas is equal to the outflow velocity of the jet escaping from the combustion chamber taking into account the influence of the wires—electrodes according to the Eq. (2.64) Along the distance and the y-section of the jet with the stipulated calculating step, the gas velocity is changing according to the momentum Eq. (2.55). It is followed by its correction according to the equation of velocity change on the axis, and on the section by

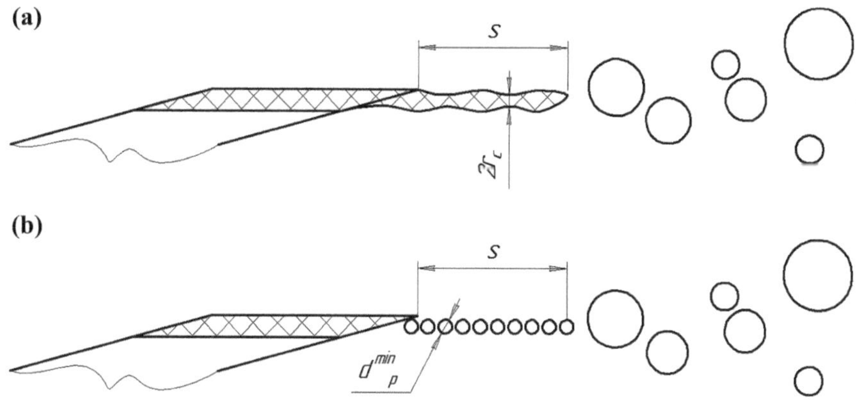

Fig. 2.24 Estimated traffic scheme of drops movement when they are being teared off from the electrode face-end

Eqs. (2.54, 2.61–2.63). It is considered that the gas velocity is equal to the velocity of the environment at the boundary of the jet: $Wg = Ws = 0$ m/s. Gas density is changing in temperature at each step of the calculation.

Aerodynamic resistance coefficient (C_d) was determined according to the calculation of Reynolds equation computed at the previous calculating step (2.70, 2.71, 2.72, 2.74). All other equations were rejected by the following reasons. Stokes formula (2.68) and the equation for the free flow regime (2.69) do not meet the requirements of turbulent nature of the jet. The results of calculations on the dependency differ from the experimental data that were made by Abramovich for large values of Re [53]. Furthermore, the surface tension stands up against a strong deformation of the particles ($f = 1.4–1.5$) as it is shown in paragraph 2.4. When $Re < 0.2$ aerodynamic resistance coefficient becomes unrealistically large. To avoid this, an assumption was made in the calculations. It was assumed that if $Re \leq 0.2$ then $Cd = 126$. This corresponds to $Re = 0, 2$ according to the Eq. (2.70).

Due to the turbulent nature of the jet, let us assume that the distribution of the specific gas consumption on the cross-section is close to uniform. In this case, it is proportional to the square of the j-circular cross-section

$$G_{gj} = G_g \cdot \frac{\left(y_{j+1}^2 - y_j^2\right)}{\delta^2} \tag{2.75}$$

Let us assume that the total mass flow rate corresponds to the maximum for nozzle (position 2, Fig. 2.21) which is expressed in terms of the gas-dynamic functions according to [14]:

$$G_g = \frac{m \cdot P \cdot F \cdot y(\lambda) \cdot 3600}{\sqrt{T_g \cdot \tau(\lambda)}}, \tag{2.76}$$

where $P = 0.981 \times 10^5$ N/m^2; is the static pressure in the flow,

$$m = \sqrt{k \cdot \left(\frac{2}{k+1}\right)^{\frac{k+1}{k-1}}} \cdot \sqrt{\frac{g}{R}} = 0.035\,\mathrm{m}^{-1}\,\mathrm{s}\,\mathrm{K}^{0.5},$$

where λ, $y(\lambda)$, $\tau(\lambda)$ are the gas-dynamic functions:

$$\lambda^2 = \left(\frac{k+1}{2} \cdot M\right) \Big/ \left(1 + \frac{k-1}{k+1} \cdot M^2\right), \tag{2.77}$$

$$y(\lambda) = \left(\frac{k+1}{2}\right)^{\frac{1}{k-1}} \cdot \lambda \Big/ \left(1 - \frac{k-1}{k+1} \cdot \lambda^2\right), \tag{2.79}$$

$$\tau(\lambda) = 1 - \frac{k-1}{k+1} \cdot \lambda^2, \tag{2.80}$$

where $k = c_v/c_p$ is the adiabatic index, taken as $k = 1.262$ for air at $T = 2100$ K [9].

Mach number (M) was determined experimentally (see below).

In the calculations of cross-sectional area of the nozzle, the presence of a two wires in it with diameter d_{np} merging at cross-sectional angle α_{np} was taken into account, see Fig. 2.21.

$$F = F_c - F_{np} = \pi \cdot \left[r_0^2 - 2 \cdot \left(\frac{d_{np}}{2 \cdot \cos \alpha_{np}}\right)^2\right], \tag{2.81}$$

where F_c is the total area of the nozzle; F_{np} is the cross-sectional area of the wire.

To determine the specific (kg/m^2/s) consumption of the particles flow over the cross-section g_j, we assume that it is distributed according to normal—curricular law

$$g_j = g_0 \cdot e^{-k_c \cdot y_j^2}, \tag{2.82}$$

where g_0 is a specific consumption on the jet axis.

For the maximum radius of the spraying material pattern (r), we take the radius of the jet. According to the experimental data, nearly 10 % of the particles are flown off the jet borders of angle (α). At the experiment, the coating on the disk was weighed. The disk of the radius (δ) is placed on the characteristic spraying distance. It allows calculating the concentration coefficient k_c in each section of the annular section of the jet:

$$0.1 \cdot g_0 = g_0 \cdot e^{-K_c \cdot r^2}$$

$$K_c = -\ln 0.1/r^2 = 2.303/r^2 \tag{2.83}$$

Integrating of the specific discharge on the cross-sectional area (F) will give the total mass flow of the particles:

$$G_p = \int_F g dF \tag{2.84}$$

Let us expand the integrand:

$$G_p = \int_F \text{g} \text{d}F = \int_0^\infty g \cdot 2 \cdot \pi \cdot r \cdot \text{d}r = \int_0^\infty g_0 \cdot e^{-k_c \cdot r^2} \cdot 2 \cdot \pi \cdot r \cdot \text{d}r = \frac{\pi}{k_c} \cdot g_0$$

Accordingly, the fuel particles consumption for each interval $\Delta\alpha_j$ in each section x_i is:

$$G_{pj} = \int_{y_j}^{y_{j+1}} g_0 \cdot e^{-k_c \cdot r^2} \cdot 2 \cdot \pi \cdot r \cdot \text{d}r = G_p \cdot \left(e^{-k_c \cdot y_j^2} - e^{-k_c \cdot y_{j+1}^2} \right) \tag{2.85}$$

The nature of the Eq. (2.85) shows that the largest consumption occurs at the cross-sections which are distanced from the jet axis on the relative distances: $y/\delta \approx 0.4$ (Fig. 2.25).

To select the initial data for calculations and confirm its accuracy, some of the parameters have been determined experimentally. The following are the measurement of the initial velocity of the gas stream, the gas distribution, particle velocity, granulometric size of the spraying jet.

Measurement of the initial velocity of the gas stream flow was carried out using pneumometric head. Its shape was designed for supersonic flow (Fig. 2.26). The initial velocity of a supersonic flow was determined by the Mach number (M) and the sound velocity ($a_{зв}$)

$$W_g^0 = M \cdot a_{зв} \tag{2.86}$$

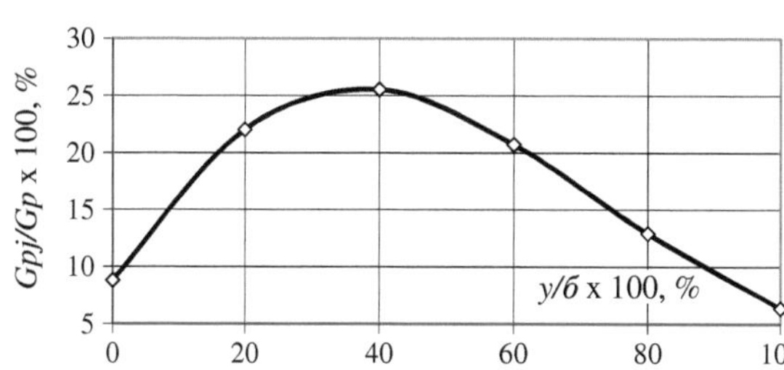

Fig. 2.25 Distribution of relative consumption of particles(G_{pj}/G_p) on the cross-sections of the jet

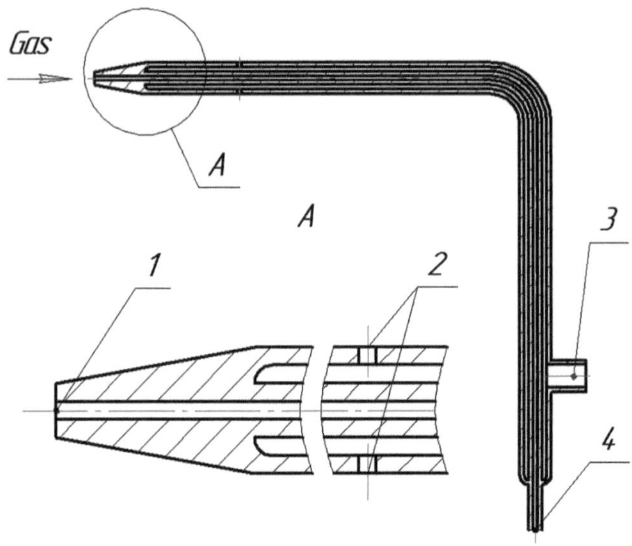

Fig. 2.26 Scheme of pneumometric head for measuring the gas flow rate [14, 60]. *1* Inlet orifice; *2* side orifice; *3* outlet tube for static pressure measuring; *4* outlet tube for total pressure measuring

The Mach number is calculated by the method of successive approximations on the Rayleigh formula [14, 60]:

$$\frac{P_1^*}{P_H} = \left(\frac{k+1}{2}\right)^{\frac{k+1}{k-1}} \cdot \left(\frac{2}{k-1}\right)^{\frac{1}{k-1}} \cdot \left(M^{\frac{2 \cdot k}{k-1}}\right) \Big/ \left(\left(\frac{2 \cdot k}{k-1} \cdot M^2 - 1\right)^{\frac{1}{k-1}}\right) \quad (2.87)$$

where P_1^*, $P_н$ are total and static pressure of the incoming flow, MPa. When calculating the absolute sound velocity according to the Eq. (2.13), the combustion temperature of propane–air mixture was taken for the temperature of the stream: $T_g = 2112$ K [5]. Adiabatic index of combustion products $k = 1{,}262$ is taken as for air at $T = 2100$ K [9].

To eliminate the influence of the nozzle design on the results of measurements of the opening half-angle of receiving hole $(\omega_н)$ is made than limit angle of the gas flow (ω), (Fig. 2.27):

$$\omega_H < \omega \quad (2.88)$$

Neglecting the difference between bevel front-section in the form of a head and clinch-section which is taken into account when calculating the velocity with the Rayleigh formula (2.87), we can assume that the oblique shock is occurred when supersonic gas flowing around the nozzle [53]. Velocity vector $\bar{W}_g$ and the bevel bow shock front form an angle α before jet with the head reach their meeting point.

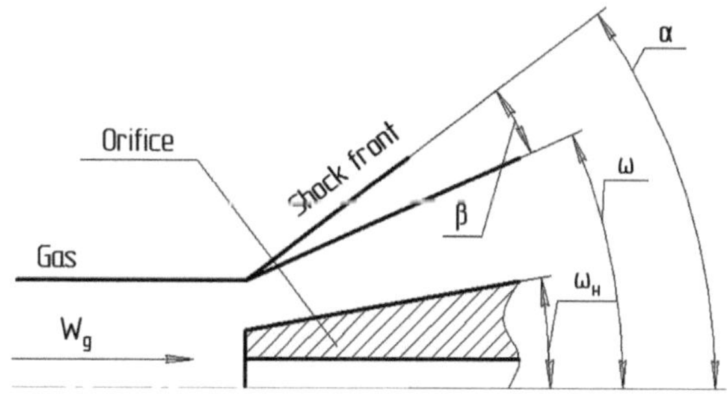

Fig. 2.27 The scheme of a pneumometric head for gas flow velocity measurement

Surface of a cone with an apex angle α is the envelope of spherical waves. The angle α is defined by:

$$\sin\alpha = 1/\mathrm{M} \tag{2.89}$$

The deflection angle of the jet after crossing the front of the oblique shock ω is:

$$\omega = \alpha - \beta \tag{2.90}$$

According to [53], the angle between the velocity vector and the shock front β is determined by the equation:

$$tg\beta = \frac{k-1}{k+1}\cdot\left(1+\frac{2}{k-1}\cdot\frac{1}{M^2\cdot\sin^2\alpha}\right)\cdot tg\alpha \tag{2.91}$$

Expected Mach number is in the range of 1–1.5. On the basis of the calculations by Eqs. (2.88–2.91), it was assumed that $\omega_н = 10°$.

Characteristic dimensions of the head (Fig. 2.28) are selected from the following conditions [60]:

- side holes 2 are positioned at a distance of not <4–6 head diameters from the front edge. In this case, the pressure P_2 in the tube 3 equals to the static pressure of the oncoming stream $P_н$. Since for the head with $\omega_н < \omega$, the slowing down of

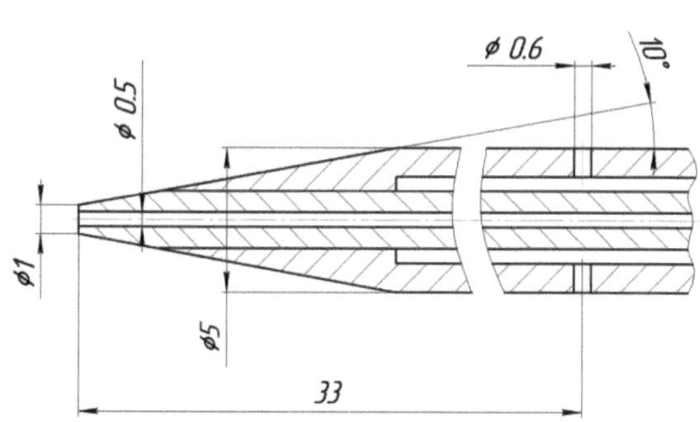

Fig. 2.28 View of the working part of the head

the gases jet takes place without friction and eddies, we can assume that in the tube 4, pressure P_1 is formed. The latter is approximately equal to the total pressure of the incoming flow P_1^*. Thus, the pressures $P_н$ и P_1^* are measured by pressure gauges attached to the head output tubes 3 and 4 respectively;

- the minimum size of the central hole 1 is made of in order to reduce measurement errors.

In this case, the following relation is correct

$$P_1^*/P_H = P_1/P_2 \tag{2.92}$$

According to the measurement and calculation results for the Eqs. (2.86 and 2.87), the initial velocity of supersonic was 950 m/s, when the combustion products of propane–air mixture were flowing out of the nozzle.

Float-type meters were used to determine the gas flow rates (Table 2.13).

Table 2.13 The results of the consumption of compressed air and propane being fed into the combustion chamber

Excess oxidizer coefficient, α_{OK}			0.8	1.0	1.2
Combustion chamber pressure, MPa			0.32	0.30	0.29
Air	Pressure, MPa		0.42	0.42	0.42
	Density, kg/m^3 [62]		4.94	4.94	4.94
	Consumption	Volume, m^3/h	1.6	1.6	1.7
		Mass, kg/h	7.9	8.1	8.2
Propane	Pressure, MPa		0.40	0.38	0.35
	Density, kg/m^3 [9]		7.67	7.28[a]	6.70[a]
	Consumption	Volume, m^3/h	0.085	0.069	0.057
		Weight, kg/h	0.65	0.50	0.38
Gas mixture consumption, kg/h			8.57	8.57	8.58

[a]Obtained by the linear approximation of table data

Determination of the particles velocity was carried out using the velocity rate meter for the luminous objects ИSSO-1. Device operation (Fig. 2.29) is based on the principle of a rotating mirror. Mirror rotation axis is parallel to the axis of the gas jet. In this case, a perpendicular component that is proportional to the frequency of the mirror rotation applied to the projection of the particle velocity. In the case of the particle motion at a constant velocity, its track is a straight line. The values of the particle velocity are defined as:

$$W_p = V \times ctg\alpha, \tag{2.93}$$

where V is the liner rotating velocity of the mirror cylinder, m/s, α *is* an angle between the jet axis and the particle track.

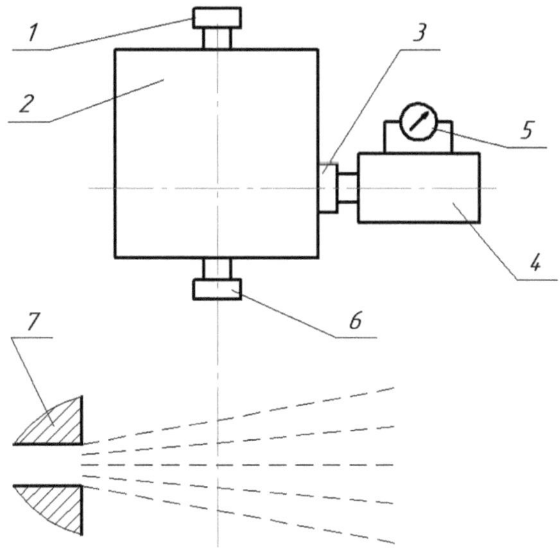

Fig. 2.29 Operating scheme of the device "ИSSO-1" [61]. *1* is an eyepiece; *2* is a registration module; *3* is a mirrored cylinder; *4* is an engine tachometer; *5* is an indicator of the rotational speed; *6* is lens; *7* is a nozzle

By varying the rotation speed of the mirror cylinder in the device, we can achieve the parallelism of luminous particles tracks and control lines with fixed angles, visible through the eyepiece of the instrument. Then, the magnitude of the velocity is counted on the dial indicator. Particle's velocity measurement was performed for the two models of arc spraying guns: with and without combustion chamber that is installed in the nozzle area. Modes: $I_Д$ = 210 A; $U_Д$ = 30 V; combustion chamber pressure equals 0.32 MPa; the pressure in the spraying head is between 0.22 and 0.23 MPa; wire material is steel: X20Cr13∅1, 6 mm. The measurement results are shown in Table 2.14. The scatter of these values is related to the difference in velocities of the particles of different diameters.

Particle size distribution at AS with steel solid wires was done according to the spraying into the water scheme. The developed device for collecting the particles (Fig. 2.30) comprises a vessel 10 into which the water 7 is poured. Two-phase "particles-gas" jet 2 from the spraying gun 1 passes through the central hole of the

Table 2.14 The particles velocity measurements for arc spraying, m/s

The distance from the nozzle face, mm		50	100	150
Spraying head with a combustion chamber	Velocity at the upper border of the interval, m/s	134	125	120
	Velocity at the lower border of the interval, m/s	110	100	100
Spraying head without a combustion chamber	Velocity at upper border of the interval, m/s	62	75	70
	Velocity at the lower border of the interval, m/s	42	52	46

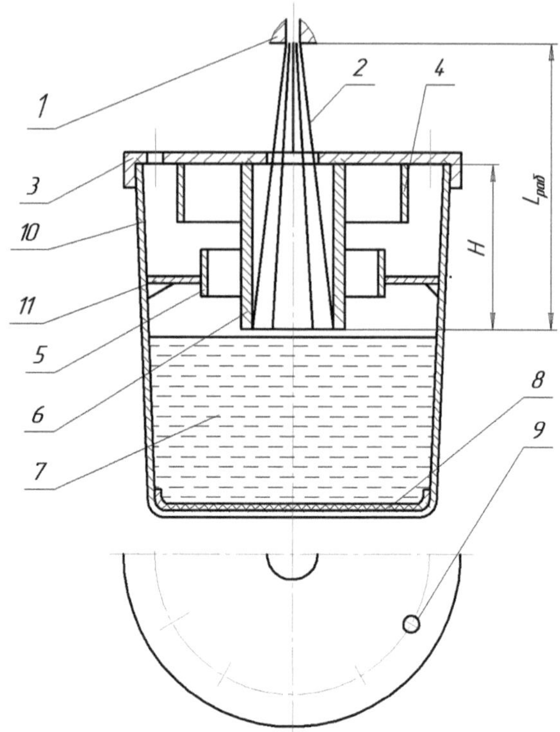

Fig. 2.30 The droplets collecting device for measuring its fractional composition. Explanations are followed in the text

cover 3, and then through the tube 6 it gets into the water. Sufficient diameter of the hole ensures achieving water all of the sprayed metal droplets in water. Height (*H*) of tube 6 is chosen so that the nozzle face of the spraying gun is at spraying working distance ($L_{раб}$) from the water surface.

In order to prevent the splashing out of water under the influence of air jet, the inside cover 11 with the sleeve 5 and the sleeve 4 were installed. They are forming a system of labyrinths to extinguish the kinetic energy of water. To release excessive air pressure inside the container 10 on the cover 3 there are special holes 9. When falling into the water, the particles cool down and settle on the tray 8 which is made from stainless steel. Thus, under the influence of air jet, the water is suspended inside the container 10 and therefore the metal droplets are subjected to a slight deformation due to their contact with it, while the conditions of heat dissipation into water are improved.

The obtained powder that was collected on the tray/pallet during spraying, is being dried for 2 h at a temperature of 200 °S. Fractional composition of the particles was determined by sieve analysis method by weight share based on GOST 18318-73 (Table 2.15).

In the diameter range 20–100 µm, the pitch mesh sizes was 10–15 µm. Particles of larger size fractions of 100–630 mm have been sown through sieves with a bigger step. No evidence of particles with a diameter bigger than 315 μm has been found. The experiments were performed using AS gun with the combustion chamber, $I_Д = 200$ A, $U_Д = 30$ V. The spraying distance was 100 mm and the steel wire X20Cr13 was used as a feedstock.

Calculations were made according to the accepted model of AS processes for the average diameters at each fraction interval as in Table 2.15. Comparison of experimental data on the velocity of the particles with the calculated ones is presented in Fig. 2.31. The curves of the experimental data correspond to the points listed in Table 2.13 when using a spray head with a combustion chamber. The diameters of the particles with a particular velocity changes rates were used for calculations. The graph shows the average calculated cross-section velocities of particles. Average velocities of the particles decrease with an increase of their

Table 2.15 Fractioned composition of particles according to experiments

Particle sizes intervals, microns											
0–20	20–35	35–40	40–50	50–63	63–71	71–80	80–100	100–125	125–160	160–200	200–315
Fraction weight content, %											
7.0	6.6	6.6	8.2	7.6	10.0	11.8	12.0	10.0	7.8	6.0	6.4
Area share of fraction, %											
38.9	10.4	3.9	6.6	7.7	4.8	4.9	8.1	6.5	4.1	2.3	1.9

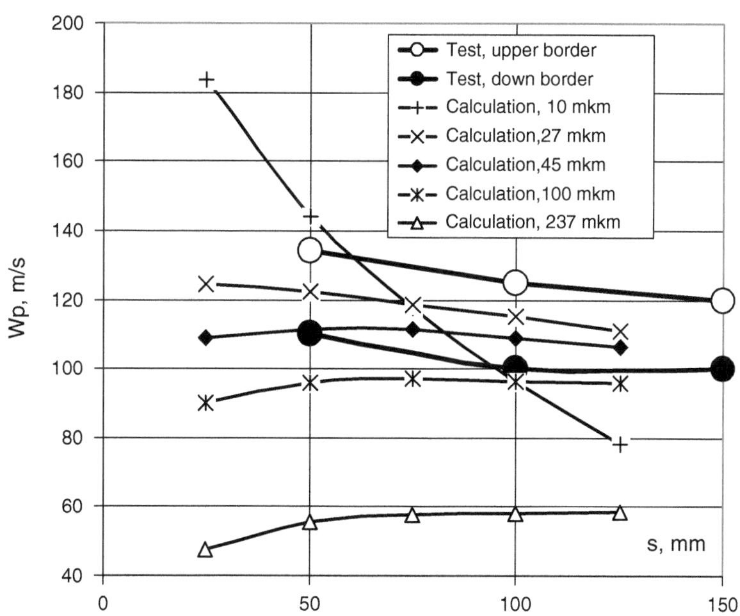

Fig. 2.31 Comparison of experimental data on the velocity of the particles with the calculated ones

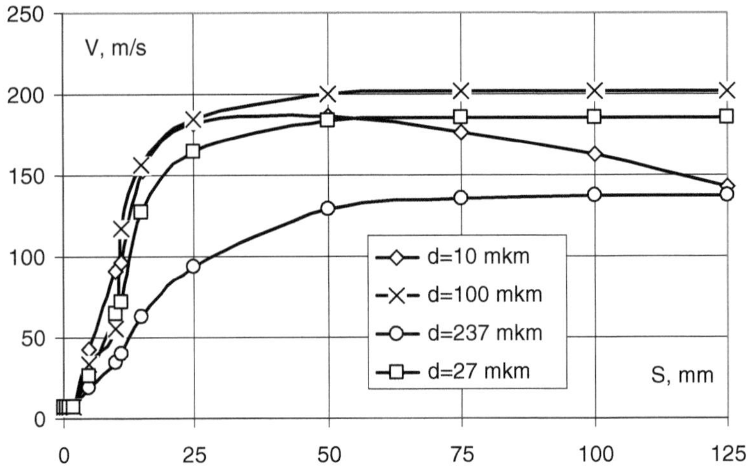

Fig. 2.32 Changing of velocity particles with different diameters on spraying distance. The axis of the jet

diameters. For particles at the diameter range of 20–125 μm, and the proportion of which accounts for about 75 wt%, a satisfactory agreement between the calculated velocities with the experimental was noted. Deviation is <10 %. The calculated velocities of the particles of larger diameter differ from the experimental ones by 20–50 %. This probably is related to the experimental procedure. Gas stream energy may not be sufficient to attract particles of a larger diameter, which are scattering across the main direction of the jet when breaking the "tongue." This leads to the fact that the tracks of those particles do not enter the objective lens of the device and therefore are not recorded therein.

Particle velocities on the jet axis are approaching the maximum value at the distance of about 50 mm, (Fig. 2.32). At a working distance of 100–125 mm, velocity varies slightly for the most fractions. Only the particles d_p = 10 μm reach the maximum velocity at the distance of ~25 mm, and the velocity decrease later. In the range of particle diameters in between 27 and 100 μm, the ratio of the weight to the particle cross-section is such, that the velocity of larger diameter particles exceeds the numbers for the smaller ones. Hereafter, for the range of diameters in between 100 and 237 μm, the dependence is reversed. The velocity of bigger particles (in diameter) is lower due to its greater inertia. In the cross-section of the cone corresponding to $y/\delta > 0.8$, the velocity of particles is in the range of 50–80 m/s, but on the jet axis the velocity of particles is 160–200 m/s.

Gas and particle velocities on the jet axis are compared at a distance of 75–100 mm. Subsequently, the gas velocity becomes lower (Fig. 2.33).

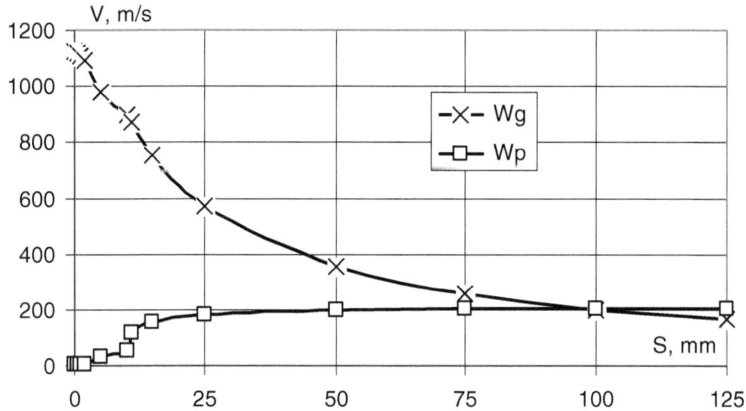

Fig. 2.33 Changing of gas and metal droplets velocities for the spraying distance. Axis of the jet

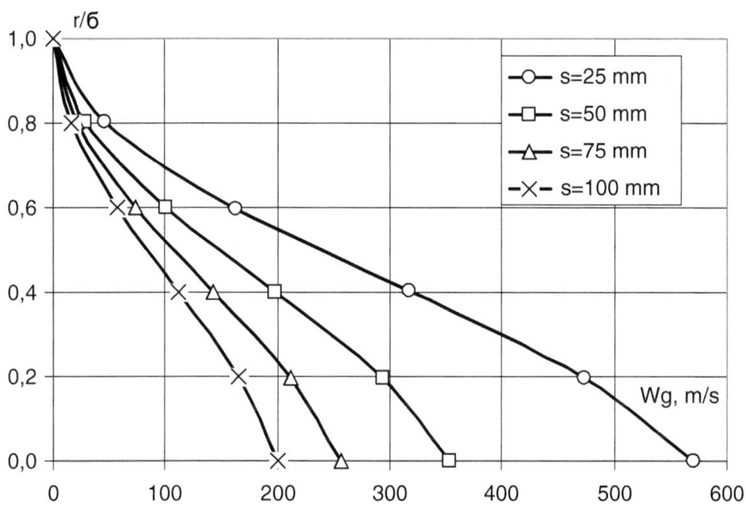

Fig. 2.34 Changing the gas velocity on the jet axis at a various distance of spraying when the diameter of particles is 100 μm

On the jet cross-section, the velocity of droplets and gas decreases drastically, whereas the gradient of the velocity drop goes down when the distance from the nozzle gets bigger (Figs. 2.34 and 2.35).

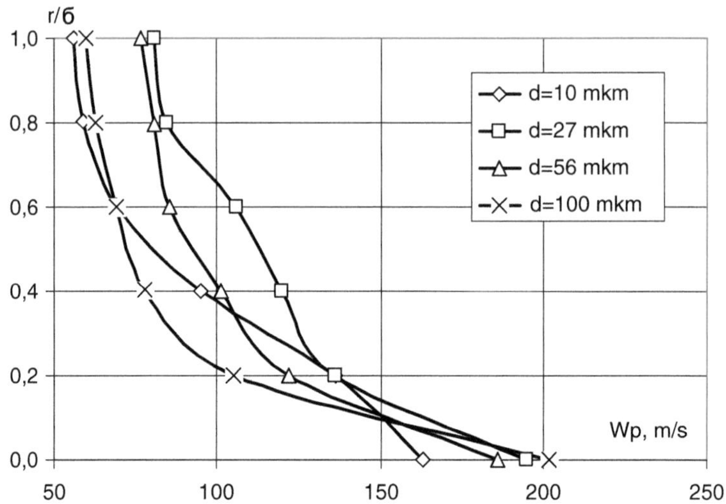

Fig. 2.35 Changing the particles velocity on the cross-section of the jet at a distance of 100 mm

2.6 Temperature Calculation of the Two-Phase Flow

The heat balance of sprayed droplets for AS can be represented as a typical scheme adopted for the TS processes (Table 2.16).

The choice of the heat exchange limiting stage "gas-drop" is made by the criterion Bio:

$$\mathrm{Bi} = \alpha_T \cdot d_p / \lambda_p, \tag{2.94}$$

where α_m is the heat transfer coefficient, W/(m^2 K); d_p is the particle diameter, m; λ is the thermal conductivity of the particle material, W/(m K).

When Bi > 1 the heat transmission is limited by heat transferring into the depth of the particle (materials with low thermal conductivity). From the data presented by Kudinov (Table 2.17), it is obvious that for metallic materials used in AS, Bi $\ll$ 1 and the loss on the internal heat transferring is negligible. Studies made by other authors can confirm this statement [11, 35, 59].

As plasma flows at a temperature about 10^4 K refractory materials lose significant amount of its heat. In the plasma process models, these losses are always taken

Table 2.16 The heat balance of sprayed droplets for TS [16, 63]

Gas		Metal droplets		
Heat transfer to the environment	Heat exchange with the particles	Heat exchange with the gas	Irradiation	Evaporation

Table 2.17 Bio criteria in plasma of various gases while heating the particles of 100 μm in diameter [63]

Material		ZrO_2	Al_2O_3	MgO	BN
λ, W/(m K)		2.39	5.86	5.86	9.2
Bi	Ar	0.18	0.07	0.07	0.05
	N_2	0.63	0.26	0.26	0.16
	NH_3	3.5	1.4	1.4	0.9
	H_2	5.5	2.2	2.2	1.4
Material		SiC	Ni	Fe	W
λ, W/(m K)		41.9	18	40	100
Bi	Ar	0.01	0.02	0.01	0.004
	N_2	0.04	0.08	0.04	0.015
	NH_3	0.2	0.5	0.2	0.08
	H_2	0.3	0.7	0.3	0.1

In the calculations the following values were taken: $Nu = 2$; the mass-average plasma temperature: for argon it is 10,000 K; for nitrogen it is 5000 K; for hydrogen and ammonia it is 3500 K. The thermal conductivity coefficient of the particles is taken for the temperature equals the half value of the melting point of its material

into account [16, 17, 35]. However, the temperature level for AS is lower, than for plasma spraying. Therefore, we are to estimate the necessity to account the irradiation losses.

Let us calculate these losses, based on the methodology proposed by Erokhin for the droplets transfer in arc welding [4], using the following system of differential equations:

for droplets thermal conductivity:

$$dQ_p = -c \cdot m_p \cdot dT \tag{2.95}$$

for irradiation losses

$$dQ_p = \xi \cdot \sigma_0 \cdot F \cdot \left(T_p^4 - T_s^4\right) \cdot dt \tag{2.96}$$

It is more convenient from the point of integration to solve the equation for T beforehand:

$$c \cdot m_p \cdot dT = -\xi \cdot \sigma_0 \cdot F \cdot T_p^4 \cdot dt,$$

where ξ is the emittance coefficient; σ_0 is a constant value of irradiation; T is temperature; F, t refer to the surface and the time of irradiation; indices "p", "s" refer to the particles and the environment.

To estimate the maximum heat losses, it was assumed that the surface temperature $T_s = 0$. Separating the variables we obtain:

$$dT/T^4 = -A \cdot dt \text{ где } A = \xi \cdot \sigma_0 \cdot F/(c \cdot m_p)$$

$$\int_{T_0}^{T} T^{-4} \cdot dT = -A \cdot \int_0^t dt,$$

$$1/T^3 - 1/T_0^3 = 3At,$$

$$T = T_0 \cdot (1 + 3 \cdot A \cdot T_0^3 \cdot t)^{-1/3},$$

Let us substitute the value of T into the Eq. (2.96):

$$dQ_p = -\xi \cdot \sigma_0 \cdot F \cdot T_0^4 \cdot (1 + 3 \cdot A \cdot T_0^3 \cdot t)^{-4/3} \cdot dt,$$

$$Q_p = -\xi \cdot \sigma_0 \cdot F \cdot T_0^4 \cdot \int_0^t (1 + 3 \cdot A \cdot T_0^3 \cdot t)^{-4/3} \cdot dt,$$

Let us bring the expression to the following form:

$$Q_p = \xi \cdot \sigma_0 \cdot F \cdot T_0^4 \cdot (1/3 \cdot A \cdot T_0^3) \cdot \int_0^t (3 \cdot A \cdot T_0^3) \cdot (1 + 3 \cdot A \cdot T_0^3 \cdot t)^{-\cdot 4/3} \cdot dt$$

According to the theorem on the invariance of integration and taking into account that

$$(3 \cdot A \cdot T_0^3) \cdot dt = d(1 + 3 \cdot A \cdot T_0^3),$$

let us integrate:

$$\begin{aligned}
\int_0^t (3 \cdot A \cdot T_0^3) \cdot (1 + 3 \cdot A \cdot T_0^3 \cdot t)^{-\cdot 4/3} \cdot dt &= \int_0^t (1 + 3 \cdot A \cdot T_0^3 \cdot t)^{-\cdot 4/3} \cdot d(3 \cdot A \cdot T_0^3) \\
&= -3 \cdot (1 + 3 \cdot A \cdot T_0^3 \cdot t)^{-\cdot 1/3} \Big|_0^t \\
&= -3 \cdot \left[(1 + 3 \cdot A \cdot T_0^3 \cdot t)^{-1/3} - 1\right] \\
&= 3 \cdot \left[1 - (1 + 3 \cdot A \cdot T_0^3 \cdot t)^{-1/3}\right]
\end{aligned}$$

Substituting the value of the integral, we obtain:

$$Q_p = (\xi \cdot \sigma_0 \cdot F \cdot T_0/A) \cdot \left[1 - \left(1 + 3 \cdot \xi \cdot \sigma_0 \cdot F \cdot T_0^3 \cdot t/(c \cdot m_p)\right)^{-1/3}\right]$$

or

$$Q_p/m_p = c \cdot T_0 \cdot \left[1 - \left(1 + 3 \cdot \xi \cdot \sigma_0 \cdot F \cdot T_0^3 \cdot t/(c \cdot m_p)\right)^{-1/3}\right] \quad (2.97)$$

The irradiation losses χ_p represent the fraction of the heat coming on to the irradiation from the enthalpy of droplets:

$$\chi_p = (Q_p / m_p)/\Delta h_{ж} \quad (2.98)$$

Let us compare the value of these losses for AS with arc welding. For the processes, the values of the droplets specific surface area and the emission time (t) differ in order of 1–3 steps. The parameters are taken based on the following:

- The value of t is assumed to be equal to the frequency of the droplets discharge in the combustion zone of the arc, where T_0 = 2500 K; and the droplets flying time at a distance, when T_0 = 1809 K;
- According to [4], ξ = 0.4 for the clean steel surface (combustion arc zone), ξ = 0.8 for droplets coated by slag (flight at a distance);
- The droplets diameter in the burning zone of the arc for AS is taken according to the idea of "tongue" formation. The latter is blown away from the electrode end face, the thickness of which is equal to the minimum size of the particles;
- For the flying of droplets at a distance, the average droplet size is taken according to the particle size analysis (Table 2.15).
- Specific enthalpy of the molten iron $\Delta h_{ж}$ is calculated according to [4]

$$\Delta h_{ж} = 1319 + 0.84 \cdot \left(T_p - 1807\right).$$

- The values of the specific heating capacity (s_p) for the combustion zone of the arc and for the arc welding are taken according to [4]; for the flight distance the value is taken according to the data [13].
- The calculation of the droplets number (n) and the total droplets surface area (F) is made for a conventional productivity of spraying/welding of 8 kg/h.

From the presented data calculations, there are the following conclusions (Table 2.18):

(1) The main share of the irradiation losses, >80 %, occurs in the burning zone of the arc.
(2) Due to the small droplet lifetime at AS, the losses are lower than in arc welding, although the total irradiating surface of the droplets is significantly higher.

Table 2.18 Loses by radiation in arc welding and AS

Parameters	AS		Arc welding [4]	
	Arc burning zone	Droplet flight at a distance	Metal drop transfer	Fine spray metal transfer
Emissivity, ξ	0.4	0.8	0.4	0.4
Stefan-Boltzman constant, σ_0, W/(kg K)	5.67×10^{-8}			
Specific heat capacity, c_P, W/(kg K)	840	740	840	840
Initial metal temperature, T_0, K	2900	1809	2500	2500
Droplet separation period, t, s	5.73×10^{-4}	1.70×10^{-3}	0.18	0.018
Droplet diameter, $d_p \times 10^6$, m	10	100	1740	420
Metal density, ρ, kg/m^3	7200	7200	7200	7200
Droplet mass, m_p, kg	3.8×10^{-12}	3.8×10^{-9}	2.0×10^{-5}	2.8×10^{-7}
Droplet surface, s, m^2	3.1×10^{-10}	3.1×10^{-8}	9.6×10^{-6}	5.6×10^{-7}
f/m, m^2/kg	83	8.3	0.48	2.0
Number of droplets, pcs.	2.1×10^{12}	2.1×10^9	4.0×10^5	2.8×10^7
Total surface area, F, m^2	660	66	3.8	16
Q_p/m, kJ/kg	41	6.9	97	42
Specific heat enthalpy of iron, Δh, kJ/kg	1901	1321	2069	2069
Irradiation loses, χ_p, %	3.3	0.5	3.8	1.6

(3) The value of the irradiation losses for AS is 0.5–3.3 % of the droplets enthalpy.

As a result of the high temperatures at plasma spraying, the major role in the heat balance of metal droplets belongs to evaporation. For instance in plasma spraying, the spraying radius of tungsten particles in the plasma jet is reduced by 78 % due to evaporation (diameter of 20 μm) and by 9 % (diameter of 80 μm) at a temperature of about 10,000 K [64].

The metal loss evaluation due to evaporation for AS was conducted in the research [65]. From the energy balance equation for the melting process of the electrodes with respect to one cycle of the arc, it was derived that for evaporated metal, the χ

$$\chi = (E - M \cdot H(T))/(M \cdot (H_u - H(T))), \tag{2.99}$$

where E is the energy released in the active spots of the electrodes within one splash/discharge; $H(T)$ is the enthalpy change from room temperature to $T_{пл}$ during heating, whereas phase transitions is considered; H_u is the enthalpy change during evaporation; $M = \pi \cdot d_p^2 \cdot v \cdot \rho_p/(2 \cdot \tau)$ is the amount of molted metal breaking away from two electrodes at a drop separation frequencyτ, and wire feeding velocity v.

The calculation of the average mass temperature T was conducted taking into consideration the derived equation of temperature distribution at the end of the electrode.

$$T(l) = T_{\kappa un} \cdot L \cdot [\exp(a \cdot l/L) - 1]/(a \cdot l), \tag{2.100}$$

where $a = ln(T_{nn}/T_{\kappa un})$, L is the thickness of the liquid layer at the end of the electrode, l is the thickness of the evacuated part of the liquid metal at the end of the electrode.

The estimated share of the evaporated metal for the close to characteristic modes (I = 270 A, U = 25 V) is in the range of 0–8 % depending on the assumed value of fractured layer (l). The further increase of the arc power according to the estimation from [65], leads to evaporation losses up to 20 %.

This is comparable with the systemized by Erokhin data when applying to arc welding. While using bare the estimation from [65 leads to evaporation losses up to 6–25 % [4]. However, the metal heat losses on evaporation are not determined by the total amount of evaporated metal, but only by that part which was not condensed in the spray screen due to deviation of the vapor stream.

Erohin has analyzed the experiments conducted by Potapevskiy at impulse welding, [66] together with his own data at MIG welding. Besides, he used calculation results about share of the evaporated metal from the electrodes ИМЕТ–4 due to the amount of metal which has passed into the slag in the form of oxides.

Based on these data, he has concluded that the metal loss during evaporation can take ≤0.5 %. A similar value of the evaporation loss is given for the iron arc melting by Zamoruev [67].

Accordingly, for the arc spraying that is similar to the arc welding process it was assumed that the heat losses of the flying particles on evaporation can be neglected in the subsequent calculations.

Since it is assumed, that the gas temperature decrease in the presence of particles is due to the heat transfer from gas to particles, the equation of heat exchange "gas-metal droplets" can be written as follows:

$$\mathrm{d}\left(c_p^p \cdot m_p \cdot T_p\right) \Big/ \mathrm{d}x = -\mathrm{d}\left(m_g \cdot h_g\right)/\mathrm{d}x, \tag{2.101}$$

where, $\alpha_T = Nu \cdot \lambda_g \cdot d_p$ is the heat transfer coefficient, W/(m^{2}K); h_g is the gas enthalpy, J/kg.

Gas heat exchange with the environment was studied by Abramovich [53]. It was assumed that the turbulent jets of the dimensionless temperature field in a gas jet are proportional to the curves of the dimensionless velocity values:

$$\left(T_{gx}-T_{gs}\right)/\left(T_{g0}-T_{gs}\right)=0.86\cdot W_{gx}/W_{g0}, \quad (2.102)$$

$$\left(\left(T_{gy}-T_{gs}\right)\right)/\left(\left(T_{gx}-T_{gs}\right)\right)^{1/\mathrm{Pr}_T}=W_{gy}/W_{gx}, \quad (2.103)$$

where, T_g is the temperature of the gas jet, K; additional indexes "0", "x", "y", "s" refer to the parameters of the initial section: on the axis, at the distance y from the axis and in the surrounding space, respectively; $Pr_T = \nu/a$ is the turbulent Prandtl number; ν is the kinematic viscosity coefficient, m²/s; $a = \lambda/\left(c_p \cdot \rho\right)$ is the coefficient of tubular temperature conductivity; m²/s; c_p^p is the specific heat capacity of the metal droplets at a constant pressure, J/(kg K).

Note that the volume distribution of the two-phase flow parameters is not usually considered in the calculations for AS. Hence, it reduces the accuracy of the calculations. Thus, Kretzschmar, Vadivasov considered parameters for the jet axis [34, 44]. Lyalyakin et al. calculated the parameters of the flying particles "… for the values of the gas parameters on the axis of the gas jet." [19].

Convective heat exchange, which is responsible for determination of the heat transferring in a two-phase jet, is estimated by the Nusselt number (heat):

$$Nu = q_n \cdot d_p/\left(\lambda_g \cdot \left(T_g - T_p\right)\right), \quad (2.104)$$

where, q_n is the specific heat stream onto the sphere, W/m².

For the spherical particles, the formulas like $Nu = f(Re, Pr)$ are used. During the study of individual fixed spherical particles in low temperature pressures conditions with constant thermal properties the dependencies that are mentioned by Klubnikin [11] were obtained:

$$Nu = 2 + 0.6 \cdot Re^{0.5} \cdot Pr^{0.333}, \quad \text{at}\, 0 < Re < 200, \quad (2.105)$$

$$Nu = 2 + 0.459 \cdot Re^{0.55} \cdot Pr^{0.333}, \quad \text{at}\, 0.1 < Re < 2 \times 10^5, \quad (2.106)$$

Tsvetkov proposed the following expression for Nu [35]:

$$Nu = 2 \cdot \lambda_{гп}/\lambda_г + 0.6 \cdot \mathrm{Re}^{0\,5} \cdot \mathrm{Pr}^{0:333} \cdot \left(\rho_g \cdot \mu_g/(\rho_{гп} \cdot \mu_{гп})\right)^{0.2}, \quad (2.107)$$

where μ is the coefficient of dynamic viscosity, Pa s; index «гп» refers to the properties of the gas at the particle surface temperature; index "g" refers to properties of the gas that has no contact with the particle.

Byrov reported that the experimental data about heat exchange between the particles moving in the stream can be determined by the formula [68]:

$$Nu = 0.186 \cdot Re^{0.8} \quad (2.108)$$

From the stated above dependencies, the most preferable for determination of the Nusselt number (2.105–2.108), is the equation of Tsvetkov (2.107). In Eq. (2.108) the deviation from the ideal gas properties (Prandtl number) is not considered. In comparison to the Eqs. (2.104–2.106), where a correction due to heat transfer conditions depending on the parameters of the boundary layer on the particle is added.

So due to above mentioned analysis lost on irradiation and evaporation of the droplets should be neglected. In the heat exchange equation of the balance of the sprayed droplet at AS, we will take into account only the thermal conductivity:

$$\frac{\mathrm{d}T_p}{\mathrm{d}x} = \frac{6 \cdot \alpha_T}{\rho_p \cdot c_p^p \cdot d_p} \cdot \frac{\left(T_g - T_p\right)}{W_p} \tag{2.109}$$

The other balance equation relates to the heat exchange "gas-metal droplets":

$$\frac{\mathrm{d}\left(c_p^p \cdot m_p \cdot T_p\right)}{\mathrm{d}x} = -\frac{d\left(m_g \cdot H_g\right)}{\mathrm{d}x} \tag{2.110}$$

$$\mathrm{d}H_g = c_p^g \cdot dT_g \text{ and } m_p / m_g = G_p / G_g$$

Let us do the same as in Sect. 2.5 and use the numerical method for solving the differential Eqs. (2.109 and 2.110). In this case, within the step limits (Δx), the values c_p^p and m_p may be considered as constant. Since $\mathrm{d}H_g = c_p^g \cdot \mathrm{d}T_g$ and $m_p / m_g = G_p / G_g$ the Eq. (2.110) should be transformed as follows:

$$\frac{dT_g}{dx} = -\frac{c_p^p}{c_p^g} \cdot \frac{G_p}{G_g} \cdot \frac{dT_p}{dx}, \tag{2.111}$$

where, c_p^p, c_p^g is the metal heat exchange of the particles and gas at the constant pressure; G_p, G_g are the mass consumption/flow of particles and gas.

Equations (2.109, 2.110) take the form:

$$T_p^{i+1} = T_p^i + \frac{6 \cdot \alpha_T^i}{\rho_p \cdot c_p^{pi} \cdot d_p} \cdot \frac{\left(T_g^i - T_p^i\right)}{W_p^i} \cdot \Delta x \tag{2.112}$$

$$T_g^{i+1} = T_g^i - \frac{c_p^{pi}}{c_p^{gi}} \cdot \frac{G_p}{G_g} \cdot \left(T_p^{i+1} - T_p^i\right) \tag{2.113}$$

To calculate the gas heat exchange with the surrounding space, let us use tubular jet dependencies (2.102, 2.103).

Thus, to determine the temperature in the selected section at each step the calculation is made successively to the Eqs. (2.112, 2.113, 2.102 and 2.103).

We assume that the temperature at the end of the electrode 2500 K, which corresponds to the values defined Erokhin and Pokhodnya at arc welding [4, 69]. According to our proposed scheme, in the "tongue" there is a further increase in temperature due to the appearance of necks. To evaluate the processes in the arc burning, it is assumed that in this zone the initial temperature of droplets is 2900 K with subsequent changes because of heat transfer due to "gas-droplet".

We assume that the droplets temperature after lowering to the melting point remains constant until the complete solidification of metal (the super cooling of the liquid metal is neglected). In the calculations, it is treated as follows. At each step of the calculation, the droplet temperature is compared with the temperature of its melting point (T_m). Upon reaching the later, the droplet temperature at each successive point remains constant. Moreover, the heat lost by the particle at all subsequent points is summed.

$$\sum_{i=1} Q^i = \sum_{i=1} c_p^i \cdot m_p \cdot \left(T^{i+1} - T^i\right) \tag{2.114}$$

This heat at each step is compared with a melting heat of the particles:

$$Q_{nn} = \Delta h_{nn} \cdot m_p, \tag{2.115}$$

where, $\Delta h_{пл}$ is a specific enthalpy of melting, J/kg.

When the following condition $\sum_{i=1} \mathrm{Q}^i > \mathrm{Q}_{nn}$ is met, the temperature fall of the solid particle is taken into account on the further steps of calculation.

Strictly speaking, the calculated temperature of the particle is mass average and is not constant over the cross-section of the droplets. However, the criteria $Bi \approx 0.04 \ll 1$ taking into account the large value of the thermal conductivity of the sprayed material (steel, $\lambda \approx 40$ W/(m K)) and the small diameter of particles. Therefore, the internal heat transfer, core and periphery temperature of droplets should be neglected in calculation.

The analysis carried out by the proposed calculation model revealed the following. On the axis of the jet, metal droplet temperature decreases to the melting point at a distance of 15–25 mm, (Fig. 2.36.) In the droplets of the majority of fractions, the melting heat $Q_{пл}$ is high enough to make this the temperature remained unchanged throughout the spraying distance and to keep the droplets in the liquid state. Only for fractions of 10 μm or less quantity of heat release exceeds $Q_{пл}$ at a distance of 100 mm. After that the temperature begins to drop below the melting point. Hence, the droplets solidify.

Gas and droplets temperatures are compared at a distance of 15–25 mm. Subsequently, droplets temperatures exceed the gas temperature, Fig. 2.37.

To analyze the interaction of droplets with the substrate, it is necessary to assess the state of the drop when approaching the sprayed surface. Let us introduce the concept of the solidification degree $R_{отв}$, representing the ratio of the total heat loss of the droplet $\sum_{i=1} Q^i$ to the heat of droplet melting $Q_{пл}$:

$$R_{отв} = \left(\sum_{i=1} Q^i \cdot 100\right) \Big/ Q_{пл} \tag{2.116}$$

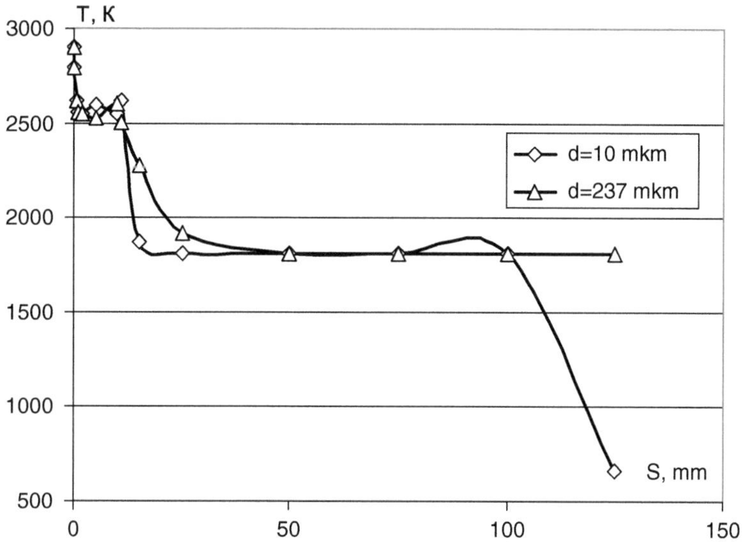

Fig. 2.36 Changing of the metal droplets temperature of different diameters along the spraying distance. The axis of the jet is considered

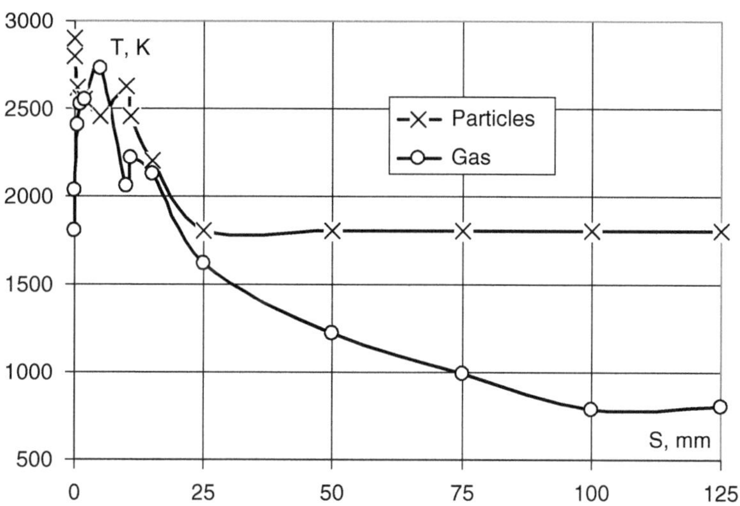

Fig. 2.37 The changing of the gas and metal temperatures along the spraying distance. The axis of the jet is considered. The diameter of droplets is 100 μm

Depending on the solidification degree, we can determine the droplet state. From the structure of the formula, it is clear that when $R_{отв} \leq 100\ \%$ the droplet is in a solid–liquid state. If heat losses exceed the heat of the droplet melting point, i.e., $R_{отв} \geq 100$ %, the droplet is in the solid state, and its temperature begins to drop below $T_{пл}$. Calculation of the solidification degree for the characteristic droplet diameters are shown in Table 2.19.

The solidification degree at the jet axis for the majority of the fractions is <20 % and varies in inverse proportion to the diameter of the droplets (see Fig. 2.38). Although, only the fraction of 10 μm is solidified completely at a distance of 100 mm. At the jets boundary, the solidification degree is higher. Fraction drops of 10 μm solidify at the distance of ~35 mm. The rest of the factions reach the solidification degree of 5–75 %, at a distance of 125 mm, Fig. 2.39.

Due to changes in the temperature and composition of the gas jet at the spraying distance is changed, the changes of thermophysical properties of the gas: density, viscosity, coefficient of diffusion, and others have occurred. On the axis of the jet

Table 2.19 Droplets solidification degree at the axis and the jet boundary

Droplets diameter	Cross-section point	*s*, mm				
		25	50	75	100	125
d_p = 10 μm	Axis	2.37	19.67	45.95	79.14	101.61
	Boundary	52.12	120.38	125.00	125.11	147.43
d_p = 27.5 μm	Axis	0.27	3.22	6.43	10.85	16.18
	Boundary	3.65	18.38	34.78	53.06	73.25
d_p = 56.5 μm	Axis	0.12	1.57	3.07	4.12	5.52
	Boundary	1.97	7.16	13.20	19.66	26.64
d_p = 100 μm	Axis	0.01	0.44	0.93	1.31	1.70
	Boundary	0.70	2.62	4.81	7.26	9.82

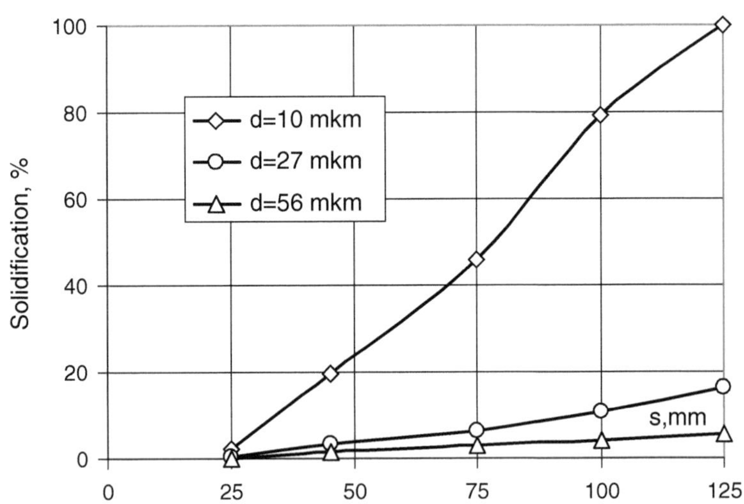

Fig. 2.38 The variation of the solidification degree of droplets of different diameters on the jet axis

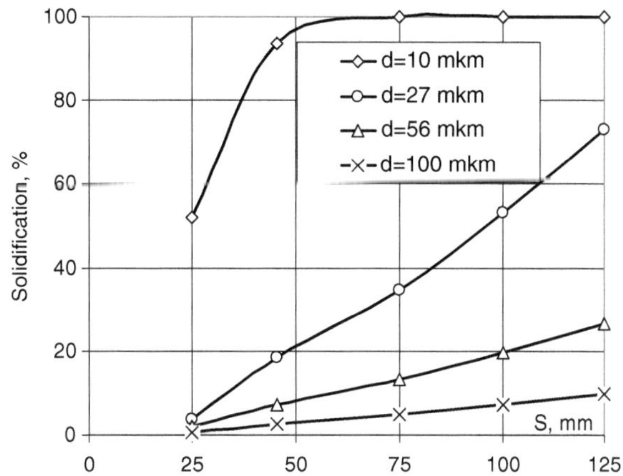

Fig. 2.39 The variation of the solidification degree of droplets of different diameters on the jet border

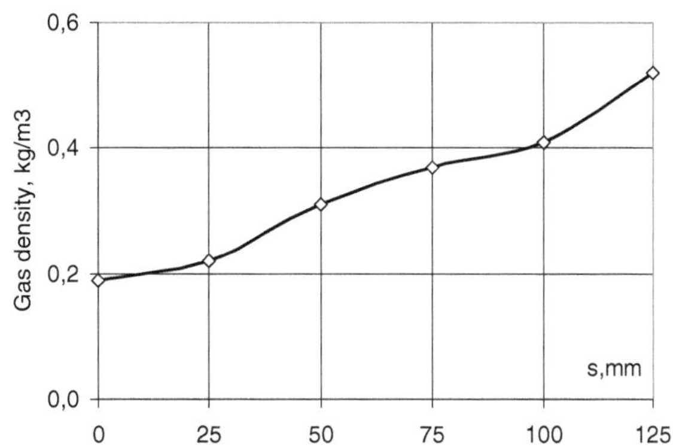

Fig. 2.40 Changing in the density of the gas on the jet axis

gas density increases by two times at a distance of 100 mm if compared to the nozzle face, Fig. 2.40.

The density of the gas is even higher along the cross-section of the jet. At the boundary of the jet, the gas density is 3–6 times higher than at the axis depending on the spraying distance, Fig. 2.41. This affects the angle of the velocity changing curves.

Changing of the number of characteristics is evaluated by the criterion dependences. The features of the flow are determined by the Reynolds number. For the case of external flowing around the flow is transferred to the turbulent regime at $Re_{\kappa} = 20\text{–}30$ [72].

According to calculations, the smaller diameter particles move in the laminar regime within the spraying distance on the jet axis. Particle having diameter of 45 μm and further up to the distance 50 mm move in the turbulent regime. Moreover, the turbulence is higher, the bigger the diameter is.

At the end of the initial section of the jet, where a sharp drop in the gas velocity begins, a turbulence peak for all diameters occurs. With the further increase of the

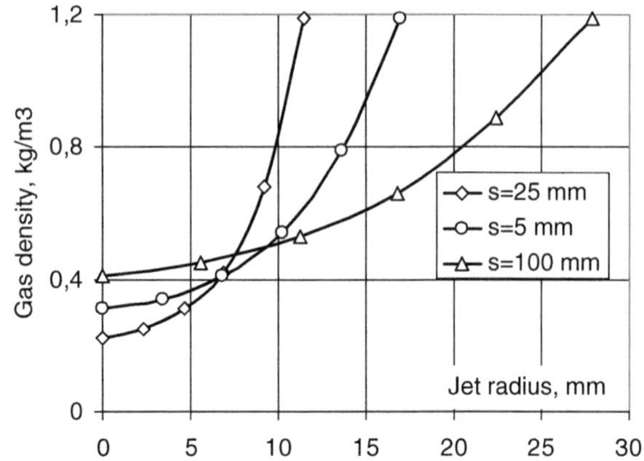

Fig. 2.41 Changing of the gas density at the cross-section of the jet at a different spraying distance

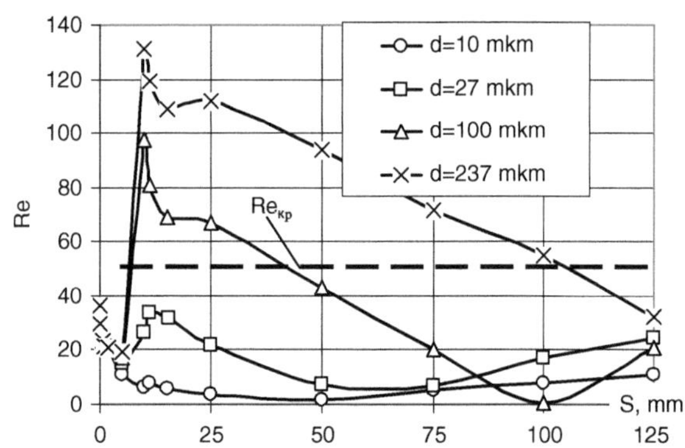

Fig. 2.42 Changing of the Reynolds number for the different diameters droplets at a spraying distance. The axis of the jet is considered

distance up to 125 mm on the jet axis, the flow regime of all the droplets is coming closer to laminar, Fig. 2.42. While distancing from the axis, the changing of *Re* has a complicated character that is determined by the value of the relative velocity of the droplets. Although, its value is greater than the critical $Re_{\kappa} = 20–30$.

Nusselt number characterizing the heat and the mass transferring is related to the Reynolds number that is reflecting the dynamic parameters of the gas and to Prandtl number characterizing the physical properties of the gas.

Calculation shows that $Pr = 0.7–0.8$. This corresponds to the data of Abramovich, who is taking the value of $Pr_T = Pr_D = 0.75$ for axisymmetric turbulent jet [53]. Figure 2.43 shows that the curve of the Nusselt number repeats the changes Re. However, these changes are expressed less drastically. Note also that $8 > Nu > 2$ is true for all factions and along the whole distance. Hence, it shows the deviation from the ideal case of Langmuir (heat and mass transfer toward the ball in the absence of air flow, $Nu = 2$).

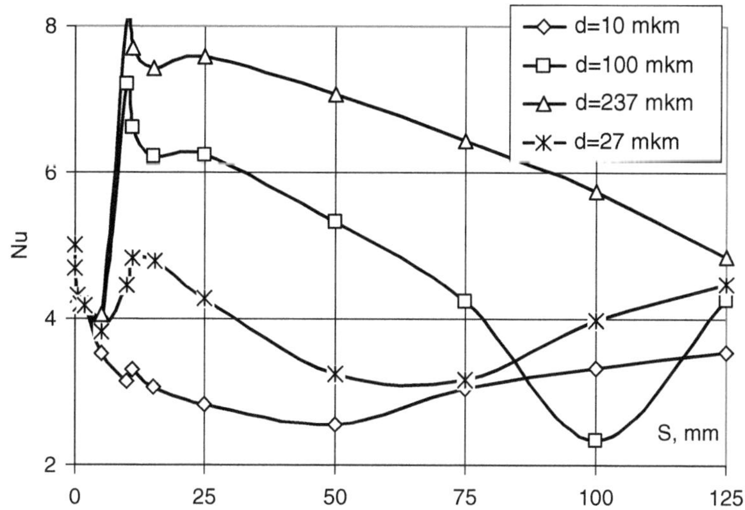

Fig. 2.43 Changing of the Nusselt number for the different diameters droplets at a spraying distance. The axis of the jet is considered

2.7 Chemical Composition of the Gas Jet

Despite the difference in the nature of high-temperature metallurgical processes and substances participating in gaseous atmosphere formation, the composition of the latter turns out to be of the same type. At temperatures below 2500 K, gas includes three groups of components: products resulting from the complete interaction with oxygen (CO_2, $H_2O_{(steam)}$, SO_3); product resulting from incomplete interaction with oxygen, from dissociation, from carbon elimination and gas removal from the metal (CO, H_2, CH_4, SO_2, O_2, N_2); inactive components (Ar and others.) [18]. The composition of such a gaseous atmosphere can be described on the basis of thermodynamic analysis of a few chemical reactions.

Arc spraying scheme use either compressed air or neutral and reducing gases as carrier medium. Compressed air spraying provides a stable atomization of the melting metal; nevertheless, the quality of the coating is not very high due to a large number of pores and oxides [57]. The use of noble gases for thermal spraying [70] is not technologically efficient because of their high price. Promising is the use of atmosphere "compressed air-carbohydrates". That will bring additional heat when the mixture is burning; and will decrease the oxygen penetration in metal drops (particles) in case of reducing gas jet. Following this variant, the use of methane mixed with air is suggested [71]. VNIITUVID "Remdetal" suggests the use of kerosene as carbohydrates [19]. Baranovskiy suggests the use of products resulting from the combustion of propane–air mixture as the carrier gas [72]. However, these papers either completely miss any quantitative calculations [70, 71] or contain some approximate ones [19, 72]. It does not allow the realization of the advantages of the selected scheme fully.

Equipment and technology improving requires more comprehensive understanding of the delivery of gases and the nature of their interaction. When investigating gas and plasma jets used for coating application, there is usually introduced an assumption regarding the consistency of the gas jet composition [73–75]. Such calculations are characterized by a low precision. In order to increase their reliability, it is desirable to calculate the composition of the gas in each point of the jet.

The local composition of gases in each point of the jet at every moment of time can be taken as equilibrium one due to the high rate of chemical reactions. The legitimacy of this assumption has been experimentally approved for fast-running processes in metallurgy [76], rocket engine design [77]. In this case, the composition of the gas mixture is determined on the basis of the balance of chemical reaction components taking into account blending of atmospheric oxygen and nitrogen to the jet beyond the initial part.

The necessity of calculation of the equilibrium gas composition when spraying was pointed out by Tsvetkov [35], nevertheless, any ad hoc solution was not given. Karp with colleagues made such a calculation for the process of plasma spraying in their work using the minimum of thermodynamic potential with additional logarithm of the system energy equation [78]. As at temperatures 4000–12,000 K typical for plasma processes, high gas ionization is highly possible, these calculations take into account atom dissociation.

At arc spraying, according to the estimation (2.2), the dissociation is thermodynamically unlikely. In respect to arc spraying using combustible mixtures, the calculation of equilibrium composition of combustion gases of current mixtures "combustible gas-air" was given by Lialiakin [19]. However, it does not take into account the decomposition of CO_2 and H_2O. It does not match, for example, the thermodynamic analysis of carbon dioxide dissociation according to reaction

$$2CO_2 = 2CO_2 + O_2,$$

which states the dissociation degree of 6.6 % at temperature 2000 K [43].

Besides changing of the concentration of components resulting from chemical reactions, the composition of the jet changes due to continuous blending of air from the atmosphere. In order to estimate oxygen blending into the plasma jet, Gershenzon used the regression dependence [79] based on partial experimental data [80], which narrows the application area of the calculation results. The analytical account of atmospheric gases blending into the gas jet in papers considering process simulation in two-phase jet at arc spraying does not occur.

This is the law of mass action that is used to determine the gas composition at arc spraying. In contrast to similar calculations for plasma spraying [78], the equations of equilibrium constants are considered as the closure equations instead of the energy equation. Simplifying assumptions are to be taken:

1. Main products resulting from the combustion of propane–air mixture are CO, CO_2, H_2, H_2O, N_2, O_2. Atom dissociation of O, N, H is to be neglected due to not very high temperatures (<2300 K). The presence of carbohydrates, which

completely dissolve at temperature above 1000 °C [81], can also be neglected. Pohodnia analyzed the possibility of NO formation at welding [69]. According to different citations, at temperature 2500 K the equilibrium concentration of NO is 2–6 mol%. However, due to the high energy of activation of NO formation reaction (527 kJ/mole), the equilibrium concentration time (τ_{NO}) goes up sharply along with decrease in temperature from 2.2×10^{-3} s at $T_g = 2600$ K to 0.27 s at $T_g = 2100$ K. Gas residence time in the arc burning zone during arc spraying is an order of magnitude less and NO formation is unlikely.
2. The composition of gas at the nozzle exit is close to the equilibrium composition corresponding to the temperature at this point. This assumption is not absolute because combustion products also contain over-equilibrium quantities of H, OH, and other radicals due to the chain mechanism of combustion [76].

In the combustion chamber propane burns according to the summary reaction

$$C_3H_8 + 5O_2 = 3CO_2 + 4H_2O \tag{2.117}$$

If volume ratio of oxidant (air) and propane is L_B, then air is assumed to contain 79 % of nitrogen and 21 % of oxygen, and stoichiometric coefficients of the reaction (2.117) is equal 23.8. Combustion completeness is characterized by excess oxidant ratio:

$$\alpha_{ox} = V_O / V_g \cdot L_B^{st}, \tag{2.118}$$

where V_O—oxidant volume in mixture; V_g—combustible gas volume in mixture.

Oxidant (air) amounts required for propane combustion at different degrees of oxidation are given in Table 2.20. Where n—ratio of oxygen and propane number of moles. It is necessary to take into account that in practice, due to imperfection of mixers, the complete combustion is achieved when $\alpha_{ox} \geq 1.05$ [82].

To improve the quality by decrease of the degree of oxidation of particles when spraying, it is desirable to neutralize the action of air oxygen. Nevertheless, in case of considerable decrease of α_{ox}, a part of heat from combustion is used for heating of the excessive, according to stoichiometry, component. It leads to a decrease in temperature of combustion products. Besides the decrease in thermal contribution of the gas jet, it reduces the velocity of the jet passing through the supersonic nozzle because the sound velocity (a_{3B}) is related to the gas temperature by the dependence:

$$a_{3B} = \sqrt{k \cdot g \cdot R \cdot T_g} \tag{2.13}$$

Table 2.20 Relative air volume required for propane combustion at different degrees of oxidation

α_{ox}	0.6	0.7	0.8	0.9	1.0	1.05	1.1	1.2
n	3	3.5	4	4.5	5	5.25	5.5	6
L_B	14.28	16.66	19.04	21.42	23.8	24.99	26.18	28.76

The velocity of particles goes down, respectively, and the degree of their oxidation can increase due to a longer residence in the gas jet even despite lower oxygen content in it. The summary effect can be determined only as the result of detailed calculations.

When the reducing mixture is supplied, the initial oxygen content is lower than according to stoichiometry (2.117) and at outlet from the nozzle there will be products of propane decomposition according to the reaction:

$$C_3H_8 = 3C + 4H_2 \tag{2.119}$$

When there is free carbon, oxygen content in the gas phase will be determined by the equilibrium of the reaction:

$$2C + O_2 = 2CO \quad 1g\ k_{PO} = 11672/T + 9.16 \quad [76] \tag{2.120}$$

When there is no carbon, it is sufficient to consider the equilibrium of two reactions only:

$$CO_2 = CO + 0.5O_2 \quad 1g\ k_{P1} = -14765/T + 4.54 \quad [76] \tag{2.121}$$

$$H_2O = H_2 + 0.5O_2 \quad 1g\ k_{P2} = -13160/T + 3.05 \quad [76] \tag{2.122}$$

Let us consider, first, the method of solving of this relatively simpler problem. The reaction equilibrium constants (2.121), (2.122) can be noted as:

$$k_{P1} = \left(P_{CO} \cdot P_{O_2}^{0.5}\right) / P_{CO_2} \tag{2.123}$$

$$k_{P2} = \left(P_{H_2} \cdot P_{O_2}^{0.5}\right) / P_{H_2O} \tag{2.124}$$

The summary number of gas moles is equal to:

$$\sum n_i = n_{CO_2} + n_{CO} + n_{O_2} + n_{H_2O} + n_{H_2} + n_{N_2} \tag{2.125}$$

The material balance equations take the form of:

$$n_C^u = n_{CO_2}^u + n_{CO}^u = n_{CO_2} + n_{CO} \tag{2.126}$$

$$n_H^u = 2 \cdot n_{H_2O}^u + 2 \cdot n_{H_2}^u = 2 \cdot n_{H_2O} + 2 \cdot n_{H_2}, \tag{2.127}$$

$$\begin{aligned} n_O^u &= 2 \cdot n_{O_2}^u + 2 \cdot n_{H_2}^u = 2 \cdot n_{H_2O} + 2 \cdot n_{H_2} \\ &= 2 \cdot n_{O_2} + 2 \cdot n_{CO_2} + n_{CO} + n_{H_2O} \end{aligned}, \tag{2.128}$$

$$n_{N_2}^u = n_{N_2}, \tag{2.129}$$

where $n^u_{N_2}, n^u_C, n^u_H, n^u_O$ are initial numbers of nitrogen moles, atomic carbon, hydrogen, and oxygen, respectively. From (2.125–2.129), we have

$$\sum n_i = n^u_C + 0.5 \cdot n^u_H + n_{N_2} + n_{O_2}, \tag{2.130}$$

$$n^u_O = 2 \cdot n_{O_2} + n^u_C + n_{CO_2} + n_{H_2O}, \tag{2.131}$$

The partial pressures of the components included in the Eqs. (2.123) and (2.124) are equal to:

$$P_{CO} = P \cdot n_{CO} \Big/ \sum n_i = P \cdot \left(n^u_C - n_{CO_2}\right) \Big/ = P \cdot \left(n^u_C - n_{CO_2}\right) \sum n_i, \sum n_i, \tag{2.132}$$

$$P_{CO_2} = P \cdot n_{CO_2} \Big/ \sum n_i, \tag{2.133}$$

$$P_{O_2} = P \cdot n_{O_2} \Big/ \sum n_i, \tag{2.134}$$

$$P_{H_2} = P \times n_{H_2} \Big/ \sum n_i = 0.5 \times P \times \left(n^u_H - n_{H_2O}\right) \Big/ \sum n_i, \tag{2.135}$$

$$P_{H_2O} = P \cdot n_{H_2O} \Big/ \sum n_i. \tag{2.136}$$

Inserting (2.123–2.136) into (2.123), (2.124) and eliminating the unknowns taking into account (2.125–2.131) we have:

$$n_{CO_2} = \frac{n^u_C \cdot B}{k_{P1} + B}; \quad n_{H_2O} = \frac{n^u_H \cdot B}{2 \cdot (k_{P2} + B)}; \quad n_{O_2} = \frac{B^2 \cdot A}{P - B^2} \tag{2.137}$$

$$-n^u_O + n^u_C + \frac{n^u_C \cdot B}{k_{p1} + B} + \frac{n^u_C \cdot B}{2 \cdot \left(k_{p2} + B\right)} + \frac{2 \cdot B^2 \cdot A}{P - B^2} = 0, \tag{2.138}$$

where

$$B^2 = P \cdot n_{O_2} \Big/ \sum n_i, \tag{2.139}$$

$$A = n^u_C + n^u_{N_2} + 0.5 \cdot n^u_H \tag{2.140}$$

Having solved the Eq. (2.138) for B, n_{O_2} can be found, then, from (2.125–2.131), mole fractions of the other gas mixture components and the mixture composition can be determined consequently. The Eq. (2.138) is a polynomial of degree four, which is solved by the method of successive approximations with magnitude B constraints resulting from the physical content of the Eq. (2.139): $B \leq P^{0.5}_{O_2}$.

Calculations beyond the initial part of the jet, take into account the blending of atmospheric oxygen and nitrogen into the jet. Fields of dimensionless concentrations and temperature by the Eqs. (2.102 and 2.103) are proportional to the curves of dimensionless values of velocity [53]:

$$\gamma_x = (C_x - C_s)/(C_0 - C_s) = 0.86 \cdot W_{gx}/W_{go}, \tag{2.141}$$

$$\gamma_y = (C_y - C_s)/(C_x - C_s) = \left(W_{gy}/W_{gx}\right)^{\mathrm{Pr}_D}, \tag{2.142}$$

where $C = G_g/\left(G_g + G_{air}\right)$—the additives concentration, which is the ratio of additives weight to the total weight of the mixture in the same volume in an arbitrary point of the jet per unit time. In particular, for the case of the gas flow through the nozzle into pure immobile air; $\mathrm{Pr}_D = \nu/D$ is the Prandtl diffusion number characterizing the physical properties of the gas; D is the coefficient of eddy diffusion of the gas.

If we denote the quantity of blended air moles by (z) and the quantity of gas moles by ($\sum n_i$) we can express dimensionless values of velocities as follows:

$$\gamma = \sum n_i/\left(\sum n_i + z\right), \tag{2.143}$$

where $\gamma = \gamma_x \cdot \gamma_y$.

One can see from the Eq. (2.143) that:

$$z = \sum n_i \cdot (1 - \gamma)/\gamma \tag{2.144}$$

Consequently:

$n_{N_2 \mathrm{blended}} = 0.79 \cdot z$ is the quantity of blended nitrogen moles
$n_{O_2 \mathrm{blended}} = 0.21 \cdot z$ is the quantity of blended oxygen moles

If we divide the jet into intervals Δx and each jet section—into intervals Δy, inside which the quantity of blended air moles can be considered as constant, we have the following equations for any section Δx and Δy:

$$n^j_{N_2} = n^{j-1}_{N_2} + 0.79 \cdot \sum n^{j-1}_i \cdot (1 - \gamma)/\gamma \tag{2.145}$$

$$n^j_{O_2} = n^{j-1}_{O_2} + 0.21 \cdot \sum n^{j-1}_i \cdot (1 - \gamma)/\gamma \tag{2.146}$$

$$n^j_C = n^{j-1}_C \tag{2.147}$$

$$n^j_H = n^{j-1}_H \tag{2.148}$$

Let us consider the case when the coefficient of oxidant excess $\alpha_{ox} < 1$, that is, the atmosphere is reducing. As noted above, when propane is completely decomposed, the gas will contain elemental carbon, for the reason it is necessary to consider the equilibrium of the reaction (2.120). Besides that, the balance equation of carbon in the gas (2.126) will change as follows:

$$n_{\mathrm{C}}^{\mathrm{u}} = n_{\mathrm{CO}_2} + n_{\mathrm{CO}} + n_{\mathrm{C}} \tag{2.149}$$

where n_{C}—is the quantity of elemental carbon moles.

It complicates the calculation because it requires establishing of a new more complex system of equations. However, specific conditions of the problem under solution allow the use of a simpler method. The calculation starts by the maximum values $\alpha_{\mathrm{ox}} > 1$, when the gas is a priori free from carbon. The calculation continues by steady decrease in α_{ox} with checking of carbon occurrence conditions in each cycle, which results from the equilibrium of reaction (2.120):

$$\mathrm{P}_{\mathrm{CO}}^2 / \mathrm{P}_{\mathrm{O}_2} \geq k_{P0} \tag{2.150}$$

when free carbon occurs, due to large values of equilibrium constants of its interaction with O_2, H_2O, CO_2 (10^5–10^8, according to [52]), the gas in equilibrium will contain almost only H_2, CO, and N_2. In this case:

$$n_{\mathrm{CO}} = n_{\mathrm{O}}^{u}, n_{\mathrm{N}_2} = n_{\mathrm{N}_2}^{u}, \quad n_{\mathrm{H}_2} = 4 \cdot n_{\mathrm{C}_3\mathrm{H}_8}^{u} \tag{2.151}$$

Consequently

$$\sum n_i = n_{\mathrm{O}}^{\mathrm{u}} + n_{\mathrm{N}_2}^{\mathrm{u}} + 4 \cdot n_{\mathrm{C}_3\mathrm{H}_8}^{u} \tag{2.152}$$

and

$$\mathrm{P}_{\mathrm{CO}} = P \cdot n_{\mathrm{O}}^{u} \Big/ \sum n_i; \quad \mathrm{P}_{\mathrm{H}_2} = P \cdot 4 \cdot n_{\mathrm{C}_3\mathrm{H}_8}^{u} \Big/ \sum n_i \tag{2.153}$$

Partial pressures of H_2O and CO_2 are found from the Eqs. (2.123) and (2.124).

If necessary, the calculation can be repeated until the required precision of values n_{H_2} and n_{CO} is achieved by the Eqs. (2.126 and 2.127) calculating the equilibrium constants by the Eqs. (2.120–2.122). The initial quantity of moles of components in terms of reactions of propane combustion and decomposition is equal to:

$$n_i^u = G_g^i \cdot 10^3 / M_i, \mathrm{mol/h} \tag{2.154}$$

where G_g^i, M_i is the consumption and the molecular weight of the i—component.

For specific components of the gas mixture the temperature dependences of components properties were obtained by the regression analysis of the reference data (Table 2.21).

In order to account the influence of components on the properties of the gas mixture there are used different methods. Thus, while determining the heating of

Table 2.21 Temperature Dependences of the thermophysical properties of mixture components, T = 298–2500 K, P = 0.1 MPa [9, 62, 83]

	CO	CO_2	H_2	H_2O	N_2	O_2
Heat capacity $c_P = A + B_1 \cdot T + B_2 \cdot T^2 + B_3 \cdot T^3 + B_4 \cdot T^4$, kJ/(kg K)						
A	0.309727	0.401553	17.364959	0.370223	0.985015	0.800472
B_1	0.001003	1.639687	−0.006933	5.513101	0.000142	0.000416
B_2	0	−0.595574	0.000004	−5.250974	0	0
B_3	0	−0.502355	0	1.690904	0	0
B_4	0	0.288831	0	0	0	0
Dynamic viscosity $\eta = A + B_1 \cdot T + B_2 \cdot T^2$, kg/(m s)						
A	53.470470	90.821075	32.467186	−30.86506	74.161049	83.962339
B_1	0.452274	0.345442	0.204200	0.407145	0.388818	0.470983
B_2	−0.000110	−0.000042	−0.000035	0	−0.000066	−0.000072
Thermal conductivity, $\lambda = A + B_1 \cdot T + B_2 \cdot T^2$, W/(m K)						
A	5.0661668	−10.05457	91.495468	15.147091	11.160771	3.392105
B_1	0.070140	0.092531	0.325269	0.001003	0.058519	0.077837
B_2	−0.000009	−0.000014	0.000034	0.000084	−0.000003	−0.000010
Density, $\rho = A/T$, kg/m^3						
A	347	530	24.1	216	348	376

particles in the plasma jet, first calculated is nitrogen as plasma-formating gas, then the results are transferred on other gases [75]. In other cases, there are used average parameters and additive dependences are taken for them proportionally to mole fractions of components [73, 84]:

$$g_m = \sum g_i \cdot x_i \tag{2.155}$$

where g_m is a mixture property; g_i, x_i are a property and mole fraction of component, respectively.

This is correct for calculations of density and heat capacity of gas mixtures. Nevertheless, the additive determination of kinetic characteristics of gas mixtures—thermal conductivity, viscosity, diffusion coefficient—gives significant inaccuracy due to the occurrence of the potential interaction between molecules besides the kinetic one. Regarding the gas mixture at plasma spraying, Strongin accounts the thermal conductivity depending on the interaction potential, at the same time, he takes the viscosity of the gas mixture as a constant value, which does not depend on the temperature [85].

To account the influence of intermolecular forces, the interaction integral Ω is calculated according to the value of the potential interaction energy $\Psi(r)$. The value Ω is equal to 1 if the intermolecular attraction does not exist. $\Psi(r)$ represents the potential molecule interaction energy at the distance r from each other. The ratio of Ψ and r is expressed as the interaction potential. The most often used interaction potential is Lennard–Jones potential

$$\Psi(r) = 4 \cdot \varepsilon \cdot \left[(\sigma/r)^{12} - (\sigma/r)^{6} \right], \tag{2.156}$$

where σ is the collision diameter, the distance between molecules when $\Psi(r) = 0$; ε is the characteristic energy corresponding to the minimum of the interaction curve.

The available data regarding the mutual influence of gas components are systematized in the monograph [86]. They allow the calculation of kinetic properties of a gas mixture.

The mixture viscosity is determined according to the Chapman-Enskog kinetic theory through the equation:

$$\eta = {}^{5}\!/_{16} \cdot (\pi \cdot M \cdot R \cdot T)^{0.5} \Big/ \left((\pi \cdot \sigma^2) \cdot \Omega \right) \tag{2.157}$$

The computational solution of this equation is approximated by the series

$$\eta_{\mathrm{CM}} = \sum_{i=1}^{n} x_i \cdot \eta_i \Big/ \sum_{i=1}^{n} \left(x_j \cdot \Phi_{ij} \right), \tag{2.158}$$

where Φ-parameter is determined by Vilke approximation:

$$\Phi_{ij} = \left[1 + \left(\eta_i / \eta_j \right)^{0.5} \cdot \left(M_j / M_i \right)^{0.25} \right]^2 \Big/ \left[8 \cdot \left(1 + M_i / M_j \right) \right]^{0.5}, \tag{2.159}$$

$$\Phi_{ji} = \eta_j \cdot M_i \cdot \Phi_{ij} / \left(\eta_i \cdot M_j \right) \tag{2.160}$$

For example, for the binary system the Eqs. (2.158–2.160) will be as follows:

$$\eta_{\mathrm{CM}} = x_1 \cdot \eta_1 / (x_1 + x_2 \cdot \Phi_{12}) + x_2 \cdot \eta_2 / (x_2 + x_1 \cdot \Phi_{21}),$$
$$\Phi_{12} = \left[1 + (\eta_1/\eta_2)^{0.5} \cdot (M_2/M_1)^{0.25} \right]^2 \Big/ \left[8 \cdot (1 + M_1/M_2) \right]^{0.5},$$
$$\Phi_{21} = \eta_2 \cdot M_1 \cdot \Phi / (\eta_1 \cdot M_2)$$

According to Vilke, while making calculations according to the Eqs. (2.158–2.160), the average deviation of η_{SM} from experimental data is <1 %.

The thermal conductivity of the mixture is not usually the linear function of the composition as well. If the molecules of components have very different polarities, the thermal conductivity of the mixture is higher than the values obtained following the rule of additivity. For non-polar molecules, the trend is converse. Similar to the theoretical ratio for the viscosity, the thermal conductivity of the gas mixture is expressed in the form of the equation of Vasilieva:

$$\lambda_{CM} = \sum_{i=1}^{n} x_i \cdot \lambda_i \Big/ \sum_{j=1}^{n} \left(x_j \cdot A_{ij}\right) \tag{2.161}$$

For A-parameter, there is used the Lindsay-Bromley modification obtained on the basis of the Sutherland gas model

$$A_{ij} = \frac{1}{4} \cdot \left\{ 1 + \left[\frac{\eta_i}{\eta_j} \cdot \left(\frac{M_j}{M_i} \right)^{3/4} \cdot \frac{T + S_i}{T + S_j} \right]^{1/2} \right\} \cdot \frac{T + S_{ij}}{T + S_i}, \tag{2.162}$$

$$A_{ji} = \frac{1}{4} \cdot \left\{ 1 + \left[\frac{\eta_j}{\eta_i} \cdot \left(\frac{M_i}{M_j} \right)^{3/4} \cdot \frac{T + S_j}{T + S_i} \right]^{1/2} \right\} \cdot \frac{T + S_{ji}}{T + S_j}, \tag{2.163}$$

where S_i, S_j are the Sutherland constants found by the concordance of the Sutherland equation with experimental data, dependence $S_i = 1.5\mathrm{T_{boil}}$ (except He, H_2, Ne, for which $S_i = 79$ K) is taken empirically; $S_{ij} = 0.73 \cdot \left(S_i \cdot S_j\right)^{0.5}$ is the Sutherland interaction constant.

The values of the boiling point ($\mathrm{T_{boil}}$) are given in Table 2.22 [9]. An average calculation inaccuracy in the Eq. (2.161) is 1.9 %.

The component diffusion coefficients can be calculated like for a binary mixture if one of the components is taken as the main one. To calculate the diffusion coefficient for the binary mixture "A–B" the following equation is used:

$$D_{AB} = 1,858 \cdot 10^{-7} \cdot T_g^{1.5} \cdot \left[(M_A + M_B)/(M_A \cdot M_B)\right]^{0.5} \Big/ \left(P \cdot \sigma_{AB}^2 \cdot \Omega_D\right), \tag{2.164}$$

where

$$\sigma_{AB} = (\sigma_A + \sigma_B)/2 \tag{2.165}$$

Table 2.22 Auxiliary parameters for calculation of the thermal conductivity and diffusion coefficient in the gas mixture (carbohydrate-air)

	Mixture components						
	CO	CO_2	H_2	H_2O	N_2	O_2	
T_{boil}, K	81.50	194.65	20.40	637.00	77.35	90.18	
σ, Å	3.690	3.941	2.827	2.641	3.798	3.467	
ε_{AB}/k, K	91.7	195.2	59.7	809.1	71.4	106.7	
Equation (2.166) coefficients							
A_1	A_2	A_3	A_4	A_5	A_6	A_7	A_8
1.06036	0.15610	0.19300	0.47635	1.03587	1.52996	1.76474	3.89411

is the characteristic distance,

$$\Omega_D = \frac{A_1}{T^{*A_2}} + \frac{A_3}{\exp A_4 \cdot T^*} + \frac{A_5}{\exp A_6 \cdot T^*} + \frac{A_7}{\exp A_8 \cdot T^*} \tag{2.166}$$

is the collision integral for the diffusion,

$T^* = k \cdot T_g / \varepsilon_{AB}$ is the dimensionless temperature,

$$\varepsilon_{AB} = (\varepsilon_A \cdot \varepsilon_B)^{0.5}$$

The values of auxiliary parameters are given in Table 2.22. The calculation inaccuracy in the Eq. (2.164) is 7–12 %.

Based on data found in the scientific literature, the density, heat capacity of gas mixtures were determined proportionally to the mole fraction of components according to the additive dependences. The kinetic characteristics of the gas mixture—thermal conductivity, viscosity, diffusion coefficient—were found taking into account the mutual influence of the components according to the additive dependences, Eqs. (2.158, 2.161 and 2.164).

The calculation results were analyzed for the jet core, where the influence of the initial content of carrier gas is maximal. Figure 2.44 shows the change in the content of oxygen-containing components of the gas along the distance when the jet is initially reducing. It is seen that due to intense atmospheric air blending beyond the initial part ($x > 25$ mm), the jet has almost the same composition as the air. When the initial gas compositions are different, oxidizing or reducing, sharp

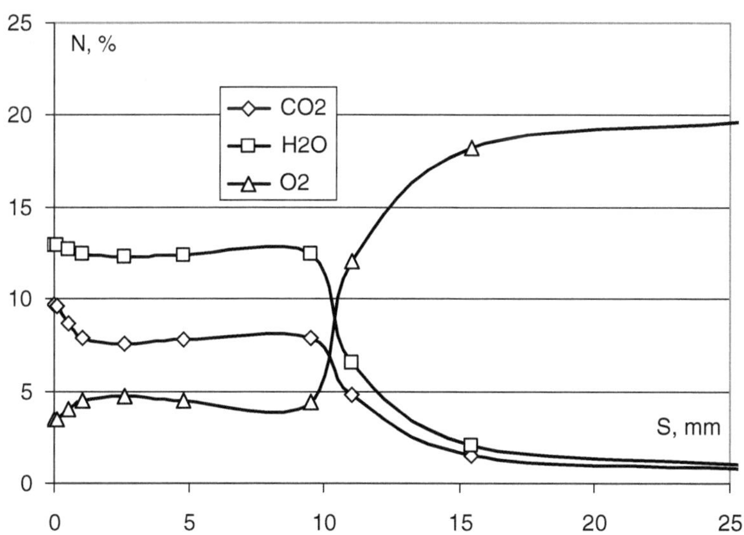

Fig. 2.44 Oxidant content in gas. Jet Core. $\alpha_{ox} = 0.8$

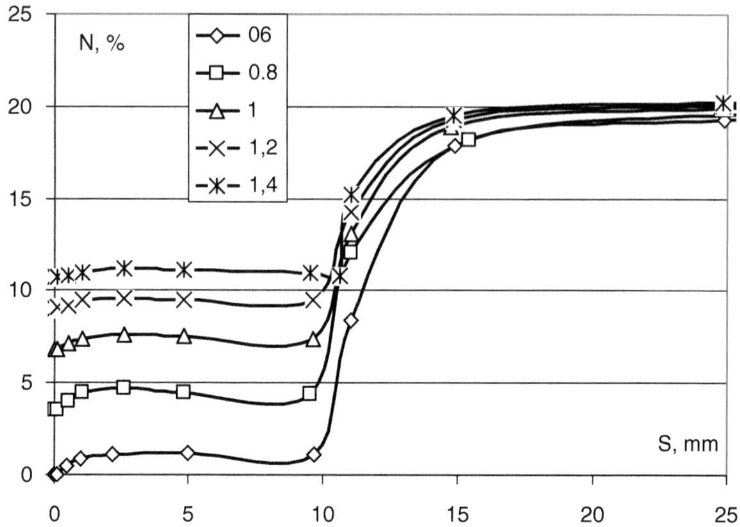

Fig. 2.45 Oxygen content in gas with different initial coefficients of oxidant excess. Jet Core

differences in oxidant content from the surrounding atmosphere are only in the initial part of the jet. The behavior of one of the oxidants—oxygen—is shown in Fig. 2.45. The calculation does not show any significant difference of jet composition in the rest of the distance.

2.8 Kinetic of Interaction of the Sprayed Metal with Oxygen

The quality of coatings obtained by arc spraying depends, in particular, on the quantity of oxygen solved in the metal of particles and contained in oxides. The process of oxygen delivery to the metal drops is important from the point of view of the final properties of coating. The principals of this process allow the prediction of properties of the coatings obtained depending on the modes of spraying or, by solving the converse problem, the selection of wire composition in order to obtain a coating of the required composition.

2.8.1 Physical Pattern of the Process

Based on the idea of the character of the interaction between melted metal and gas at arc spraying, as well as taking into account experimental data about slag behavior on metal drops within the gas jet, the physical pattern of oxygen delivery to the sprayed metal can be presented as follows.

First, all oxygen delivered is solved in the liquid metal until the saturation point. Then, slag forms on "metal-gas" interface. This process involves not only the oxygen coming from the gas phase but also the oxygen discharged by the supersaturated solution as the result of metal cooling from the overheat temperature, 2500–2900 K at the end face of the electrode and in the arc burning zone, to the melting point when drops are flying.

The reaction surface at the end face of the electrode is renewed in cycles due to the removal of liquid metal film with the frequency of drop discharge.

After film removal, the liquid metal surface opens. Oxygen content in it is equal to the initial solid metal, which is significantly below the saturation point at the temperature on the end face.

Experimental data show that the surface of flying drops is irregularly covered with slag.

In the paper of Borisova and colleagues, there are microphotos of particles obtained during the arc spraying of wire 65Г (like steel ASTM 1566) [46]. It is noted that all particles, as a rule, are covered with a layer of iron oxides, which can crack or partially pitted. There occur particles representing a complete oxide casing submitted to a partial destruction, Fig. 2.46.

In case of crucibleless melting of metal in high-frequency magnetic field, Yavorskiy and his colleagues described their observation of metal drops formed as the result of melting of specimens in the inductor [87]. The drops were blown with inert gas flow. Then, they added oxygen to the gas blowing a drop during a certain period of time. The observation showed that when carbon content is $C < 1$ wt%, there are formed particles of oxide phase on the drop surface, which are visible due to their different, compared to the metal, color, and brightness. The slag film does not cover the entire metal surface but floats on it in the form of tiny islands, which flow to the lower part of the specimen, where they gather into a larger structure, which breaks away from the metal drop. It was often observed that the metal stirred by electromagnetic field caught these slag islands, and their reverse descending on the surface.

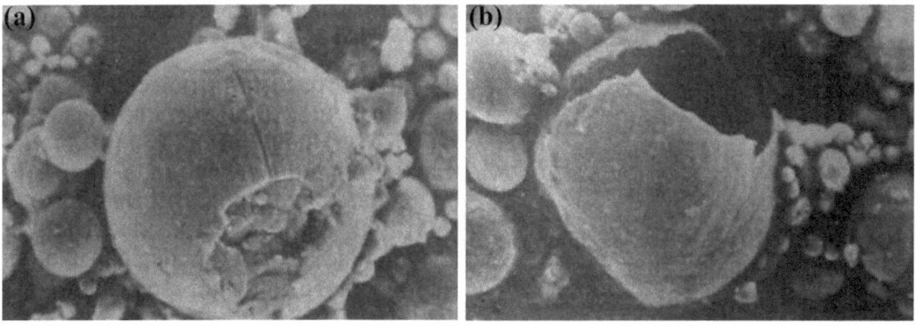

Fig. 2.46 View of particle of steel 65Г caught from spraying jet. **a** Cracking and partial pitting of the oxide casing on a particle, x550; **b** partially destroyed empty oxide casing, x600

Our investigations of the structure and properties of coatings have shown the presence of individual layers in the coating consisting of iron oxides [88].

In order to give a quantitative description of processes taking place, the idea of the continuous change of the area covered with slag is to be replaced by a model, according to which the spherical particle is filmed with slag. The area of the slag film is constant or changes under a certain law. The slag film is pushed aside under the influence of the gas flow. According to the calculations made before, at the most part of the spraying distance the gas velocity is higher than the velocity of the drops that is why the rear side of the drop surface is mainly clear of the slag film. The value of the surface share clear of slag can be assessed by the parameter (*L*):

$$L = f_s / f_d, \tag{2.167}$$

where f_s, f_d are slag covered area and total area of the drop surface, respectively.

It is most likely that the value of the area clear of slag is continuously changing due to changes in drop velocity relative to the gas and existence of drop rotation because of the turbulence. The calculations assumed a conventional scheme, according to which the share of the surface clear of slag remains the same during all the life time of a drop. Just opposing the calculations with the measures of oxygen content in the coating—the end point of drop flight—it is possible to take an average value *L*.

It is found that nitrogen in the presence of an arc discharge can pass into liquid metal in quantities several times larger than its equilibrium solubility under the similar conditions in the absence of an arc discharge [89]. However, the formation of supersaturated solution is less likely for oxygen. More likely, when the temperature is going down, oxygen will discharge from the solution and interact with iron forming FeO.

Thus, there are two types of reaction surfaces on a drop, which show different behavior while interacting with oxygen:

- Surface clear of slag. Oxygen from the gas phase comes on this surface and interacts with liquid metal forming slag. The latter, under the influence of the gas flow, is pushed aside from this surface opening the metal surface again for the interaction with the gas phase.
- Surface filmed with slag. Interdependent reactions between the components of the liquid metal and slag take place on the interphase interface “metal-slag”.

The total quantity of oxygen (m_O) having come into the drop of the *i*—fraction will be equal to the sum of oxygen solved in the liquid metal and contained in the slag:

$$m_O = m_{\mathrm{Me}} \frac{[\mathrm{O}]}{100} + m_w \left(\sum_{\mathrm{i}} \frac{(\ni_{\mathrm{m}} O_n)_i \cdot M_O}{100 \cdot M_{(\ni_{\mathrm{m}} O_n)}} \right), \tag{2.168}$$

The degree of oxidation of particles in this fraction r_i will be:

$$r_i = (m_O \cdot 100)/(m_{Me} + m_w) \tag{2.169}$$

The general degree of oxidation R is summed according to the distribution of drops by the weight fraction:

$$R = \left(\sum (r_i \cdot dm_i)\right) \cdot 100, \tag{2.170}$$

where for the drops of the i—fraction: m_{Me} is the weight of the liquid metal, kg; $m_{ш}$ is the weight of slag, kg; dm_i is the weight fraction of a component; M is the molecular weight of a component, kg/mole; $(Э_m O_n)_i$ are concentrations of the i—component of slag, wt%.

2.8.2 Principles of Thermodynamic and Kinetic Calculations of Oxygen Delivery

Let us consider the features of oxygen delivery to the clear and slag-covered surfaces.

2.8.2.1 Free of Slag Surfce

Macrokinetics divides the conditions of reactions into three areas: kinetic, diffusion, and transition ones.

A reaction takes place in the kinetic area if it is limited by the rate of the chemical transformation itself. In this case, the rate of the chemical transformation is proportional to the concentration of the reacting substances according to the law of mass action:

$$w = kc_1^{\nu_1} c_2^{\nu_2}, \tag{2.171}$$

where k is the constant of the reaction rate depending on the temperature in accordance with the Arrhenius equation; ν_1, ν_2 are stoichiometric coefficients;

$$k = Ze^{-\frac{E}{RT}}, \tag{2.172}$$

Z is the factor, which reflects the probability of the transformation; T is the temperature; R is the gas constant; E is the activation energy (the level of energy when a molecule or another particle becomes reactive).

Along with the temperature growth, the rate of the chemical transformation goes up quickly, whereas the velocity of delivery and taking off chemical agents in liquid media changes much more slowly. Consequently, under high-temperature conditions inherent to arc welding processes, gas passes into metal in the diffusion area; that is, the rate of the reaction between metal and gas is limited by transport agent (molecular or convection diffusion) [4, 69].

Let us introduce the basic concepts of the mass transfer—processes of chemical agent transfer to the phase interface (reaction zone) and reaction product take off from this interface [4, 56, 90].

The specific diffusion flow j, kg/(m^2 s), which express the velocity of mass transfer, is equal to the amount of the substance transferred through a unit of surface per unit time:

$$j=\beta(c_{об}-c_{пов}), \tag{2.173}$$

where β is the coefficient of mass transfer, m/s; $s_{об}$, $s_{пов}$ are the concentrations of a substance in the liquid or gas volume and on the phase interface, respectively, kg/m^3.

If a heterogeneous reaction takes place in the diffusion area, there should be supported an equilibrium on the phase interface: $s_{пов} = s_p$. In this case the Eq. (2.173) can be written as:

$$j=\beta(c_{об}-c_{пов})=\beta\Delta c \tag{2.174}$$

In physical–chemical hydrodynamics, it is assumed that a substance in the liquid or gas volume is distributed uniformly in consequence of convective stirring and all changes in concentration related to its transfer take place in the thin interface, where the main role is played by molecular diffusion. The thickness of this layer δ is conventionally represented as the diffusion path. Then, in compliance with the first Fick's law, when the linear change in interface is admitted, the diffusion flow is equal to:

$$j=D\cdot\Delta c/\delta \tag{2.175}$$

Comparison of the Eqs. (2.174) and (2.175) gives:

$$\beta=D/\delta \tag{2.176}$$

The transfer of the substance in the interface is also influenced by convection diffusion, thus, δ must be considered as a nominal value determined by the relation (2.176).

At different spraying process stages, the relation of values of the diffusion flows in gas and metal can change. So, in Gershenzon's paper [79] concerning plasma spraying, the delivery of oxygen into drops following the mechanism of external diffusion was explained without taking into account the kinetics of interaction in the system "metal-slag". According to the calculation made in this paper, the amount of oxygen delivered to a particle will be:

$$m_O = \int_0^\tau j \cdot F \cdot d\tau = \pi \cdot d^2 \int_0^l \frac{j}{V_p} dl, \quad (2.177)$$

where l—spraying distance, m; d, F, V_p—diameter, m, surface area, m^2, and velocity of a particle, m/s; j—specific diffusion flow, kg/(m^2 s).

Taking that the oxygen concentration in a particle is equal to 0, it was found:

$$j = \beta_g \cdot c_O = Nu \cdot D \cdot c_O / d, \quad (2.178)$$

where β_g—coefficient of mass transfer in gas, m/s; c_O—concentration of oxygen in gas, kg/m^3; D—coefficient of diffusion in gas, m^{2}s.

Considering the diffusion in gas only without considering the diffusion in metal like it was done by Gershenzon leads to the fact that the calculated degree of drop oxidation is insensitive to the chemical composition of the starting materials. This differs from the data obtained by Novozhilov for the similar, as far as the temperature level is concerned, welding process. His experiments showed that the amount of oxygen having reacted with metal during welding of austenitic steel 12X18H9T (like steel ASTM 302) was about two times larger than the amount of oxygen having reacted during the welding of pearlitic steels 30XГCA, Cт.3, and Cт.3кп (like steels 30CrMnSiA and A57036, correspondingly) [91]. Besides that, Gershenzon did not take into account the eventual oxygen saturation of the liquid metal.

Correspondingly, the following slag formation and the kinetics of slag interaction with metal were not taken into account as well. This is in contradiction with the results of the experimental studies. Besides that, according to the data obtained by vacuum melting, the oxygen content in the coating in case of arc spraying is 2–6 % [88, 92], which is an order of magnitude higher than the oxygen saturation limit in liquid iron at the melting point, $[O]_{Fe}^{sat} \approx 0.17\,\%$. This also confirms the presence of slag in the coating and, consequently, in drops forming the coating.

To choose the limiting stage for arc spraying, let us estimate possible limiting diffusion flows in gas $J_g^{\lim it}$ and in metal $J_{Me}^{\lim it}$:

$$J_g^{n} = \beta_g \cdot c_g = \beta_g \cdot P_{O_2} / (2 \cdot R \cdot T), \quad (2.179)$$

$$J_{Me}^{npeд} = \beta_{Me} \cdot c_{Me} = \beta_{Me} \cdot \rho_{Me} \cdot [O]_{Fe}^{Hac} / (100 \cdot M_O) \quad (2.180)$$

where β_g, β_{Me}—coefficients of mass transfer in gas and metal, m/s; s—oxygen concentration, wt%; P_{O_2}—partial pressure of oxygen, MPa; T—temperature, K.

The results of calculations of limiting diffusion flows for different process stages are given in Table 2.23. β_{Me} value for the arc zone is taken according to the estimation [4, 89], for the flying stage—like for the melt at plasma-hydrogen deoxidation [89].

Table 2.23 Estimation of the limiting diffusion flows of oxygen in gas and metal

Stages	1. End face of electrode	2. Arc burning zone	3. Distance, mm				
			10	25	50	75	100
Tg, K	2400	2900	2102	1266	968	784	712
$P_{O_2} \times 10^{-5}$, Pa	0.0485	0.0713	0.21	0.21	0.21	0.21	0.21
$V_г$, m/s	1127	1127	889	571	351	254	199
V_p, m/s	0.045	7.63	69	92	105	111	114
Re	238	19.5	273	269	221	184	127
Nu	10.3	4.0	11.1	10.9	10.2	9.2	8.1
$D \times 10^4$, m^2/s	6.6	9.1	3.6	2.1	1.3	0.9	0.7
$\nu \times 10^4$, m^2/s	4.8	5.7	2.7	1.6	1.0	0.7	0.6
$\beta_{Me} \times 10^2$, m/s	0.25	0.25	0.025	0.025	0.025	0.025	0.025
$\beta_g \times 10^2$, m/s	6844	36298	4224	2562	1469	924	673
I_g^n, mole/($m^2 \cdot s$)	0.13	55	26	26	21	15	11
I_{Me}^n, mole/(m^2 s)	14	41	0.2	0.2	0.2	0.2	0.2

The coefficient of mass transfer in gas was estimated according to Nusselt criterion, which was determined in compliance with criterial dependences according to [35]:

$$\beta_g = \text{Nu} \cdot \text{D/d}, \tag{2.181}$$

$$Nu = 2 + 0.459 \cdot Re^{0.55} \cdot Pr^{0.75},\ Re = W \cdot d/\nu,\ Pr = \nu/D \tag{2.182}$$

where d—characteristic size, m.

Diffusion and viscosity coefficient values in gas (D, ν) were calculated taking into account the mutual influence of the mixture components. The corresponding gas composition at the nozzle exit was calculated for the case of metal spraying by products of propane–air mixture combustion, $\alpha_{ox} = 0.8$. As d value there were taken: the gap between electrodes—for the end face of the electrode, the thickness of the "tongue"—for the burning zone, drop diameter—for the flying stage. According to the estimation of the gap size between the electrodes made in the paper [22], we take $d = 0.1–1$ mm for the end face of the electrode. The thickness of the tongue is taken as the smallest drop diameter of the sprayed metal, $d = 10 \times 10^{-6}$ m, according to the data obtained from the sieve test. As for the flying stage, the most likely diameter of sprayed particles $d = 90 \times 10^{-6}$ m is taken.

When comparing the calculated values of the limiting diffusion flows, it is seen that at the end face of the electrode we deal with the inner problem of the mass transfer and for stages 2 and 3 the outer problem takes place. The value of the mass transfer coefficient in gas is more than an order of magnitude higher than at welding or plasma-arc melting [89]. It can be explained by small size of drops transferred and high relative velocity of the drops and gas compared to the above mentioned processes. The order of magnitude β_g at the distance 25–125 mm corresponds to the values of this parameter for plasma-arc melting given by Lakomskiy [89]. β_g is several times higher at the starting segment of the jet, where temperatures are higher and drops are smaller. If taken are drops from the side fractions, their diameters are 10 and 237 μm, in this case the mass transfer in gas is about two times different along the distance (Fig. 2.47).

The calculations given show that at the end face and in the arc burning zone, at stages 1 and 2, the limiting segment of the oxidation process is the external diffusion and the diffusion in metal along the spraying distance.

It can be noted the specific character of the drop transfer in arc spraying in comparison with the data at arc welding given by Pokhodnya at the same values of current and electrode diameter [69]. The specific area of drops in arc spraying is 1–2 orders higher, Fig. 2.48. On the other hand, the time of liquid metal drop interaction with air is 100–1000 times shorter. It is evident that there is no direct similarity with the welding process according to these parameters. For the purposes of the welding process estimation Pokhodnya introduced the generalized kinetic index of interaction, which is the characteristic of transfer under electrode melting:

$$\omega = F_K t_K / V_K, \tag{2.183}$$

where F_d, V_d—drop surface area, m^2, and volume, m^3; $\tau_k = m_{av}/v_{av}$—mass-averaged time of interaction, s; m_{av}—drop mass, kg; v_{av} = average melting velocity, kg/s.

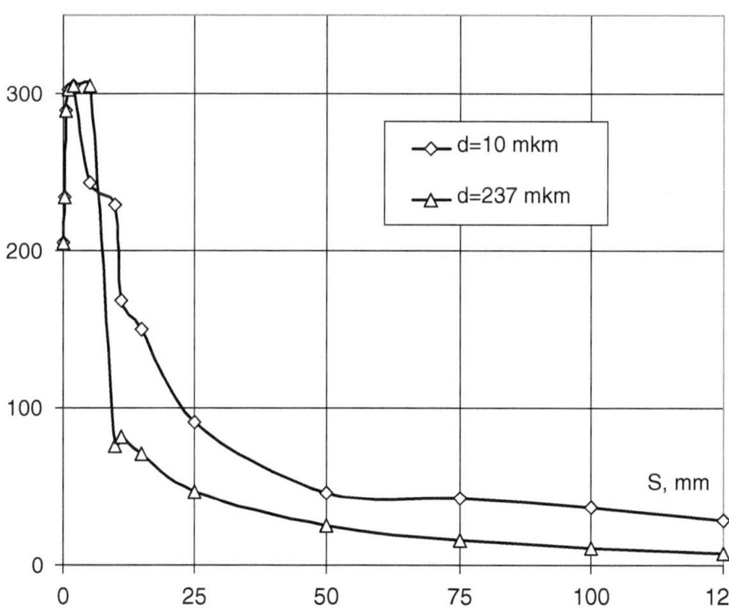

Fig. 2.47 Change of mass transfer coefficient in gas for drops with different diameters along the distance. Jet axis

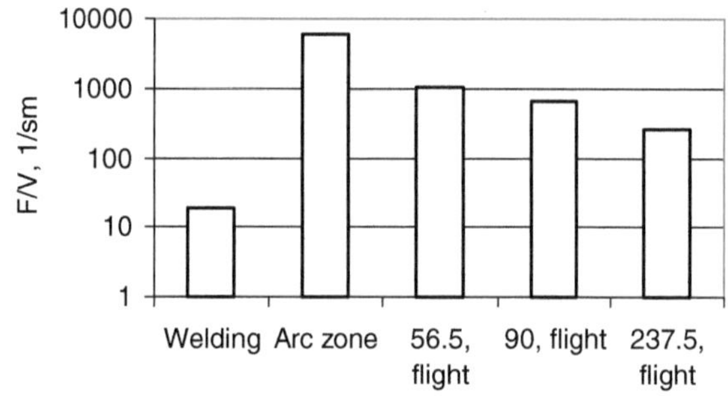

Fig. 2.48 Comparison of the specific area of drops at welding and arc spraying

When ω values are compared for welding and arc spraying (Fig. 2.49), it is seen that the intensity of the mass transfer at arc spraying is about 30 % higher for the arc burning zone.

The possibility of mass transfer in flight goes down sharply for the drops belonging to the main range of diameters and becomes an order of magnitude lower than for welding in drop transfer. Despite the sharp difference of individual indices of interaction, the generalizing index for both welding and arc spraying has the same order of values. This justifies the application of welding parameters of transfer in arc spraying process.

Erokhin [4] studied regarding the metal melting on the end face of the electrode under arc welding the kinetics of the interaction process of liquid metal with surrounding atmosphere. For an open system, where reacting agents are delivered and taken off, the mass transfer balance was described by the equation of the fist-order reaction, when the process rate is proportional to first-degree concentrations:

$$\frac{dm_O}{d\tau} = \frac{g_{пл} \cdot [\mathrm{O}]_0}{100} - \frac{g_{омв} \cdot [\mathrm{O}]}{100} + \frac{\beta_{\mathrm{Me}} \cdot \rho \cdot F}{100} \cdot ([\mathrm{O}]' - [\mathrm{O}]), \tag{2.184}$$

where $m_O = \rho \cdot V \cdot [\mathrm{O}]$—amount of oxygen in liquid metal, kg; ρ, V—density, kg/m^3, volume, m^3, of liquid metal; $g_{пл}$, $g_{омв}$—mass rates of metal melting and

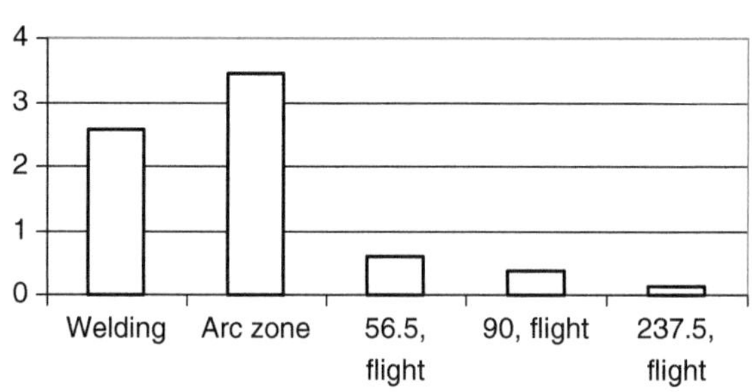

Fig. 2.49 Kinetic index of interaction at welding and arc spraying. *At Welding* drop diameter is 3 mm, $\tau_d = 0.14$ s. *At Arc spraying* in the arc burning zone the conventional drop diameter is 100 µm, $\tau_d = 5.74 \times 10^{-4}$ s; when flying, $\tau_d = 0.001$ s

taking off, kg/s; β_{Me}—coefficient of mass transfer in metal, m/s; F—reaction surface area, m^2; $[O]_0$, $[O]$, $[O]'$—starting, current and phase interface concentrations of oxygen, wt%.

This approach can be successfully applied in calculations if the mutual influence of simultaneous reactions is not significant and if inhibitions due to internal diffusion mechanisms are not important. It should be also noted that the equilibrium oxygen concentration in the phase interface ($[O]'$) in the Eq. (2.184) was not analytically determined by Erokhin. This is related to the lack of data regarding the influence of melting conditions on the equilibrium and due to non-isothermal conditions of melting. As for plasma-arc melting, Erokhin determined equilibrium concentrations by experiment [93].

As noted before, at first all delivered oxygen is dissolved in liquid metal until the saturation point. The thermal dependence of solubility, according to Ryzjonkov [18], is:

$$lg[O]_{Fe}^{sat} = -6320/T + 2.734 \qquad (2.185)$$

where $[O]_{Fe}^{sat}$ is the oxygen saturation point in iron, wt%; T—melt temperature, K.

Unlike hydrogen and nitrogen, which comply with Sieverts' law, oxygen discharge from liquid iron into gaseous atmosphere is not likely. Let us show that, the oxygen solvability in iron is described by the equation:

$$0.5\{O_2\} = [O]_{Fe} \qquad (2.186)$$

When general air pressure is 3×10^{-3} Pa, in other words, under low vacuum, oxygen partial pressure in gas phase at 1809 K will be $\leq 6 \times 10^{-4}$ Pa [43]. This value is not proper to the arc spraying process.

2.8.2.2 Slag-Covered Surface

For quantitative description of the interaction within the system "metal—slag" Boronenkov and colleagues proposed kinetic analysis method of simultaneous reactions taking place between metal and slag under diffusion mode. It takes into account the mutual influence of reactions and the diffusion of all components in metal and slag [94].

The theoretical ground of the method includes two principles:

(1) under the diffusion mode of all reactions, the relation of concentrations in the interface "metal—slag" is close to equilibrium for each reaction;
(2) velocity of transfer of each agent in metal and slag to or from their interface is proportional to the difference of concentrations in the phase volume and on the interface "metal—slag".

Let us consider an example, which has a practical importance, of oxidation of iron impurities by slag:

$$\frac{n}{m}[\ni_i] + (\mathrm{FeO}) = \frac{1}{m}(\ni_{in}\ \mathrm{O}_m) + [\mathrm{Fe}], \tag{2.187}$$

where $[\ni_i]$—are chemical agents (Mn, Si, Cr, P, Ti, and etc.).

Under the diffusion mode, when reactions take place in phase interface, the relation of concentrations near the border "metal–slag" is determined by the expression:

$$K_i = \frac{a^{1/m}_{\ni_{in}\mathrm{O}_m} \cdot a_{\mathrm{Fe}}}{a^{n/m}_{\ni_i} \cdot a_{\mathrm{FeO}}} = \frac{\gamma^{1/m}_{\ni_{in}\mathrm{O}_m} \cdot \gamma_{\mathrm{Fe}}}{\gamma^{n/m}_{\ni_i} \cdot \gamma_{\mathrm{FeO}}} \cdot \frac{N^{1/m}_{\ni_{in}\mathrm{O}_m} \cdot N_{\mathrm{Fe}}}{N^{n/m}_{\ni_i} \cdot N_{\mathrm{FeO}}} \tag{2.188}$$

where K_i includes the activities of corresponding components or, at normalizing of reaction products by infinitely weak solution, coefficients of activities and coefficients for calculation of concentration in their weight equivalents; a_i, γ_I, N_i—activities, respectively, mole fractions of agents $Э_i$.

This implies:

$$\frac{(\mathrm{FeO})'}{[\mathrm{Fe}]'} = \frac{1}{K_i}\frac{(\ni_{in}\ \mathrm{O}_m)'^{1/m}}{[\ni_i]'^{n/m}}, \tag{2.189}$$

where index (′) corresponds to the parameters on the phase interface.

Under quasi-stationary mode the mass transfer velocity I_j, mole/(m^2 s) of any agent j to the interface Me-slag (or from it) is equal to:

$$I_j = \pm k_j^{\partial}\left(C_j - C_j'\right), \tag{2.190}$$

which implies:

$$\mathrm{C}_\mathrm{j}' = \mathrm{C}_\mathrm{j}\left(1 \pm \mathrm{I}_\mathrm{j}\big/\mathrm{I}_\mathrm{j}^{\Pi}\right) \tag{2.191}$$

where κ^{∂}—mass transfer constant, m/s; S, S' are concentrations in the volume and on the interface, mole/m^3; $I^{\Pi} = \kappa^{\partial}\ s$—limiting diffusion flow, mole/(m^2 s); (−)—for initial agents, (+)—for reaction products.

Substituting molar concentrations in the Eq. (2.191) by weight concentrations and putting them into the expression (2.189) we have

$$x = \frac{(\mathrm{FeO})'}{[\mathrm{Fe}]'} = \frac{1}{K_i}\frac{(\ni_{in}\ \mathrm{O}_m)^{1/m} \cdot \left(1 + I_{\ni_{in}O_m}\big/I^{\Pi}_{\ni_{in}\mathrm{O}_m}\right)^{1/m}}{[\ni_i]^{n/m} \cdot \left(1 - I_{\ni_i}\big/I^{\Pi}_{\ni_i}\right)^{n/m}} \tag{2.192}$$

The stoichiometry of the reaction (2.187) implies:

$$I_{\mathrm{FeO}} = I_{\mathrm{Fe}}; \quad I_i = \frac{m}{n} I_{\ni_i} = m I_{\ni_{in}\mathrm{O}_m} \tag{2.193}$$

Besides that, it should be taken onto account that I_i is a part of the total flow of iron oxide (I_{FeO}), which is consumed for the oxidation of additive i. Consequently:

$$I_{\mathrm{FeO}} = I_1 + I_2 + \cdots + I_i + \sum I_i = \sum \frac{m}{n} I_{\ni_i} \tag{2.194}$$

Let us express I_{FeO} and $I_{\ni_i}$ in terms of x using the Eqs. (2.192 and 2.193):

$$I_{\mathrm{FeO}} = \left(\frac{(\mathrm{FeO})}{[\mathrm{Fe}]} - x \right) \Big/ \left(\frac{x}{I^n_{\mathrm{Fe}}} + \frac{(\mathrm{FeO})}{[\mathrm{Fe}] \cdot I^{\Pi}_{\mathrm{FeO}}} \right), \tag{2.195}$$

where $n = 1$

$$I_{\ni_i} = \left(x^m - \frac{1}{K_i^m} \frac{(\ni_i \, \mathrm{O}_m)}{[\ni_i]} \right) \Big/ \left(\frac{x^m}{I^{\Pi}_{\ni_i}} + \frac{(\ni_i \, \mathrm{O}_m)}{K_i^m [\ni_i] \cdot I^{\Pi}_{\ni_i \mathrm{O}_m}} \right), \tag{2.196}$$

where $n = 2$

$$I_{\ni_i} = I^{\Pi}_{\ni_i} \left[1 + b_i \frac{I^{\Pi}_{\ni_i}}{4 \cdot I^{\Pi}_{\ni_{i2} O_m}} - \sqrt{\left(1 + b_i \frac{I^{\Pi}_{\ni_i}}{4 \cdot I^{\Pi}_{\ni_{i2}\mathrm{O}_m}} \right)^2 - 1 + b_i} \right], \tag{2.197}$$

where $b_i = \frac{(\ni_{i2}\mathrm{O}_m)}{K_i^m \cdot x^m \cdot [\ni_i]^2}$

The limiting diffusion flows of components on the interface metal—slag will be, according to [90]:

$$I^{\Pi}_i = \kappa_i^{\partial} \cdot C_1 = \frac{D_i}{\delta_i} \cdot C_i = \beta \cdot D_i^{\alpha} \cdot C_i = \beta \cdot D_i^{\alpha} \frac{\rho}{100 \cdot M_i} \cdot c_i, \tag{2.198}$$

where K_i^{∂} is mass transfer constant, m/s; δ_i—diffusion layer thickness, m; $\alpha = 0.5$, according to [52]; S_i, c_i—molar, mole/m^3, and wt%, concentrations; β—convection constant, s$^{-1/2}$; ρ—metal or slag density, kg/m^3; M_i—molecular weight of components, kg/mole.

The convection constant is the same for all elements, is characterizes the intensity of mass transfer as the result of hydrodynamic conditions. According to calculations and experimental data obtained by Boornenkov [95], the range of most likely values is

$$\beta = (0.5 - 2.0),\ \mathrm{c}^{-0.5}.$$

Putting I_{FeO} and $I_{\ni_i}$ into the expression (2.194), we have an equation with one unknown x. Furthermore, from the expressions (2.196, 2.197), $I_{\ni_i}$ is found.

It should be taken into account that, in case of significant differences of metal and slag compositions in the volume and their interface, changes in K_i values due to, first, changes in coefficients of activity, must be accounted in the Eq. (2.189). Popel and colleagues analyzed the influence of this factor [52].

The results of calculations will not depend on which reactions of all the possible ones are taken into consideration. It is sufficient that they are algebraically independent, include all necessary components, and contain a common chemical agent. This is related to the essence of the method itself, which does not allow the determination of rates of reactions themselves, but determined the flows of agents to (or from) the phase interface, consequently, the velocities of their transfer through this interface. The question of what exactly adsorption and chemical acts give place to this process cannot be solved under the diffusion mode of reactions.

2.8.3 Example Calculation of Oxygen Delivery to a Drop

After calculations of velocity, temperature of two-phase flow, and concentrations of components of the carrier mixture, items (2.2–2.7), there are enough data to calculate the kinetics of oxygen delivery to metal. We offered a scheme of oxygen delivery to a drop at arc spraying, which takes into account processes of both external and internal diffusions [96, 97]. Particularly examined is the stage of drop separation in the arc burning zone. This stage has a significant influence on the final amount of oxygen delivered. The deliveries of oxygen into metal by solution from the gas phase and by interaction with slag are considered separately.

Let us divide the process into three consequent stages characterized by different features of the interaction between liquid metal and oxygen.

Stage 1. Metal on the End Face of the Electrode

Metal is continuously delivered to the zone of interaction (to the end face) and taken off from it to a drop. The agent (oxygen) concentration in gas phase is constant. Metal temperature on the end face, according to Erokhin and Pokhodnya [4, 69], is 2400–2700 K. We take the temperature $T = 2500$ K in our calculations.

There is a restricted amount of data concerning the oxygen mass transfer coefficient in the melt (β_{Me}) for liquid metal in the arc discharge under gas flow. According to Lakomskiy [89], nitrogen in a drop of electrode metal at the end face of the welding electrode has the mass transfer coefficient in metal $\beta_{Me} = 0.29 \times 10^{-2}$ m/s. This is an order of magnitude higher than under metallurgical processes in melts during steel production, where

$$\beta_{\mathrm{Me}} = (0.8 - 2.9) \times 10^{-4} \mathrm{m/s}.$$

It is evidently explained by higher values of diffusion coefficient at the drop temperature. At the same time, the interface becomes thinner due to active mixing of the drop metal. If assumed that the character of the change β_{Me} of oxygen is similar, it is likely that β_{Me} = (0.1–0.3) × 10^{-2} m/s. According to Erokhin's experiments, where the mass transfer coefficient is determined for drops at shielded metal arc welding, β_{Me} = (0.19–0.25) × 10^{-2} m/s [4]. Let us take β_{Me} = 0.25 × 10^{-2} m/s for our calculations.

Stage 2. Metal in the Arc Burning Zone

Let us take, in calculations, the scheme proposed when the features of drop formation were considered. The metal is in the form of tongue in the arc burning zone. In necks formed as the result of gas pressure drops in places of local perturbations, there take place breaks of the liquid metal jet. Joule heat discharges more intensively here than in other segments of the conductor due to higher ohmic resistance. It is assumed that this leads to increase in temperature in necks up to 2900 K and the process of jet breakage intensifies sharply due to joint processes of liquid metal boiling and collisional broadening of gas near the necks. The value of the oxygen mass transfer coefficient in metal was taken like for Stage 1.

The shape of the tongue is continuously changing under joint influence of hydrodynamic and electro-dynamic forces [48, 98] and it seems to be difficult to describe it analytically. Let us a simplification taken for the description of the initial velocity of metal drops, according to which, there are spherical metal drops in the arc burning zone with a nominal diameter equal to the "tongue" thickness, that is d_{conv} = 10 × 10^{-6} m. The dimension of the arc burning zone was estimated from the condition of the equality of the volumes of the broken away layer and drop, according to the data obtained by oscillography of drop breaking away at arc spraying. Its length, s = 6.3 × 10^{-3} m, is quite accurate because it corresponds to the data obtained by high-speed shooting [20, 98]. At the end of the arc burning zone, the fraction composition of drops corresponds to data obtained by grain-size analysis, (Table 2.15). This simplification allows making calculations though the physical pattern of the process has more complicated character. Furthermore, in compliance with experimental data on oxygen amount in the coating parameter d_{conv} can be corrected.

Stage 3. Flight along the Spraying Distance

Let us give a short characteristic of parameters in accordance with the calculations of velocities and temperatures having been made before. Approximately, after 15–25 mm of the spraying distance the drop temperature goes down to $T_{nл}$ = 1809 K and then it stays unchanging until the end of the distance due to a significant enthalpy of melting. When fraction decreases from 237 × 10^{-6} to 10 × 10^{-6} m, average in section drop velocities on the spraying distance increase 2.5–3 times, Fig. 2.31.

Consequently, the time of drop interaction with oxygen is different. Gas temperature drops to 600–800 K. Due to intense blending of the gas jet with atmosphere the gas jet composition does not differ from air composition after the end of the initial part of the jet.

Figure 2.50 shows the physical model of oxygen delivery to a drop at arc spraying that we propose.

When the surface oxygen concentration $[O]'$ is lower that the saturation point $[O]_{Fe}^{sat}$, oxygen solves in liquid metal, which is described by the Eq. (2.185).

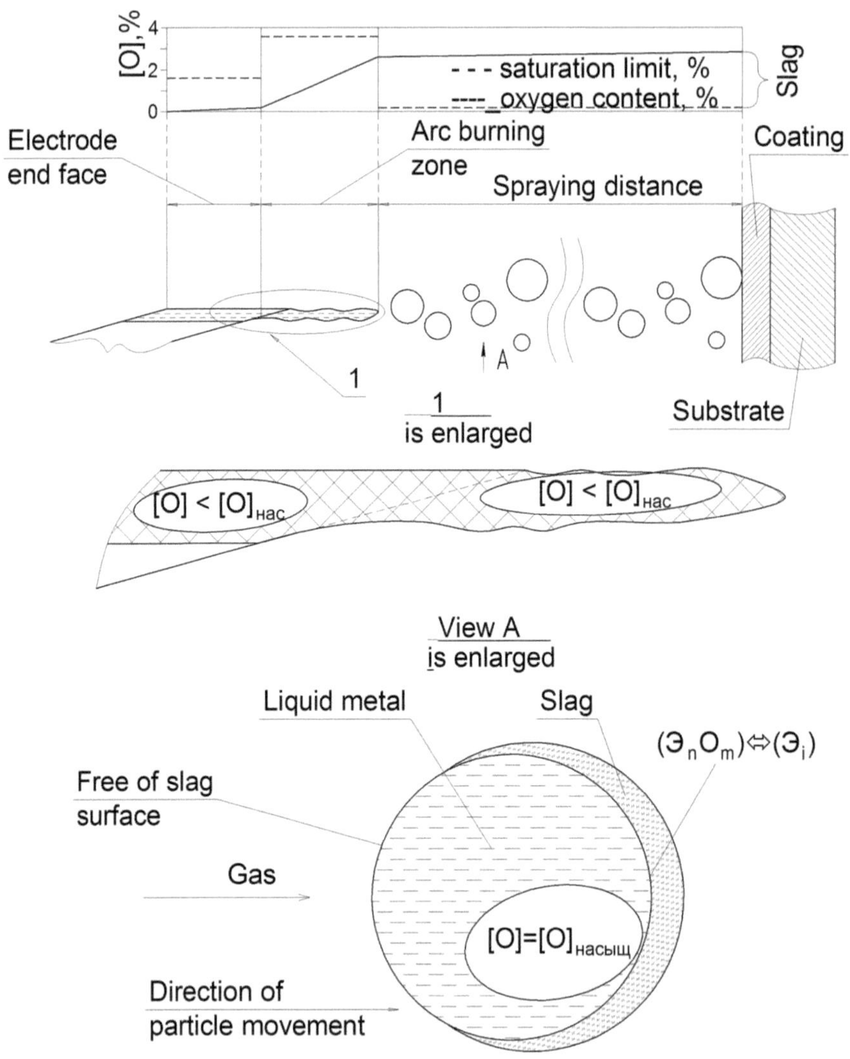

Fig. 2.50 Scheme of oxygen delivery at different stages of arc spraying process

Table 2.24 Calculation results of solubility limit of oxygen in liquid metal according to the Eq. (2.185)

T, K	1809	2500	2900
$[O]_{Fe}^{sat}$, wt%	0.174	1.607	3.587

Due to high temperatures at Stages 1 and 2, that is at the end face of the electrode and in the arc burning zone, the solubility limit of oxygen in metal is significantly higher than during drop flight before the distance, Table 2.24.

At these stages, all oxygen delivered or, at least, most of it solves in liquid metal. At the same time, the specific area of drops in the arc burning zone is much larger than the area of the end face of the electrode. It leads to the creation of the most favorable conditions for oxygen delivery to metal in the arc burning zone. At the third stage, in flight, the drop temperature decreases to $T_{пл}$ = 1809 K. That leads to a decrease in solubility limit (see Table 2.24) and, correspondingly, to oxygen discharge from the supersaturated solution.

Oxygen discharged from the solution is fixed in oxides forming slag film on the surface of drops. Due to the action of the carrier gas flow, this film is pushed aside from the rear side of the drop leaving a part of surface open for oxygen delivery from outside (Fig. 2.50 view A). Oxygen from outside is used for slag film formation, which is also pushed to the front side. Here, the elements are redistributed between slag and metal.

Let us consider an example. Welding wire Св08Г2С (wire like ER70S-6) is taken as the feedback. It contents, wt%: 0.110 C, 1.92 Mn, 0.79 Si, 0.020 P, 0.021 S.

The total amount of oxygen (m_O) delivered to the drop of the i-fraction will be equal to the sum of oxygen solved in liquid metal and contained in slag:

$$m_O = m_{\mathrm{Me}} \frac{[\mathrm{O}]}{100} + m_{sl} \left(\frac{(\mathrm{FeO}) M_\mathrm{O}}{100\, M_{\mathrm{FeO}}} + \frac{(\mathrm{MnO}) M_\mathrm{O}}{100 M_{\mathrm{MnO}}} + \frac{(\mathrm{SiO_2}) M_\mathrm{O}}{100 M_{\mathrm{SiO_2}}} \right) \quad (2.199)$$

The degree of oxidation of particles in this fraction r_i will be:

$$r_i = (m_\mathrm{O} \cdot 100)/(m_{\mathrm{Me}} + m_{\mathrm{sl}}) \quad (2.200)$$

The total degree of oxidation R is summed according to the distribution of drops in the weight fraction:

$$R = \left(\sum (r_i \cdot dm_i) \right) \cdot 100, \quad (2.201)$$

where for drops belonging to i-fraction: m_{Me}—liquid metal weight, kg; m_{sl}—slag weight, kg; dm_i—weigh fraction; M—molecular weight of a component, kg/mole; (FeO), (MnO), (SiO_2)—slag component concentrations, wt%.

The investigations of metallurgical interaction within the system "liquid metal–slag–gas" at welding [99–101] and steel production [52] show that the influence of reaction of slag with gas phase can be neglected. The exception is provided by reactions with hydrogen and nitrogen participation [69], which are not considered here. Based on these papers, we assume that the process of oxygen adsorption takes place in the diffusion mode.

The kinetics of oxygen delivery into melt at the stages indicated above will be considered as in paper [4], but taking onto account the diffusion not only in metal but also in gas. For an open system with the delivery and taking off agents, the description of the process will be made through the first-order equation:

$$\frac{dm_O}{d\tau} = \frac{g_{пл} \cdot [\mathrm{O}]_0}{100} - \frac{g_{отв} \cdot [\mathrm{O}]}{100} + \frac{\beta_{\mathrm{Me}} \cdot \rho \cdot F}{100} \cdot ([\mathrm{O}]' - [\mathrm{O}]), \tag{2.184}$$

An additional equation to find $[\mathrm{O}]'$ and $[\mathrm{O}]$ will be obtained from the condition of the equality of diffusion flows on the phase interface:

$$j_g = j_{\mathrm{Me}} \tag{2.202}$$

Let us make transformations:

$$j_г = \beta_g \cdot (C^0_{O_2} - C'_{O_2}) = \beta_г \cdot (P_{O_2} - P'_{O_2}) / (2 \cdot R \cdot T),$$

$$j_{Me} = \beta_{Me} \cdot (C'_{[O]} - C_{[O]}) = \beta_{Me} \cdot \rho \cdot ([O]' - [O]) / M_O,$$

$$1/2\beta_g (C_{O_2} - C'_{O_2}) = \beta_{Me} (C'_O - C_O)$$

$$\beta_g \cdot (P_{O_2} - P'_{O_2}) / (2RT) = \beta_{Me} \cdot \rho \cdot ([O]' - [O]) / (100 \cdot M_O)$$

The Eq. (2.202) will take the following view:

$$A_1 \cdot \left(\mathrm{P}_{\mathrm{O}_2} - \left([\mathrm{O}]'\right)^2 / k^2\right) = [\mathrm{O}]' - [\mathrm{O}] \tag{2.203}$$

where $A_1 = 100 \cdot \beta_g \cdot M_{\mathrm{O}} / (2 \cdot \beta_{\mathrm{Me}} \cdot R \cdot T \cdot \rho)$.

The partial pressure of oxygen on the phase interface surface ($\mathrm{P}'_{\mathrm{O}_2}$) will be determined through the equilibrium constant of the process of oxygen solution in liquid iron:

$$k = f_{[\mathrm{O}]} \cdot [\mathrm{O}]' \Big/ \left(P'_{\mathrm{O}_2} / P^{\mathrm{O}}\right)^{0.5}, \quad \lg k = 6120/T + 0.15 \quad [52], \tag{2.204}$$

where P^{O} is partial pressure of oxygen above the diffusion layer in gas; $f_{[\mathrm{O}]}$—coefficient of oxygen activity, due to its low concentration we take $f_{[\mathrm{O}]} \approx 1$.

The calculation results given in Table 2.23 show that the relation of values of diffusion flows in gas and metal changes following the stages of metal drop life. Taking onto account of only one of them distorts the pattern of the present processes. The introduction of an additional equation of equality of diffusion flows on the phase interface let us take into account both internal and external diffusions, obtain a more real idea of the kinetics of the processes.

We assume that at stage 1, on the end face of the electrode, quasi-stationary state is established. Here, the velocities of agent delivery and taking off are the same: $g_{dl} = g_{\text{take off}} = g$; and the volume of liquid layer on the face end of the electrode is constant:

$$dm_O/d\tau = 0.$$

In this case, the Eq. (2.184) will be as follows:

$$g \cdot [\mathrm{O}]_0/100 - g \cdot [\mathrm{O}]/100 + \beta_{\mathrm{Me}} \cdot \rho \cdot f \cdot \left([\mathrm{O}]' - [\mathrm{O}]\right)/100 = 0 \tag{2.205}$$

From the joint solution of the Eqs. (2.203) and (2.204):

$$[\mathrm{O}] = \left(-M/2 + \sqrt{M^2/4 + A_1 \cdot (1 + A_2)^2 \cdot N}\right) \Big/ \left(A_1 \cdot (1 + A_2)^2\right), \tag{2.206}$$

$$[\mathrm{O}]' = \left([\mathrm{O}] \cdot (1 + A_2) - [\mathrm{O}]_0\right)/A_2, \tag{2.207}$$

where $A_2 = \beta_{\mathrm{Me}} \cdot f \cdot \rho / g$,

$$M = A_2 \cdot (1 + A_2) \cdot k^2 - 2 \cdot A_1 \cdot (1 + A_2) \cdot [\mathrm{O}]_0 - (A_2 \cdot k)^2,$$
$$N = A_1 \cdot (A_2 \cdot k)^2 \cdot \mathrm{P}_{\mathrm{O}_2} - A_1 \cdot [O]_0^2 + A_2 \cdot k^2 \cdot [\mathrm{O}]_0$$

At stages 2 and 3, in the arc burning zone and in flight, we assume that a drop is a closed system, where metal is not delivered to or taken off: $g_{dl} = g_{\text{take off}} = 0$.

The third stage differs from the second one by decreasing temperatures of gas and metal, growing partial pressure of oxygen, smaller coefficient of mass transfer in gas (Table 2.23). The Eq. (2.184) for stages 2, 3 is as follows:

$$\mathrm{d}(\rho \cdot V \cdot [\mathrm{O}])/(100 \cdot d\tau) = \beta_{\mathrm{Me}} \cdot \rho \cdot f \cdot ([\mathrm{O}]' - [\mathrm{O}])/100 \tag{2.208}$$

If the process time is divided into so short time intervals that

$$\frac{\mathrm{d}[\mathrm{O}]}{\mathrm{d}\tau} \approx \frac{\Delta[\mathrm{O}]}{\Delta\tau},$$

we can note for time interval $(n + 1)$:

$$\left([\mathrm{O}]_{n+1} - [\mathrm{O}]_n\right)/\left([\mathrm{O}]' - [\mathrm{O}]_{n+1}\right) = A_3 \cdot \Delta\tau \tag{2.209}$$

where $A_3 = \beta_{\mathrm{Me}} \cdot f / V$.

From the condition of equality of diffusion flows in gas and metal $j_ә = j_{Me}$ the Eq. (2.203) for the $(n + 1)$th step of calculation will be as follows:

$$A_1 \cdot (P_{O_2} - ([O]')^2/k^2) = [O]' - [O]_{n+1} \tag{2.210}$$

Oxygen contents $[O]_{n+1}$ and $[O]'$ are defined from the joint solution of the Eqs. (2.209) and (2.210):

$$[O]' = \frac{-\frac{1}{2\cdot(1+A_3\cdot\Delta\tau)} + \sqrt{\left(\frac{1}{2\cdot(1+A_3\cdot\Delta\tau)}\right)^2 + \frac{A_1}{k^2}\cdot\left(A_1\cdot P_{O_2} + \frac{[O]_n}{1+A_3\cdot\Delta\tau}\right)}}{A_1/k^2} \tag{2.211}$$

$$[O]_{n+1} = ([O]_n + A_3\cdot\Delta\tau\cdot[O]')/(1 + A_3\cdot\Delta\tau) \tag{2.212}$$

When $[O]' > [O]_{Fe}^{sat}$, all oxygen coming from the atmosphere is consumed for slag formation. The oxygen content in metal is found from the Eq. (2.209), when $[O]' = [O]_{Fe}^{sat}$. The diffusion flow from gas into metal I_g is equal to:

$$I_g = \beta_g \cdot \left(P_{O_2} - ([O]_{Fe}/\mathrm{k})\right)/(2\cdot R\cdot T) \tag{2.213}$$

Oxygen can additionally discharge from the supersaturated solution when the drop temperature is going down after the segment of the arc burning zone. The additional diffusion flow I_{add} will be:

$$I_{add} = \beta_{Me}\cdot\rho\cdot([O]_{sat} - [O]_n)/(100\cdot M_O) \tag{2.214}$$

The amount of slag formed in the front side on the calculation step, Δm_s:

$$\Delta m_s = (I_ә \cdot L + I_{add})\cdot F\cdot\Delta\tau\cdot\cdot M_{FeO}\cdot(FeO)/100 \tag{2.215}$$

where L is the part of the surface area of the rear side free of slag.

In comparison with the previous step, the weight of liquid metal will decrease due to transfer of Fe to slag:

$$m_{Me}^{n+1} = m_{Me}^{n} - \Delta m_w \cdot M_{\ni}/M_i \tag{2.216}$$

Slag from the Eq. (2.215) will be added to the slag formed before. The content of slag components on a step after events at the rear side will be:

$$(FeO)^{n+1} = ((FeO)^n\cdot m_s/100 + \Delta m_s)/(m_s + \Delta m_s) \tag{2.217}$$

for other components of type ($\ni_K O_m$):

$$(\ni_k O_m)^{n+1} = (\ni_k O_m)^n \cdot m_w / ((m_w + \Delta m_w) \cdot 100) \tag{2.218}$$

Besides slag formation, there will take place its interaction with metal. Boronenkov exposes the principles of analytical description of kinetics of this process with colleagues [94]. Arc spraying process has a number of features, which are necessary to be taken into account in the calculation of metal—slag interaction:

- Stages (electrode end face, arc burning zone, spraying distance) have different thermal, geometrical, and hydrodynamic conditions;
- Continuous renewal of metal and slag masses in interaction on the electrode end face and in the arc burning zone;
- Simultaneity of all reactions at each stage and their mutual thermodynamic and kinetic influence;
- Influence of diffusion inhibition of all agents in metal and slag on the rate of each reaction.

There are some simplifying assumptions used in the model:

- Chemical composition of metal and slag is uniform at each stage within a phase (ideal mixing);
- Contact surface of metal with slag at each stage is uniform under kinetic aspect; that is it can be characterized by some average in time contact area and mass transfer constant;
- Influence of reactions between slag and gas phase can be neglected at the first approximation.

In case the above assumptions are taken, a complex of reactions of additive oxidation with the common agent—FeO, describes the interaction of metal and slag on the phase interface:

$$[\mathrm{C}]' + (\mathrm{FeO})' = [\mathrm{Fe}]' + \mathrm{CO} \quad \lg(\mathrm{K}_1) = -5152/\mathrm{T} + 4.8 \tag{2.219}$$

$$(\mathrm{FeO})' = [\mathrm{Fe}]' + [\mathrm{O}]' \quad \lg(\mathrm{K}_2) = -6320/\mathrm{T} + 2.734 \tag{2.220}$$

$$[\mathrm{Mn}]' + (\mathrm{FeO})' = [\mathrm{Fe}]' + (\mathrm{MnO})'; \quad \lg(\mathrm{K}_3) = 5550/\mathrm{T} - 2.34 \tag{2.221}$$

$$0.5[\mathrm{Si}]' + (\mathrm{FeO})' = [\mathrm{Fe}]' + 0.5(\mathrm{SiO_2})' \quad \lg(\mathrm{K}_4) = 7825/\mathrm{T} - 3.06 \tag{2.222}$$

In the Eqs. (2.219–2.222) constants K_i correspond to the Eq. (2.188)

Let us assume:

- Influence of CO diffusion on gas is small to negligible because in bubbles on the interface "metal—slag" $P_{CO} \sim 0.1$ MПa;
- Sulfur and phosphorus do not have any significant influence on slag composition because of their small concentrations.

As every reaction is accompanied by consumption of some amount of FeO, the summary diffusion flow of components is equal to the diffusion flow of FeO. For these components, the process is described by the equation:

$$I_{\mathrm{FeO}} - I_{\mathrm{C}} - I_{\mathrm{O}} - I_{\mathrm{Mn}} - 2I_{\mathrm{Si}} = 0 \tag{2.223}$$

According to the equation of concentration dependence on the phase interface from the concentration in the volume and diffusion flows (2.191):

$$(\mathrm{FeO})' = (\mathrm{FeO}) \cdot \left(1 - I_{\mathrm{FeO}} / I_{\mathrm{FeO}}^n\right) \tag{2.224}$$

$$[\mathrm{C}]' = [\mathrm{C}] \cdot \left(1 - I_{\mathrm{C}} / I_{\mathrm{C}}^n\right) \tag{2.225}$$

$$[\mathrm{O}]' = [\mathrm{O}] \cdot \left(1 + I_{\mathrm{O}} / I_{\mathrm{O}}^n\right) \tag{2.226}$$

$$(\mathrm{MnO})' = (\mathrm{MnO}) \cdot \left(1 + I_{\mathrm{MnO}} / I_{\mathrm{MnO}}^n\right) \tag{2.227}$$

$$[\mathrm{Mn}]' = [\mathrm{Mn}] \cdot \left(1 - I_{\mathrm{Mn}} / I_{\mathrm{Mn}}^n\right) \tag{2.228}$$

$$[\mathrm{Si}]' = [\mathrm{Si}] \cdot \left(1 - I_{\mathrm{Si}} / I_{\mathrm{Si}}^n\right) \tag{2.229}$$

$$(\mathrm{SiO_2})' = (\mathrm{SiO_2}) \cdot \left(1 + I_{\mathrm{SiO_2}} \Big/ I_{\mathrm{SiO_2}}^n\right) \tag{2.230}$$

Let us introduce a variable

$$x = \frac{(\mathrm{FeO})'}{[\mathrm{Fe}]'} = \frac{1}{K_i} \frac{(\ni_{in}\, O_m)^{1/m} \cdot \left(1 + I_{\ni_{in} O_m} \Big/ I_{\ni_{in} O_m}^n\right)^{1/m}}{[\ni_i]^{n/m} \cdot \left(1 - I_{\ni_i} \Big/ I_{\ni_i}^n\right)^{n/m}} \tag{2.231}$$

Because of high concentration of iron, its diffusion, as the inhibiting agent, can be excluded, then, from the Eq. (2.192):

$$x = (\mathrm{FeO})'$$

From the Eqs. (2.219–2.222):

$$[\mathrm{C}]' = \frac{1}{K_1 \cdot x}; \quad [\mathrm{O}]' = K_2 \cdot x; \quad K_3 = \frac{(\mathrm{MnO})'}{[\mathrm{Mn}]' x};$$

$$K_4 = \frac{\left((\mathrm{SiO_2})' \Big/ [\mathrm{Si}]'\right)^{0{,}5}}{x} \tag{2.232}$$

From the Eqs. (2.224) and (2.231):

$$I_{\mathrm{FeO}} = I^n_{\mathrm{FeO}}\left(1 - (\mathrm{FeO})'/(\mathrm{FeO})\right) = I^n_{\mathrm{FeO}}(1 - x/(\mathrm{FeO})) \quad (2.233)$$

From the Eqs. (2.219), (2.225), and (2.232):

$$I_{\mathrm{C}} = I^n_{\mathrm{C}}\left(1 - [\mathrm{C}]'/[\mathrm{C}]\right) = I^n_{\mathrm{C}}(1 - 1/(K_1 \cdot x \cdot [\mathrm{C}])) \quad (2.234)$$

From the Eqs. (2.220), (2.226), and (2.232):

$$I_O = I^n_O\left([\mathrm{O}]'/[\mathrm{O}] - 1\right) = I^n_O(\mathrm{K}_2 \cdot \mathrm{x}/[\mathrm{O}] - 1) \quad (2.235)$$

From the Eqs. (2.221), (2.227), (2.228), and (2.232):

$$K_3 = \frac{(\mathrm{MnO})'}{x \cdot [\mathrm{Mn}]'} = \frac{(\mathrm{MnO}) \cdot \left(1 + I_{\mathrm{MnO}}/I^n_{\mathrm{MnO}}\right)}{x \cdot [\mathrm{Mn}] \cdot \left(1 - I_{\mathrm{Mn}}/I^n_{\mathrm{Mn}}\right)} \quad (2.236)$$

Under the stationary mode, substances diffuse toward the interface in accordance with the stoichiometry of the reaction; that is $I_{\ni_i} = I_{\ni_{\mathrm{in}} O_m}$, or for this particular case $I_{\mathrm{Mn}} = I_{\mathrm{MnO}}$, then

$$I_{\mathrm{Mn}} = \frac{\mathrm{K}_3 \cdot x \cdot [\mathrm{Mn}] - [\mathrm{MnO}]}{\mathrm{K}_3[\mathrm{Mn}] \cdot x/I^n_{\mathrm{Mn}} + [\mathrm{MnO}]/I^n_{\mathrm{MnO}}} \quad (2.237)$$

From the Eqs. (2.222), (2.229), (2.230), and (2.232):

$$K_4 = \frac{(\mathrm{SiO}_2)^{0.5}\left(1 + I_{\mathrm{SiO}_2}/I^{\mathrm{n}}_{\mathrm{SiO}_2}\right)^{0.5}}{[\mathrm{Si}]^{0.5}\left(1 - I_{\mathrm{Si}}/I^n_{\mathrm{Si}}\right)^{0.5} \cdot x} \quad (2.238)$$

Like in the previous case, in this one $I_{\mathrm{Si}} = I_{\mathrm{SiO}_2}$, then

$$I_{\mathrm{Si}} = \frac{x^2 \cdot K_4^2 \cdot [\mathrm{Si}] - (\mathrm{SiO}_2)}{x^2 \cdot K_4^2 \cdot [\mathrm{Si}]/I^{\mathrm{n}}_{\mathrm{Si}} + (\mathrm{SiO}_2)/I^{\mathrm{n}}_{\mathrm{SiO}_2}} \quad (2.239)$$

Taking into account (2.233)–(2.239):

$$I^n_{\mathrm{FeO}}\left(1 - \frac{x}{(\mathrm{FeO})}\right) - I^n_C\left(1 - \frac{1}{K_1[\mathrm{C}]x}\right) - I^n_O\left(\frac{K_2 x}{[\mathrm{O}]} - 1\right) - \frac{\mathrm{K}_3[\mathrm{Mn}]x - [\mathrm{MnO}]}{K_3[\mathrm{Mn}]x/I^n_{\mathrm{Mn}} + [\mathrm{MnO}]/I^n_{\mathrm{MnO}}} - 2 \cdot \frac{x^2 K_4^2[\mathrm{Si}] - (\mathrm{SiO}_2)}{x^2 K_4^2[\mathrm{Si}]/I^n_{\mathrm{Si}} + (\mathrm{SiO}_2)/I^n_{\mathrm{SiO}_2}} = 0 \quad (2.240)$$

Table 2.25 Coefficient of diffusion of components [52, 102]

Component	(FeO)	[C]	[O]	(MnO)
D_0, m^2/s	5.00×10^{-4}	3.90×10^{-6}	3.34×10^{-7}	5.00×10^{-4}
E, J/mole	2.04×10^5	6.68×10^4	5.00×10^4	2.04×10^5
Component	[Mn]	(SiO_2)	[Si]	
D_0, m^2/s	1.80×10^{-7}	5.00×10^{-4}	3.20×10^{-5}	
E, J/mole	5.43×10^4	2.04×10^5	1.23×10^5	

We assumed $\beta = 1.2$, $s^{-1/2}$. For the calculation of the diffusion coefficients, we used the following temperature dependence:

$$D_i = D_0 \cdot \exp(-E/\mathrm{RT}). \tag{2.241}$$

Numerical values of parameters D_0 and E are given in Table 2.25. Drop temperature is taken according to calculation results in item 2.3.

Having determined x from the Eq. (2.240) and, correspondingly, diffusion flows in the interface metal—slag: for FeO–I_{FeO}, for components—$I_{\ni i}$; we can calculate the content of oxides and elements in a drop after the events on a step of calculation in the front side.

MnO, SiO_2 in slag:

$$\left(m^{n+1}_{(\ni_K O_M)}\right)_i = \left(m^{n}_{(\ni_K O_M)}\right)_i + I_{\ni i} \cdot (M_{\ni_K O_M})_i \cdot f \cdot (1-L) \cdot \Delta\tau \tag{2.242}$$

FeO in slag:

$$m^{n+1}_{(\mathrm{FeO})} = m^{n}_{(\mathrm{FeO})} - I_{\mathrm{FeO}} \cdot M_{\mathrm{FeO}} \cdot f \cdot (1-L) \cdot \Delta\tau \tag{2.243}$$

C, Mn, Si in a particle:

$$\left(m^{n+1}_{\ni}\right)_i = \left(m^{n}_{\ni}\right)_i - \mathrm{I}_{\ni i} \cdot M_{\ni i} \cdot f \cdot (1-L) \cdot \Delta\tau \tag{2.244}$$

Oxygen in a particle:

$$m^{n+1}_{\mathrm{O}} = m^{n}_{\mathrm{O}} + \mathrm{I_O} \cdot \mathrm{M_O} \cdot f \cdot (1-L) \cdot \Delta\tau \tag{2.245}$$

Slag mass:

$$\begin{aligned} m^{n+1}_{w} = {} & m^{n}_{w} - \mathrm{I_{FeO}} \cdot \mathrm{M_{FeO}} \cdot f \cdot (1-L) \cdot \Delta\tau \\ & + \sum \left(\mathrm{I}_{\ni i} \cdot (M_{\ni_K \mathrm{O_M}})_i \cdot f \cdot (1-L) \cdot \Delta\tau\right) \end{aligned} \tag{2.246}$$

Particle mass:

$$m_{\mathrm{Me}}^{n+1} = m_{\mathrm{Me}}^{n} + \mathrm{I_O} \cdot \mathrm{M_O} \cdot f \cdot (1 - L) \cdot \Delta\tau - \sum \left(\mathrm{I}_{\ni_i} \cdot \mathrm{M}_{\ni_i} \cdot f \cdot (1 - L) \cdot \Delta\tau\right) \quad (2.247)$$

Concentration of slag components:

$$(\mathrm{FeO}) = m_{(\mathrm{FeO})}^{\mathrm{n}+1} \cdot 100/m_{w}^{n+1} \quad (2.248)$$

$$(\ni_{\kappa}\, O_m)_i = \left(M_{(\ni_{\kappa}\mathrm{O_M})}^{n+1}\right)_{\mathrm{i}} \cdot 100/\left(m_{w}^{n+1}\right) \quad (2.249)$$

The degree of oxidation of drops belonging to the *i*-fraction is calculated through the Eq. (2.200) and the summary degree of oxidation—through the Eq. (2.201).

When comparing experimental and calculated data (Fig. 2.51), it is seen that under limited amount of oxygen in the arc burning zone (α_{ox} = 0.8–1.2, which corresponds to the partial pressure of oxygen $\mathrm{P_{O_2}}$ = 0.0035–0.009 MPa) the calculated values fit the interval of experimental data when the part of drop surface clean of slag is L = 0.4–0.5. If only compressed air is used as the carrier gas ($\mathrm{P_{O_2}}$ = 0.021 MPa), the calculated and experimental data match each other when L = (0.3–0.35), in this case the velocity of oxygen delivery gets slower.

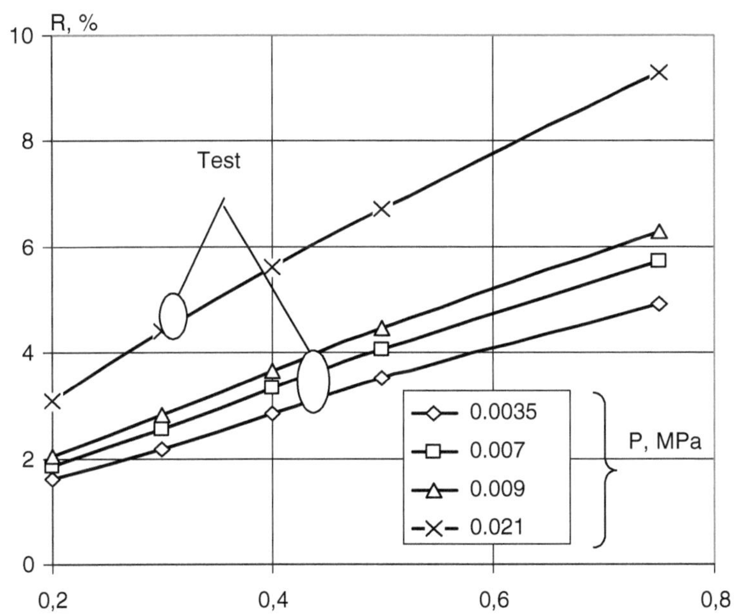

Fig. 2.51 Correspondence of experimental and calculated data according to oxidation degree (*R*) and partial pressure of oxygen (*P*) for different areas of drop surface clean of slag (*L*)

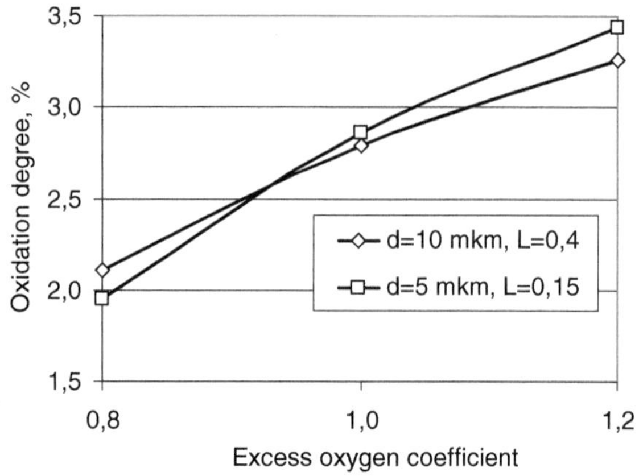

Fig. 2.52 Increase in oxygen content resulting from decrease in initial drop diameter

Thus, increase in oxygen delivery gas double action on the drop. On the one hand, the increase of oxygen content in gas leads to the growth of values of diffusion flows on the drop surface. On the other hand, decrease in surface tension of liquid metal in the presence of oxygen leads to the growth of the surface part covered by slag and, consequently, to the decrease in oxygen delivery to the drop. In case of large amount of oxygen, this inhibition is spectacular. Based on the hydrodynamic pattern of the process, a similar result can be obtained by use of alloying components increasing slag viscosity.

Decrease in surface tension of liquid metal on the end face of the electrode (by temperature increase or adding of surface-active agents) will lead to decrease in size of sprayed drops. However, to provide low content of oxygen at the same time, it requires more reliable protection against oxidation. Thus, if we assume that in the arc burning zone $d_{экв} = 5$ µm, at $L = 0.15$ the degree of oxidation is almost the same as at $L = 0.4$ for $d_{экв} = 10$ µm for all compositions of atmosphere, Fig. 2.52. Thus, the requirements to factors determining the quality of coating are opposite. Increasing the density requires fine spray of metal, and decreasing the oxygen content requires a decrease in drop diameter. Bringing these requirements closer is possible, if protective atmosphere is created.

On the end face of the electrode $[\mathrm{O}]' < [\mathrm{O}]_{\mathrm{Fe}}^{\mathrm{sat}}$ and all oxygen delivered at this stage solves in metal. On the end face of the electrodes, the amount of the oxygen delivered goes down in direct proportion to the voltage on the arc when the gap between electrodes increases. The estimation of the gap size h given in the paper [22] shows $h = 0.1–1.0$ mm. Let us show the calculation results of oxygen content on the end face of the electrode when $\alpha_{\mathrm{ox}} = 0.8$ (Table 2.26).

Decrease in gap size leads to growth of turbulence in this area, which, in its turn, improves the conditions of mass transfer in gas.

The contribution of drops belonging to small fractions to oxygen delivery $\mathrm{d}R_i = r_i \cdot 100/R, \%$ is over their weight fraction, and for drops belonging to large fractions

Table 2.26 Changes of oxygen content in liquid metal on the end face of the electrode

h, mm	*Re*	*Nu*	β_g, m/s, Eq. (2.181)	[O], %, Eq. (2.206)	[O]′, %, Eq. (2.207)
1.0	2376	31.5	20.90	0.067	0.288
0.1	237.6	10.7	68.44	0.171	0.914

Different gaps between the electrodes. Calculated data

Fig. 2.53 Relation of the contribution of fractions to oxygen delivery (*dR*, %) and their mass fractions (*dm*, %), P_{O_2} = 0.035 MPa

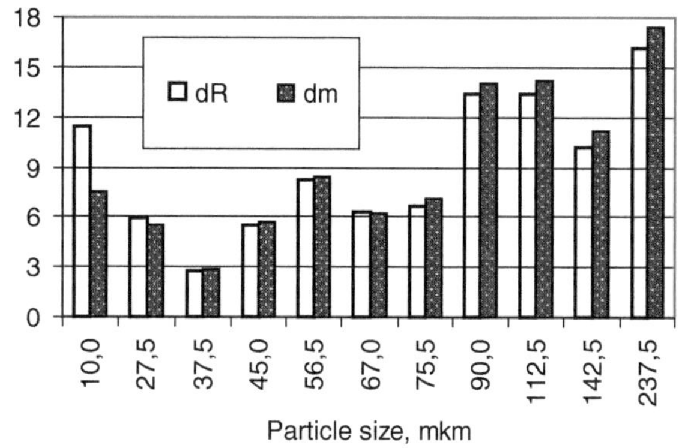

is under the corresponding weight fractions, Fig. 2.53. The contribution of drops belonging to fractions 27–75 μm to oxygen delivery is 1.5–2.5 times lower than of drops in other sizes. It is evidently related to the ratio of the specific surface $s = f/v$ and drop velocity along the spraying distance. Thus, the calculations show that the developed equipment and technologies will provide the smallest amount of oxygen delivered to coating if fraction composition of the jet is 27–75 μm.

The contribution of small fractions to oxygen delivery dR_i goes down due to increase in initial partial pressure of oxygen, Fig. 2.54.

The main oxygen delivery, 85–95 %, takes place in the arc burning zone. This is conditioned by an extremely high value of the coefficient of mass transfer in gas and large specific surface of particles. The value of oxidation degree depends very much on the coefficient of excess of gas oxidant at this stage (Fig. 2.55).

At stages 1and 2, all oxygen delivered solves in the metal of particles. At the stage of flight, as the result of sharp drop in saturation point, there takes place slag formation only due to oxygen discharging from supersaturated solution and delivered to the rear side from atmosphere. For the most likely L = 0.45 on the spraying distance, the weight fraction of slag is 5–12 % and it increases along with the growth of partial pressure of oxygen and decrease in particle diameters (Fig. 2.56).

The indicated features must be taken into account for the development of technologies of coating application and design of equipment as well.

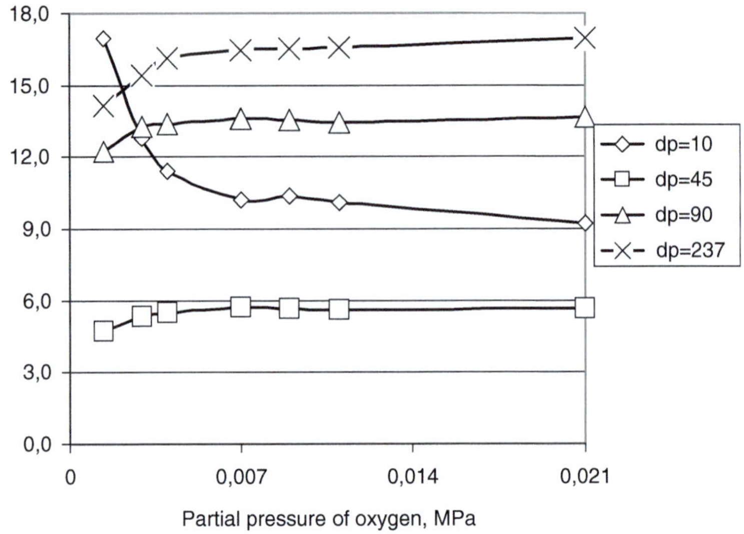

Fig. 2.54 Contribution of fractions to oxygen delivery under different initial partial pressures of oxygen

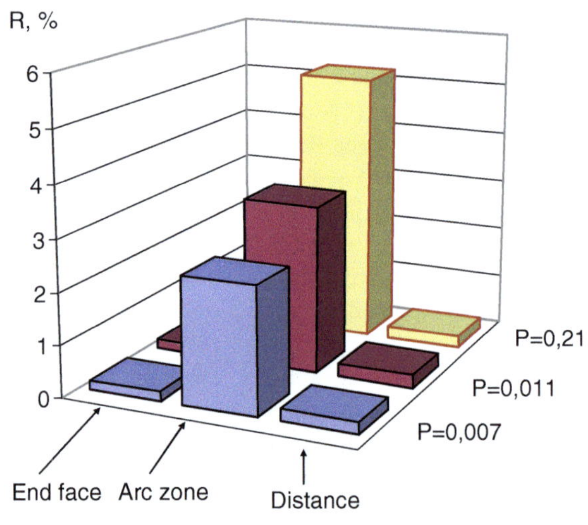

Fig. 2.55 Contribution of arc spraying process stages to oxygen delivery into drops under different initial partial pressures of oxygen, *P*, MPa

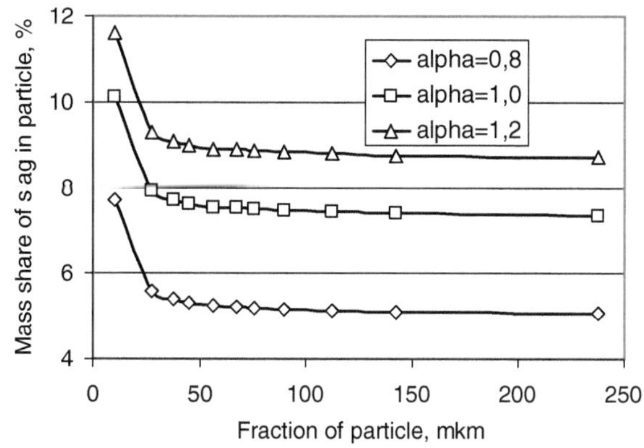

Fig. 2.56 Weight fraction of slag in a drop under different initial gas composition (alpha is oxidant excess coefficient)

References

1. Bykhovsky DG (1972) Plasma cutting. Mechanical Engineering, Leningrad
2. Leskov GI (1970) Electrical welding arc. Mechanical Engineering, Moscow
3. Lenivkin VA, Durgerov NG, Sagirov HN (1989) Technological properties of welding arc in arc shielded gases. Mechanical Engineering, Moscow
4. Erohin AA (1973) Fundamentals of fusion welding. Mechanical Engineering, Moscow
5. Nekrasov YI (1974) Gazes—acetylene substitutes. Mechanical Engineering, Moscow
6. Sokolov II (1978) Gas welding and metal cutting. High School, Moscow
7. Mikheev MA, Mikheeva IM (1977) Fundamentals of heat transferring. Energy, Moscow
8. Korobov YS, Izoitko VM, Pryadko AS et al (2000) Restoration of parts by active arc spraying method. Automobile Eng 1:23–24
9. Vargaftik NB (1972) Handbook of thermo physical properties of gases and liquids. Science, Moscow
10. Petrov GL, Tumarev AS (1977) Theory of welding processes. High School, Moscow
11. Donskoy AV, Klubnikin VS (1979) Electric plasma processes and units in engineering. Mechanical Engineering, Leningrad
12. Zhukov MF, Smolyakov VY, Urukov BA (1977) Arc heaters (plasma generators). Science, Moscow
13. Korobov YS, Boronenkov VN (1998) Calculation of parameters of movement, heating and oxidation of particles in electric arc metallizing. Welding Int. 12(9):726–730
14. Abramovich GN (1979) Applied Gas Dynamics. Science, Moscow
15. Kudinov VV (1977) Plasma coatings. Science, Moscow
16. Borisov YS et al (2000) Computer simulation of plasma spraying. Autom Weld 12:42–51
17. Petrov SV, Saakov AG (2000) Plasma of combustion products in surface engineering. Tapas, Kiev
18. Ryzhonkov DI (ed) (1989) Theory of metallurgical processes. Metallurgy, Moscow
19. Lyalyakin VP, Sablukov AS, Litovchenko NN (2000) Researching the processes of metal dynamic dispersion at arc spraying. Weld Prod 1:16–24
20. Vahalin VA, Maslennikov SB, Kudinov VV et al (1981) Process of melting and spraying of the electrode material at arc spraying. Phys Chem Mater Process 3:58–63
21. Wilden J et al (2007) Investigation about the chrome steel wire arc spray process and the resulting coating properties. In: Global coating solutions. ITSC 2007, Beijing
22. Royanov VA, Mosiyenko GA (1990) Effect of electrodes on the properties of the spray at arc metallization. Weld Prod 3:6–8

23. Asnis AE, Gutman LM, Pokladiy VR et al (1982) Welding in a mixture of active gases. Naukova Dumka, Kiev
24. BAMZ (1987) Arc Spraying Gun EM-17: Specification. Barnaul, USSR
25. Korolev NV (1996) Calculations of thermal processes in welding, fusing and thermal cutting. UPI, Ekaterinburg
26. Elliott F, Glaser M, Ramakrishna V (1969) Thermochemistry of steelmaking processes. Metallurgy, Moscow
27. Erohin AA (1975) Arc plasma melting of metals and alloys. Science, Moscow
28. Shumyakov VI, Korobov YS (2004) Development features of cored wires for activated arc spraying. In: Welding and control. Scientific conf. Perm
29. Gubin BA et al (1992) Determination of the coefficient of heat transfer from the slag to the electrode wire at electric slag welding steel. Autom Weld 6:26–29
30. Zinoviev VE (1989) Thermal properties of metals at high temperatures. Metallurgy, Moscow
31. Maslenkov SB (1991) Steels and alloys for high temperature. Metallurgy, Moscow
32. Goldfarb EM (1967) Heating engineering in metallurgical processes. Metallurgy, Moscow
33. Tillmann W, Vogli E, Abdulgader M (2007) Influence of atomization gas on the particle formation during arc spraying with cored wires In: Global Coating Solutions. Proceedings of ITSC 2007, Bejing
34. Kretzschmar E (1966) Spraying of metals, ceramics and plastics. Mechanical Engineering, Moscow
35. Tsvetkov V, Panfilov SA (1980) Low-temperature plasma in the process of recovery. Science, Moscow
36. Levich VG (1959) Physical chemical hydrodynamics. Phismathgiz, Moscow
37. Sternin LE et al (1980) Duo phases and mono phases gas flows with particles. Mechanical Engineering, Moscow
38. Volynskiy MS, Lipatov AS (1970) Deformation and fragmentation of droplets in the gas flow. Phys Eng J 5:838–843
39. Dubrovsky VV, Pidvysotsky AM, Schreiber AA (1990) Influence of aerodynamic forces changes the on the critical conditions for the destruction of droplets. Chem Technol 1:36–38
40. Petrov SV, Kotliarov OV, Mironov MV (1991) Description of dispersed particles fragmentation in the plasma jet. Chem Technol 3:46–49
41. Ageev VA, Belashchenko VE et al (1989) Analysis of methods for controlling the parameters of the particle spraying in arc metallization. Weld Prod 12:30–32
42. Popel SI (1994) Surface phenomena in fused metals. Metallurgy, Moscow
43. Linchevskiy BV (1995)Theory of metallurgical processes. Metallurgy, Moscow
44. Vadivasov DG (1958) Research of the influence of spraying process on the properties of metal coatings. SIMSKH, Saratov
45. Pohmursky VI, Student MM, Peech VS (1986) Fundamentals of protection and restoration coatings formation by arc spraying method from cored wires. Physicochem Mech Mater 6:11–16
46. Gubenko BG, Borisova AL, Kostin VA et al (1990) Structural heterogeneity of electric arc coating from steel 65G. Autom Weld 6:19–23
47. Korobov YS, Boronenkov VN (2000) Modeling of liquid metal oxidation at arc metallization. In: Mathematical modeling and simulation of metal technologies. Proceedings of MMT-2000, Ariel, pp 683–692
48. Hussary JVR, Heberlein NA (2001) Atomization and Particle-Jet Interaction in the Wire-Arc Spraying Process. J Therm Spray Technol 10:604–610
49. Sheyhaliev SM (1977) Investigation of crushing metal jets by current. Dissertation, IPI, Sverdlovsk
50. Dyatlov VI (1964) Elements of theory of electrode metal transferring in electric arc welding. In: New problems of welding equipment. Technics, Kiev
51. Bron OB (1975) Plasma flows in plasma electric arc of turning off machines. Energy, Leningrad

52. Popel SI, Sotnikov AI, Boronenkov VN (1986) Theory of metallurgical processes. Metallurgy, Moscow
53. Abramovich GN, Girshovich TA, Krasheninnikov SY et al (1984) The theory of turbulent jets. Science, Moscow
54. Marshak IS (ed) (1978) Pulsed light sources. Energy, Moscow
55. Jvorski BM, Detlaf AA (1985) Handbook of physics. Science, Moscow
56. Zhukhovitsky AA, Shvartsman LA (1987) Physical Chemistry. Metalurgy Moscow
57. Katz NB, Antoshin EV, Vadivasov DG (1966) Metallization by Spaying. Mechanical Engineering, Moscow
58. Rosenbaum RB, Todes OM (1957) Ball drop in a cylindrical tube. Pap AN USSR 115(3)
59. Gorbis ZR (1970) Heat Transfer and hydromechanics of dispersed going through flows. Energy, Moscow
60. Pogorelov YI, Sokolovsky AG (1960) Pneumometric tubes for measuring the parameters of a supersonic flow. Views High Educ, Power Eng 7:121–128
61. Shipai AK, Makarenko AG, Ivashko VS et al (1981) Methodology of studying the characteristics of plasma systems. INDMASH, Minsk
62. Predvoditelev AS (ed) (1962) Tables of thermodynamic functions of air. AN USSR, Moscow
63. Kudinov VV, Bobrov GV (1992) Theory, technology and equipment for spraying. Metallurgy, Moscow
64. Lokhov YN et al (1974) Heating and evaporation of particles in low-temperature plasma jet. Phys Chem Mater Process 6:52–56
65. Vahalin VA, Kudinov VV, Belashchenko VE (1985) Study of efficiency at heating the electrodes and deposition efficiency of feeding materials at arc spraying. Phys, Chem, Mater Process 5:65–69
66. Potapevsky AG et al (1971) Metal transferring at MAG. Autom Weld 6:1–4
67. Zamoruev VM (1960) Evaporation processes in steel production Physics and chemical fundamentals of steel production. AN USSR, Moscow
68. Burov IS (1967) Calculation of heat exchange of dispersed material particles with plasma flow. Phys, Chem Mater Process 2:70–74
69. Pokhodnya IK (1972) Gases in welding. Mechanical Engineering, Moscow
70. Yakimavichus CS (1955) A method of melted metal spraying. USSR Certificate of Authorship 107023
71. Matthäus HD, Busse K (1986) Verfahren und Lichtbogenspritzdüse zum Beschichten von Werkstückoberflächen durch Schmelzen von Drähten in einem elektrischen Lichtbogen. DE Patent 3533966
72. Baranovski VE (1986) Technology development of the process for applying anti-friction coatings with arc spraying. Dissertation, INDMASH of Minsk
73. Polak LS, Surov NS (1969) Investigation of the interaction of powder particles with the plasma flow in nozzle. Phys, Chem Mater Process 2:19–29
74. Ermakov SS, Shmakov AM (1972) Movement and the heating of the powder material during thermal spraying. Powder Metall 7:67–72
75. Sakseltsev VG et al (1972) Heating of powder materials in nitrogen plasma-jets. Mech Eng 1:127–131
76. Esin OA, Geld PV (1962, 1966) Physical chemistry of metallurgical processes. Part 1 and Part 2. Metallurgy, Moscow
77. Glushko VP (ed) (1971) Thermodynamic and transport properties of the combustion products. Moscow
78. Karp JH, Martsevoy EP, Paschenco VN (1985) Mathematical modeling of heat and motion of particles in the plasma environment. Chem Technol 6:27–33
79. Gershenson SM (1980) Investigation of oxidation and improving the technology of plasma spraying of resistant coatings based on nickel alloys. Dissertation. UPI of Sverdlovsk
80. Hasuy A (1975) Spraying technology. Mechanical Engineering, Moscow
81. Glushko VP (ed) (1973) Thermodynamic and thermo physics properties of the combustion products. Moscow

82. Vintovkin AA, Ladygichev MG, Gusovsky VL et al (1999) Burners of industrial furnaces (design and specifications). Intermet, Moscow
83. Suris AL (1985) Thermodynamics of high temperature processes. Metallurgy, Moscow
84. Zourabov VM, Puzryakov AS (1985) Mathematical model for control of plasma spraying processes. Powder Metall 11:50–53
85. Strongin MM (1989) Mathematical modeling of flows in high-temperature technology. Novosibirsk State University, Novosibirsk
86. Reed R, Prausnits J, Sherwood T (1982) Properties of gases and liquids. Chemistry, London
87. Yavoysky VI, Yavoysky AV, Sizov AM (1985) Using of pulsed-blowing for steel production. Metallurgy, Moscow
88. Korobov YS, Polyakov AM Yakovleva IL et al (1997) The structure and properties of steel coatings deposited by activated arc spraying. Weld Prod 1:4–6
89. Lakomskiy VI (1974) Plasma—arc remelting Technology, Kiev
90. Frank–Kamenetskiy DA (1967) Diffusion and heat transferring in chemical kinetics. Science, Moscow
91. Novozhilov NM (1979) Fundamentals of metallurgy of arc welding in gases. Mechanical Engineering, Moscow
92. Litovchenko NN, Radjabov GG, Denisov VV (2001) Recovery of the run-out parts with arc spraying. In: Proceedings of conference Ural Welding. Ekaterinburg
93. Erohin AA (1975) Plasma-arc fusion of metals and alloys. Nauka, Moscow
94. Boronenkov VN, Shanchurov SM, Zinigrad MI (1979) Kinetics of o multicomponent metal slag interaction in the diffusion mode. Views AN USSR 6:21–27
95. Boronenkov VN (1973) Kinetic analysis of oxidation of iron impurities of molten slag. Physical and chemical studies of metallurgical processes. UPI, Sverdlovsk
96. Korobov Ju, Boronenkov V (2000) Modeling of liquid metal oxidation at arc metallization. In: Mathematical modeling and simulation of metal technologies. Proceedings of MMT-2000, Ariel, Israel
97. Korobov YS, Boronenkov VN (2004) Kinetics of interaction of the sprayed metal with oxygen in electric arc metallizing. WeldInt 18(1):42–48
98. Korobov YS (2004) Evaluation of the forces acting on the atomized metal during its spraying. Autom Weld 7:23–27
99. Bagryansky KV, Lavryk PF (1963) Mechanism of alloying metal drops below the ceramic flux at submerged arc welding. Weld Prod 10:5–7
100. Frumin II (1961) Automatic arc surfacing. Mechanical Engineering, Moscow
101. Petrov GL (1972) Welding materials. Mechanical Engineering, Leningrad
102. Boronenkov VN, Salamatov AM (1985) The estimation of interaction kinetics of multi-component metal and slag at submerged arc welding. Autom Weld 8:19–24
103. Tillmann W, Abdulgarder M, Vogli E (2008) Asymmetric melting behavior in twin wire arc spraying with core wires. In: Thermal Spray crossing borders. Proceedings of ITSC 2008, Maastricht, The Netherlands
104. Newbery AP, Grant PS, Neisar RA (2005) The velocity and temperature of steel droplets during electric arc spraying. Surf Coat Techn 195:91–101

Chapter 3
Coatings Properties and External Influences

As mentioned previously, the internal coating properties can be represented as interrelated subsystems, defining their geometrical parameters, thermal and structural characteristics, and strength and residual strains. These properties, in turn, depend on as the characteristics of the previously discussed carrier gas, sprayed material, energy parameters of the power source, and external impacts on the sprayed process.

Let us consider internal properties of the coatings in the following sequence: formation stage, the study of the structure and residual strain. Next, we will consider the main external physical effects affecting the properties of the coatings.

Standard methods adopted for solid materials are used for studies of the internal parameters of coatings. They include methods for evaluating the geometric dimensions and roughness, methods of chemical, X-ray structure and phase analysis and mechanical testing. However, the study of the properties of coatings has some specific features associated with their small thickness, laminated structure and non-uniformity of properties regarding the coating's cross-section. The description and analysis of the basic methods of studying the internal coatings parameters are performed in this chapter.

3.1 Coatings Formation

The structure of sprayed coatings comprises layered elements, formed during high velocity impact onto the cold surface and further solidification of the particles (Fig. 3.1). Molten or heated to the plastic state particles splash on the surface engaging preliminarily by its asperities and then contacting with each other. The particles are oxidized along the spraying distance and the resulting oxide film gets into the coating. Oxides strongly influence on the properties of the coating: they may prevent diffusion of particles, affect the adhesion strength and the hardness of the coating. Due to the small size and low velocity, the single particles have time to cool below its melting point before the moment of impact on the substrate, and their kinetic energy is insufficient to deform. This leads to the fact that they remain in a globular shape within the coating. In the interval between spraying of the individual

V. Boronenkov and Y. Korobov, *Fundamentals of Arc Spraying*,
DOI 10.1007/978-3-319-22306-3_3

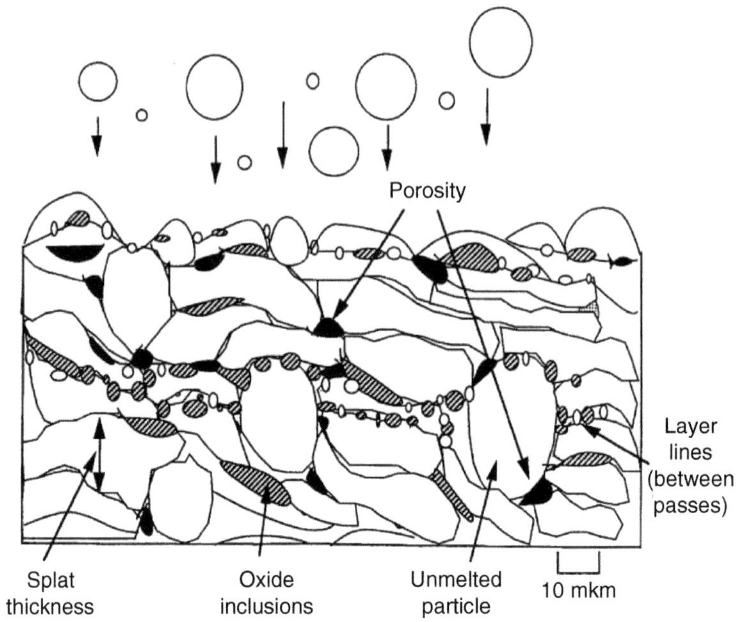

Fig. 3.1 Scheme of TS coating formation [1]

layers, the adsorption of gases and deposition of dust fractions of sprayed material or oxides occurs on the surface of the last layer. Furthermore, the particles are characterized by high rate of crystallization and flattening at impact. Due to that feature, there are cavities and defects in zone of contact with the surface of the previously deposited particles.

This leads to microcavities at the joints of the particles but since the coating is formed in an atmosphere microcavities are filled with gas. The part of the particles, which are overheated above the melting point, may evaporate and deposit as-vapored.

TS coatings are formed from discrete particles in the case of AS with diameter of 10–300 μm and being in the molten state, which are accelerated and sequentially deposited on a relatively cool substrate. Total lifetime of the particles is $(1–10) \times 10^{-4}$ s. The following events occur during that period with it: formation and breakdown from the electrode end face, flying at a velocity in the range of 40–500 m/s and the deposition on the substrate upon impact.

During this period, the particle is successively heated from ambient to a temperature above the melting point of 100–300 °C and then rapidly cooled at impact on the substrate heated typically below 150 °C. Thus, the rate of heating/cooling of particles is 10^4–10^5 °/s. In the case of metals, the possibility of growth of grains formed initially is suppressed. The size of the metal grains in the coating may be less than 1 μm (Fig. 3.2).

In comparison with TS coatings the rolled, cast and forged materials remain at elevated temperatures for a longer period of time. For example, castings in many

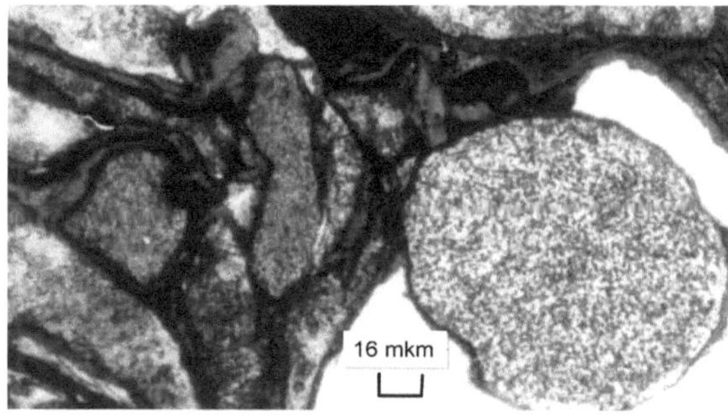

Fig. 3.2 Structure of the AS coating of steel 20X13, ×625 [2]

cases require more than 1 h solidifying. Accordingly, the cooling rate is several orders lower. Forging requires a significant contribution of time and heat to reach the required temperature of the process, more than 1000 °C. The grain size is more than several orders of magnitude, up to 1 mm because of longer holding at high temperatures and solidification time comparing to TS coatings.

Kudinov [3] shows the regularities of the coatings formation. He established a number of important propositions, which lay based on the further study of the properties of coatings.

The calculation made by Kudinov shows that the cooling rate of the molten particle in contact with the surface is faster by 2–3 orders of approach the next particle, even at maximum spraying capacity. High-speed filming shooting supports this statement with a frequency higher than 10^3 shots/s, where we can see the disappearance of particles luminosity on a substrate—they cool before imposing on them the next layer of particles.

Mutual independence of the particles crystallization on the substrate greatly facilitates the analysis of the causes of coating, practically coming down them to the study of the contact interaction of the single particles.

The formation of splashes depends on particles parameters: velocity, size, degree of melting, chemical composition and the angle of incidence to the surface. Furthermore, the shapes of the surface, its temperature and reactivity have an influence.

At impact on the surface spherical particles with diameter d are strongly deformed, harden, and take the form of a thin disk in size $D \times h$ (Fig. 3.3). High velocity of deformation and spreading of the molten particles results in a physical contact of particles material with the substrate at the interaction boundary. The pressure (P) in the contact zone has two components: P_H—head pressure, and P_u—shock pressure arising from a water hammer (or impulse pressure, due to short-term actions). The shots of high-speed filming of liquid particles impact on at the surface show that the liquid is elastically deformed in the first period of the collision, $t_u = 10^{-10}$–10^{-9} s. At the point of impact, a thin flat layer of the liquid of spreading particles is formed. Furthermore, the particle is deformed uniformly.

The formation of flat layer is associated with an elastic compression of the particle during the impact of pressure (P_u), which causes intensively spread the

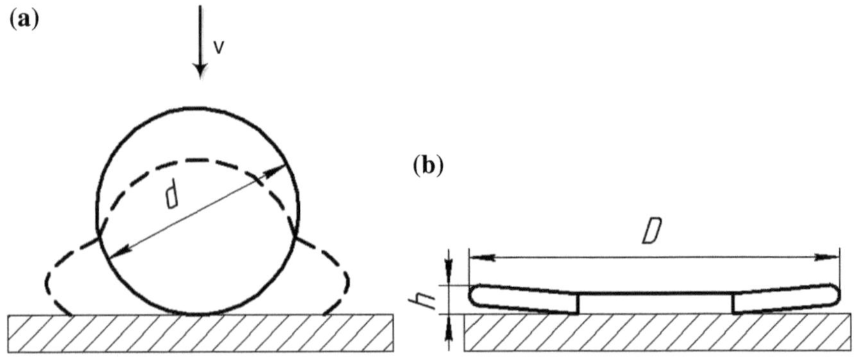

Fig. 3.3 Scheme of the deformation of a particle moving with the velocity (v), when it impacts on the flat surface of the substrate. State of the particle: **a** initial, **b** final

liquid over the surface during a pulse impact. The maximum value of the impact pressure is determined by the well-known Joukowsky equation of water hammer where factor $(\mu/2)$ is introduced to account for the spherical shape of the particle and to estimate its rigidity:

$$P_u = \frac{\mu}{2}\gamma_1 c v, \tag{3.1}$$

where μ is stiffness coefficient of the particle; γ_1—liquid density, kg/m^3; c—the velocity of sound in the metal, m/s; v—velocity of the particles, m/s.

The pulse pressure is P_u = 500–1250 MPa using data of Kudinov; $\mu = 0.5$; γ_1 = 10,000 kg/m^3; v = 100 m/s, c = (2–5) × 10^3 m/s. Such a large quantity contributes to cleaning the substrate surface in the point of impact and results in physical contact of material of particles and the substrate during period $\sim 10^{-9}$ s.

Head pressure P_H for a uniformly moving fluid flow is calculated by the Bernoulli equation. Based on the kinetics deformation of a fluid particle, it can be assumed that the P_H is applied to the area of the substrate surface, similar in size to the diameter of the particles before the impact. If we assume that the substrate is rigid body and particle—a perfect liquid, we obtain the simplest case of an impact:

$$P_H = \gamma_1 v^2 \tag{3.2}$$

The duration of pressure on the axis of the particle:

$$t = (d - h)/c \tag{3.3}$$

At the particle velocity of 100–150 m/s and the duration of 10^{-7}–10^{-5} s the head pressure is 50–100 MPa. Under its influence, the strong adhesion of the substrate and particles occurs.

To consider the elementary processes of physical and chemical interactions that determine the strength of adhesion of the particles, it is necessary to determine the temperature of the contact between the molten particle deformed by contact and a solid substrate. The problem was solved by Kudinov who applied it to plasma spraying, as the problem of the contact thermal conductivity, taking into account the phase transition [3] for the following conditions:

(1) The substrate is a semi-infinite body of a constant initial temperature (T_0) and has non-zero conductivity in the direction perpendicular to the substrate plane. In this direction, the thermal conductivity is equal to the thermal conductivity of the substrate material.
(2) The particle is melted and its initial temperature is equal to the melting temperature (T_m). Its conductivity is not equal to zero only in the direction perpendicular to the plane of contact (or substrate surface). In this direction, the conductivity is equal to thermal conductivity of the solid material of the particle. The solid phase of the particle is fixed in relation to the contact plane and there is a layer of liquid at the time of solidification above the particle.
(3) The contact of the particle to the substrate is ideal and their free surfaces are heat insulated.

The AS conditions are similar, and the following solution can be extended to this process.

The thermal cycle $T_k(t)$ of spherical particle, which is deforming due to the kinetic energy of an impact and solidifying on the substrate we can imagine using the following model (Fig. 3.4).

It consists of two stages: (1) the solidification of the particle when the contact temperature T_k remains constant and there is liquid over a movable solidification front $h(t, r)$; (2) cooling the solidified particles to the substrate temperature. The level T_0, which is the starting point for temperature readings is consistent with the

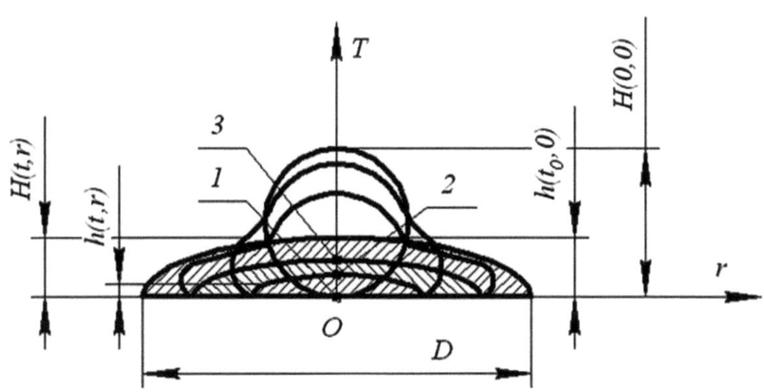

Fig. 3.4 The temperature distribution at various times $t_I < t_{II} < t_{III}$ in system "solid substrate with the temperature T_0—liquid substrate at a melting temperature T_m—rapidly deformed and simultaneously hardening particle". *h(t, r)*—the height of the movable front of crystallization at the distance (*r*) from the axis of the particle at different points in time; *H (t, r)*—the height of the free surface of the particle

initial temperature of the heating surface of the substrate in different points of the melting stain.

The problem of particle's hardening from the moment of contact between the liquid particles and the substrate prior to complete solidification of the liquid over a given point of contact is equivalent to the problem of contact between two half-spaces: liquid and solid [4].

If the contact temperature $T_\kappa(t)$ is below the melting temperature of the substrate for any t, then it has the form:

$$T_k(t) = T_0 + T_k^0, \tag{3.4}$$

$$T_k = K_\varepsilon (T_m - T_0)/(K_\varepsilon + erf(\alpha)), \tag{3.5}$$

where $K_\varepsilon = \frac{\lambda_1}{\lambda_2}\sqrt{\frac{a_1}{a_2}}$ is the criterion of thermal activity of particles regarding to the substrate, the indices "1, 2" refer to the thermal parameters of the particle and the substrate, respectively; λ is thermal conductivity coefficient, a—thermal diffusivity; erf (α)—a function of the probability integral; α—root of the equation

$$K_\varepsilon + \Phi(\alpha) = K_L \cdot e^{-\alpha^2} \Big/ \alpha,\ K_L = cT_m/(1.77L) \tag{3.6}$$

where K_L is a criterion of latent heat of material fusion (L), c—the heat, specific heat of the particle.

To determine α a nomogram was plotted (Fig. 3.5).

The duration of solidification will be

$$t_0 = h^2 \big/ \left(4\alpha^2 a_1\right) \tag{3.7}$$

The temperature of the particles and the substrate at $0 \le t \le t_0$:

$$\mathrm{x} > 0\ T_1(\mathrm{x}, t) = T_k^0 \left(1 + \frac{1}{K_\varepsilon} erf \frac{\mathrm{x}}{2\sqrt{a_1 t}}\right) + T_0, \tag{3.8}$$

$$\mathrm{x} < 0\ T_2(\mathrm{x}, t) = T_k^0 erfc\left(\frac{|\mathrm{x}|}{2\sqrt{a_2 t}}\right) + T_0. \tag{3.9}$$

For the time $\tau = t - t_0 > 0$ the temperatures $T_1(\mathrm{x}, t)$ and $T_2(\mathrm{x}, t)$ will be defined by the conditions 1–3 and the formulas (3.7–3.9) from the solution of a system of differential equations of heat conduction:

$$\frac{\partial T_1}{\partial \tau} = a_1 \frac{\partial^2 T_1}{\partial \mathrm{x}^2}, \tag{3.10}$$

$$\frac{\partial T_2}{\partial \tau} = a_2 \frac{\partial^2 T_2}{\partial \mathrm{x}^2} \tag{3.11}$$

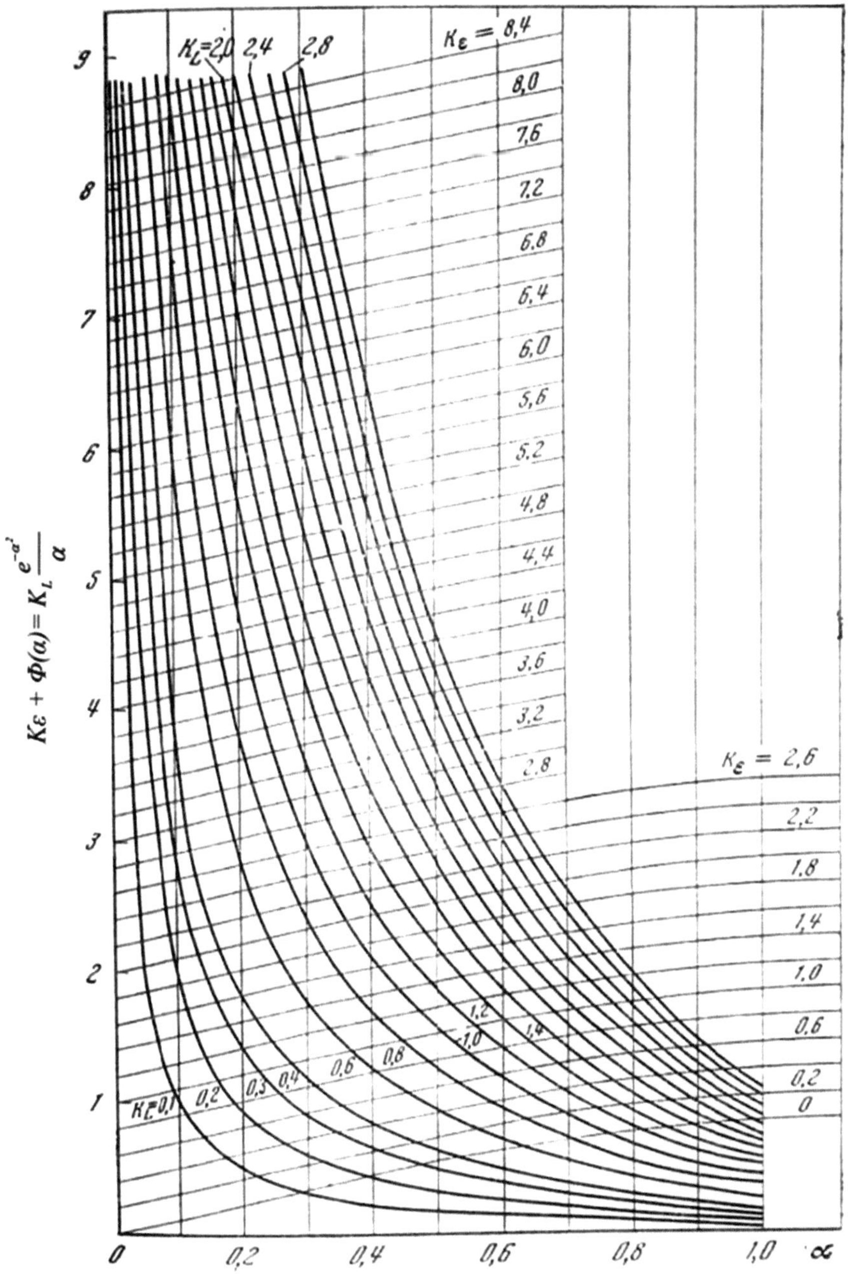

Fig. 3.5 Nomogram for determining α

with initial conditions:

$$T_1(\mathrm{x},\tau)|_{\tau=0} = T_1(\mathrm{x},t_0);\ \frac{\partial T_1}{\partial \mathrm{x}}|_{\mathrm{x=h}} = 0, \tag{3.12}$$

$$T_2(\mathrm{x},\tau)|_{\tau=0} = T_2(\mathrm{x},t_0);\ T_2(-\infty,\tau) = 0. \tag{3.13}$$

In the contact

$$T_1(0,\tau) = T_2(0,\tau);\ \lambda_1\frac{\partial T_1(0,\tau)}{\partial \mathrm{x}} = \lambda_2\frac{\partial T_2(0,\tau)}{\partial \mathrm{x}}. \tag{3.14}$$

Equations (3.4–3.14) represent a mathematical model of the particles solidification on the substrate. The calculations performed by Kudinov [3] show:

- When spraying the contact temperature for the majority of cases corresponds to the solid state of the particles and the substrate (Table 3.1);
- The substrate is heated intensively under particle but the depth of the heat-affected zone under the particle does not exceed several tens of microns.
- Heating the substrate, i.e., rising T_0, we can increase the temperature T_k in contact of sprayed particles and intensify their welding to the substrate.
- The temperature gradient in the surface layer reaches 10^5 °/s.

The given value of the temperature gradient corresponds to the lower threshold of the cooling rate at which the liquid passes into the solid amorphous state, bypassing the stage of crystallization. Amorphous structure provides improved coatings performance and the development of materials for such coatings is one of the most promising directions in the field of thermal spraying. Amorphizators (phosphorus and other non-metallic elements) [5] or spraying ultradispersed materials can help in obtaining amorphous structures.

The study of the coatings structure at spraying confirms indirectly the accuracy of the calculations of the particles cooling rate made by Kudinov. Thus, amorphous structures were not found in the coating from the wire of chromium steel wire like X20Cr13 [2]. However, Borisova et al. when studying the structure of the coatings of carbon steel like 1566 found that "… in the surface layers, and occasionally in the deeper lying layers of the coating there are small, preferably spherical, and a relatively large, generally having elongated shape, particles consisting of poor etched crystals with abnormally high hardness (H_μ = 8.5–11.7 GPa)" [6]. The authors have assumed that such areas are formed by the partial or total amorphization of the sprayed particles during their ultrafast cooling on the substrate. It states that range of cooling velocities of particles is near the velocity of structure amorphization.

Applied to the AS, the study of the deformation nature of particles on impact, depending on the substrate temperature and the impact velocity was studied by Abedini et al. [7].

Table 3.1 The temperature in the contact of the particle and the substrate

Material		Experiment			Calculation					
Particle	Substrate	$T_1 = T_m$, °C	T_2, °C	h, μm	K_ε	K_L	α	$erf(\alpha)$	$t_0 \times 10^6$, °C	T_k, °C
Ag	Ag	960	300	17	1	0.93	0.5	0.52	2.3	737
Al	Al	660	0	10	1	1.03	0.51	0.53	0.94	430
Fe	Fe	1530	300	10	1	1.48	0.63	0.63	9	1055
Ni	Ni	1455	100	18	1.18	1.47	0.6	0.6	23	1000
Al_2O_3	Al_2O_3	2050	400	10	1.16	1.2	0.54	0.56	65	1510

The polished surface of the stainless steel substrate AISI304L was coated with aluminum. The gun ValuArc 100 from Sulzer-Metco was used. The regimes were used: voltage—32.1 V; wire feed rate—7.0 m/min; spraying distance—200 mm; gas–air, the gas pressure—0.21/0.41 MPa, gas flow rate—1.13/1.81 m^3/min. Different pressure/flow rate corresponded to the impact velocity of the particles on the surface 109 ± 28 and 143 ± 36 m/s. The temperature of preheating of the substrate varied from 25 to 450 °C. To estimate the splashes during the deformation of particles the surface roughness R_a 0.01–0.04 m km was provided by polishing.

For adhesion tests, aluminum was sprayed (143 m/s) onto the flat ends of cylindrical stainless steel substrates (~50 mm long and 24.7 mm in diameter) that were grit blasted to an average R_a ranging between 7 and 8 μm. Once sprayed, the substrate was glued to a counterpart test coupon using the FM1000 epoxy film. Pull tests were conducted in accordance with ASTM C-633 standards (Standard Test Method for Adhesion or Cohesion Strength of Thermal Spray Coatings). Average value of tests was resulted from 3 to 5 specimens.

A DPV-2000 system was used to characterize size, velocity, and temperature distributions of in-flight particles.

Optical images were made of splats and coating cross-sections. Image analysis software was used to determine porosity. Deposition efficiency was determined by weighing coupons before and after depositing coatings and comparing with the mass of wire sprayed.

Figure 3.6 shows photographs of the particles impact on the surface at different modes of AS. It is evident that all the particles are splashed in the form of stars of irregular shape at the temperature of the preheating substrate (T_S) below 100 °C. The rays form is very uneven; some of them are accompanied by finger-like projections. Rapid solidification of border of flattering zone limits the spreading of liquid. It increases the liquid head pressure and causes edge splashes of spreading particles in the form of the above-mentioned finger-like projections on its boundary rim. Another sign of confined flow is the occurrence of cavities in the splashing center. Their presence leads to the formation of pores in the transition zone between splashing and the base. At T_S higher than 100 °C the cavities are reduced significantly, both in magnitude and size.

At temperatures above 100 °C changes in the structure of the splash begin. The higher the temperature is, the closer the shape of the splash disk to the regular shape. Further increase in T_S to 400 °C leads to fractures on the disk edge in the kind of splashes, obviously due to the influence of T_S on the surface tension in the solidified crust. The slower particles solidification due to increased substrate temperature enables the particles flow on a larger diameter before solidifying. The comparison of particles diameters at different velocities at the same T_S indicates that particles diameter increases with the higher impact velocity of splashing.

The transition temperature from splash to the disk of regular shape was evaluated by the transition frequency using computer image processing, Fig. 3.7. Transition

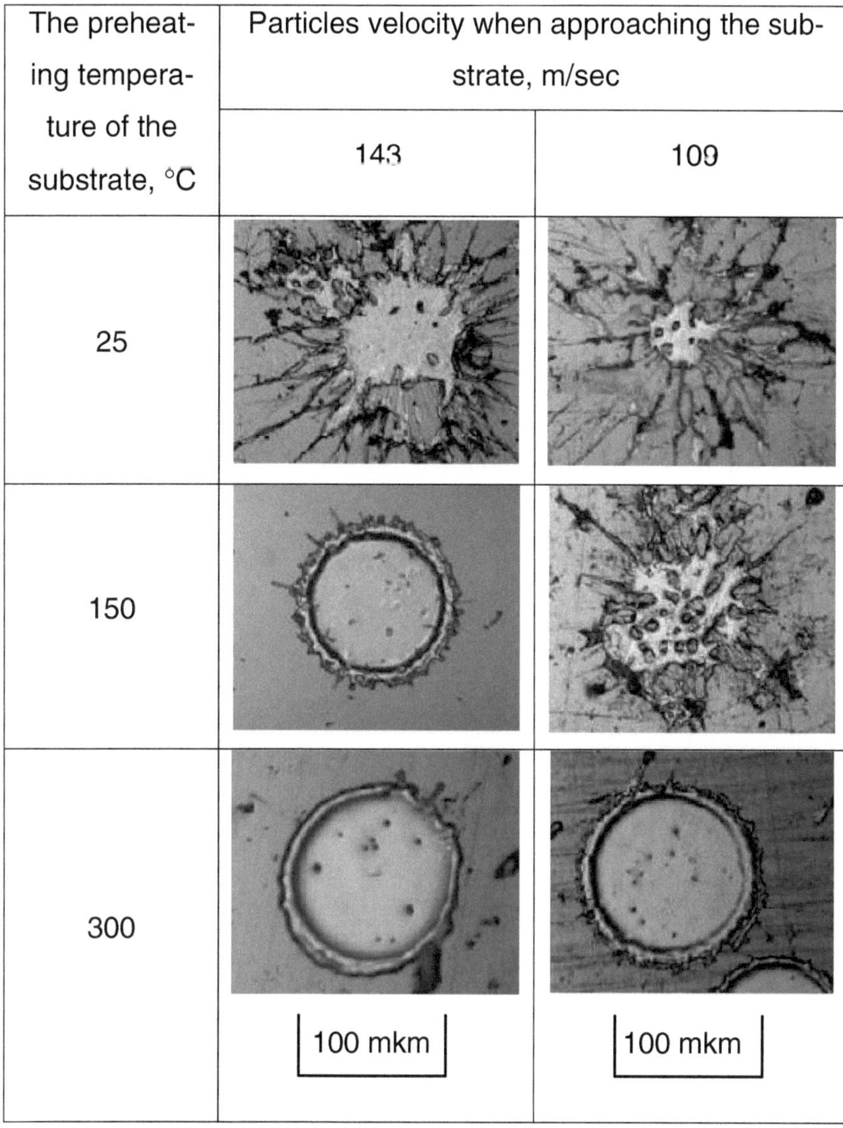

Fig. 3.6 Splash form and microstructure of AS coating of aluminum sprayed on a polished stainless steel surface, which has been preheated to different temperatures

frequency shows the proportion drops turned into a coating as regular shaped disk. For this indicator, the transition temperature is 150 and 225 °C for velocities of 143 and 109 m/s, respectively.

In addition, to quantitate the nature of the flow of a drop, the index of degree of splashing (*DS*) was introduced:

$$DS = (P^2)/(4\pi A)$$

where *P*, *A*—perimeter and splashing area.

For a perfect circle *DS* = 1.0. Its increase indicates an increase in uniformity of the particle shape during the spreading. The data analysis shows that the desired value of *DS* is less than 1.5; it corresponds to the transition temperature 150 and 230 °C for velocities of 143 and 109 m/s, respectively, Fig. 3.8. Authors showed a good agreement between the experimental results and the calculated spread factors.

The comparison between the deformation degree after the particles impact at different velocities shows that the high degree of deformation corresponds to high velocity (Fig. 3.9). This can be explained by higher kinetic energy of a droplet of higher velocity at the impact.

The porosity of the coatings decreases with increasing of the velocity of the particles and in contact with a substrate heated to 100–200 °C (Table 3.2).

For coatings on a substrate at low heating temperature, the pores on "coating-substrate" transition zone can be seen. Their quantity and size sharply decrease when heating the substrate above 100 °C, while heating the substrate above 250 °C leads to the disappearance of the pores in the transition zone. This may be due to the peculiarities of border splashing formation, as well as due to the breakage of the liquid film of the spreading droplet.

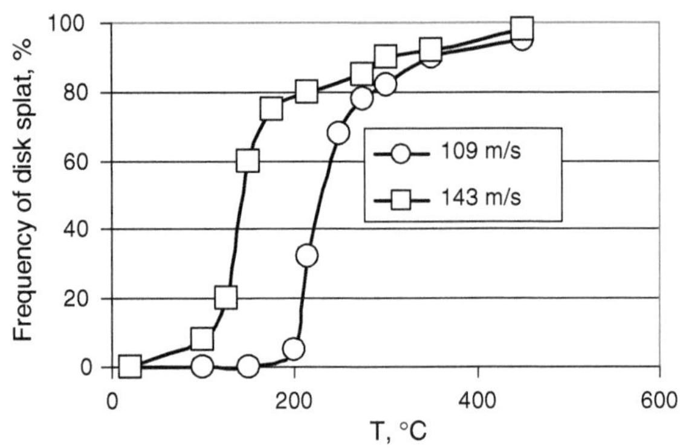

Fig. 3.7 Effect of the preheating temperature of the substrate on the frequency of splash

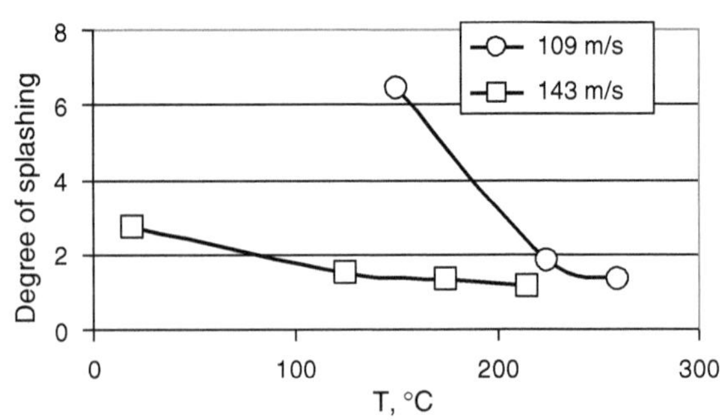

Fig. 3.8 Effect of the preheating temperature of the substrate on the degree of splashing

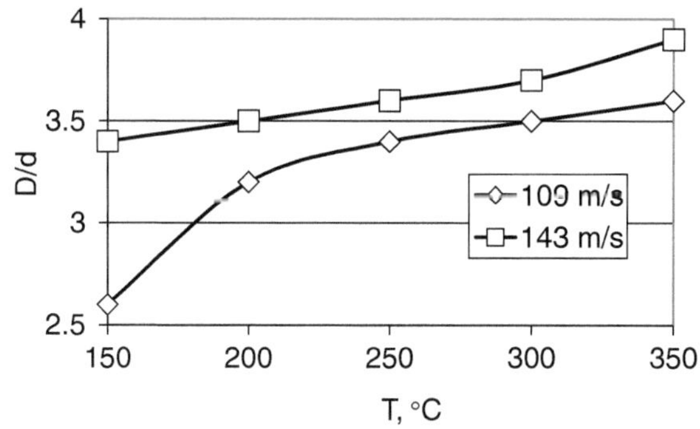

Fig. 3.9 Changing the degree of deformation of droplets with a different impact velocity. *d, D*— droplet diameters before and after impact, respectively

Table 3.2 Porosity of the coatings made at different heating temperatures of the substrate and impact velocity

Substrate temperature, °C	143 ± 36 m/s		109 ± 28 m/s	
	Porosity, %	Pores size, μm	Porosity, %	Pores size, μm
25	2.4 ± 1.1	5.3	2.6 ± 1.1	4.8
100	1.8 ± 1.7	5.8	1.7 ± 0.8	3.0
150	–	–	1.9 ± 1.2	3.3
200	1.1 ± 0.2	2.5	1.9 ± 1.7	4.1
250	1.9 ± 1.1	2.6	2.8 ± 1.0	2.8
300	2.3 ± 0.1	2.3	3.3 ± 0.8	4.2
Average	1.9 ± 0.8	3.7	2.4 ± 1.1	3.7

Our experimental data show that the magnitude of the open porosity decreases sharply with increasing the velocity and decreasing the particle size. Coatings produced by typical Arc Spraying (AS) and by Activated Arc Spraying (AAS) were compared, see [6, 8] and Table 2.15. In case of AS average particle velocity is 55 m/s, the particle size is 1–600 μm. In case of AAS, wherein the products of combustion of propane–air mixture are used as a carrier gas the average particle velocity is 140 m/s, the share of droplet diameter of 20–125 μm is about 75 %, its maximum size is 315 μm.

Analysis showed that at the coating thickness of 0.05 mm open porosity of AS and AAS coatings differs by almost one order at different spraying distances *L*, Fig. 3.10.

Obviously, this is associated with a decrease in porosity of the coating surface due to the reduction of roughness. The roughness is directly related to the size of the particles impinging on the surface. Profilograms analysis shows that the average height of the peaks on AAS-sprayed surface (spraying distance L = 100 mm, coating thickness 0.1 mm), while for AS coating this value reaches 0.03–0.05 mm under similar conditions.

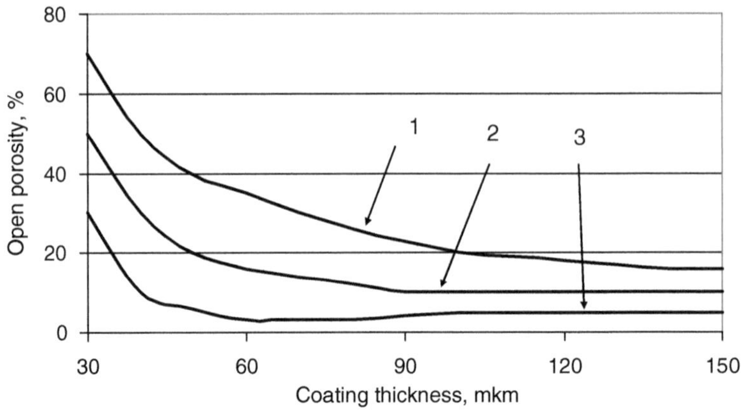

Fig. 3.10 Influence of the coating thickness on the open porosity. AS, *2, 3* AAS, *1, 3* $L = 100$ mm, *2* $L = 50$ mm

This difference in the open porosity leads to an even greater difference in the gas permeability. For AAS coating, this quantity is 1–2 orders of magnitude smaller than in the case of AS, with a coating thickness greater than 0.1 mm. With a layer thickness of 0.05–0.1 mm the difference is 4–5 orders of magnitude, Fig. 3.11.

For AS guns providing the same velocity of the particles the increase of the oxygen share in the composition of the carrier gas leads to not only an increase in the number and size of oxide layers, as well as the increase the porosity of the coating. It is especially noticeable in the surface layers (see Fig. 3.12). Increased porosity of the coating reduces its cohesive strength, since the pores are areas with near zero strength. The porosity may occur due to the following reasons: shrinkage during transition of the liquid to the solid phase, decreasing the solubility of gases with the lowering of the temperature and with the entrapment of gas by particles as they impact on the surface [8].

Due to the concepts of coating formation shown above the coating can be represented as composition of thin plates joined together along the contact surface

Fig. 3.11 Effect of coating thickness on gas permeability. *1* AS; *2, 3* AAS; *1, 3* $L = 100$ mm; *2* $L = 50$ mm

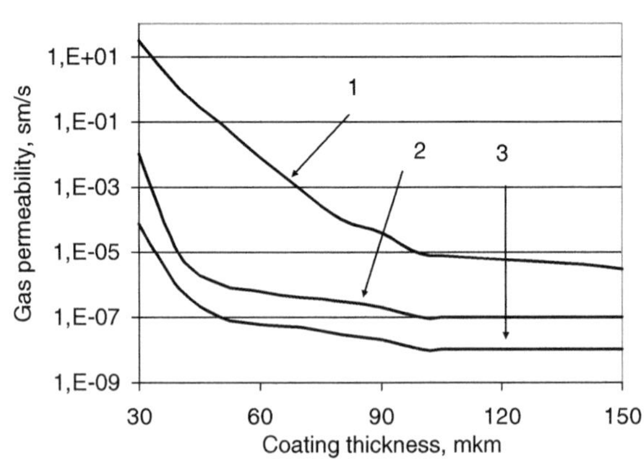

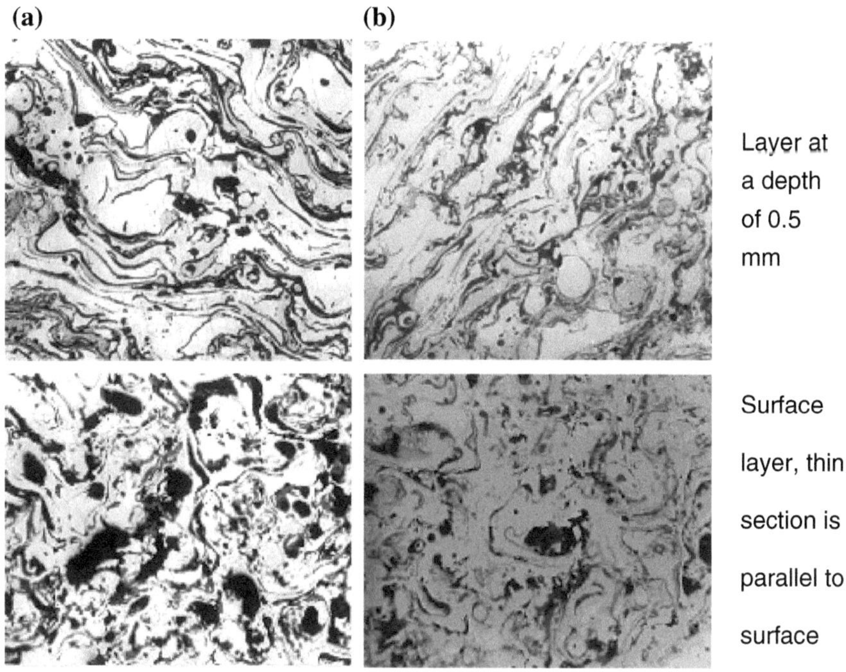

Fig. 3.12 Comparison of the coating structures made from structures of coatings from chromium steel wire like X20Cr13, ×250, with a different composition of the spraying gas jet: **a** air, **b** air and propane

by the welded section or bonded sections with area F_x, which were formed during solidification of the particles (Fig. 3.13) [3].

Welded areas do not fill the entire contact area between particles, and therefore the strength and density of the coating are lower than the strength and density of the compact material. The coating can be identified by number of structural elements that are separated by interfaces with certain properties. The interface between the coating and the substrate (1) determines the strength of adhesion between them (coating adhesion). The interface between layers (2) occurs because of the interval between spraying of the individual particles. During that interval the coating surface is contaminated, obstructing the interaction at this boundary. Also there are contact boundaries (3) between the particles in each layer. Boundaries (2, 3) determine the bonding strength in the layer which is called cohesion. Conditions for the formation of boundaries (2) between the layers differ from the conditions for the formation of boundaries (3) between the particles primarily in the time of staying in the atmosphere. The interval between the moments of the layer deposition can reach tens of seconds, which is several orders of magnitude more than pause time between the interaction of particles in the layer.

It is considered nowadays that the strength of the coatings is due to the structures of coatings from chromium steel wire like X20Cr13 by several mechanisms:

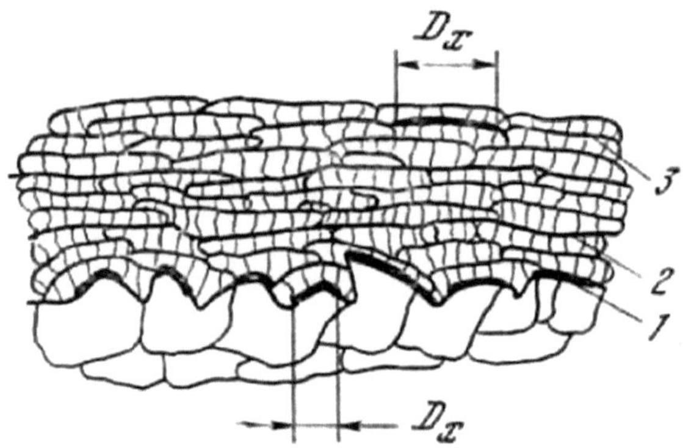

Fig. 3.13 Schematic structure of the coating. *1* boundary between the coating and the substrate; *2* the boundary between the layers; *3* boundary between the particles in each layer; D_x the diameter of the contact area where particles are welded

- Purely metallurgical bond when heat from the particle is enough for increasing the local temperature of the surface material or previously coated particles, up to the level which is necessary for the refining, diffusion/mixing or simple diffusion. This would result in chemical interaction when the particles and the substrate will have an excess of energy needed to end the process of chemical reaction—which is called as activation energy. The atoms further apart close on a distance of $\sim 3 \times 10^{-10}$ m.
- Mechanical engagement of depositing particles with a rough surface or with previously deposited particles;
- The close physical interaction between the particles under the influence of Van der Waals forces, where the atoms of sprayed material and the substrate approach to a distance of the order of the atoms size, 10^{-10} m.

Summarizing various estimates of the contribution of each form of binding, one can conclude that the share of metallurgical bond is usually small; otherwise, the tensile strength of the coating is close to the initial material.

The thermal conductivity of the coating is lower than that of a compact material. Contact temperature calculations made by Kudinov show that due to the difference of thermal conductivity the contact temperature between the coating layers is higher than that between the coating layer and the compact substrate. Moreover, the temperature of the coating during the spraying without substrate preheating is higher than the contact temperature required for strong bonding of the particles with the compact material. Therefore, in most cases the cohesive strength of the coating is higher than the adhesive one.

Barvinok [9] develops numerical methods for calculation of temperature fields in the coating—substrate system at a constant and a variable velocity of the interface crystallization. When passing from discrete particles to the layer model of a continuous buildup of coating, the thermal balance of the system can be represented as follows:

$$QT = Q\text{кин} + Q\text{пл} + Q\text{кр} + Q\text{ост} - Q\text{к} \qquad (3.15)$$

where the indices "*T*", "кин", "пл", "кр", "ост", "к" refer to following energy types: accumulated, transfer of kinetic energy into heat, the cooling from the final heating temperature to the melting point, crystallization, cooling from the melting point to the surface temperature of the deposition lost by convection and radiation, respectively.

Considering temperature relationships of thermo physical parameters the calculations for materials with low thermal conductivity (with a high content of nitrides, carbides, and nichrome) during plasma spraying showed that the coating temperature after 90–120 s reaches 600–900 °C (depending on technological parameters) and is stabilized at this level. Calculation error was 5–10 % in comparison with measurements of temperature by a chromel–alumel thermocouple installed at a depth of 1 mm from the surface of the substrate when the diameter of the electrodes tips is 0.1 mm.

The thermal conductivity of steel is considerably higher. With regard to the metal-sprayed coating of chromium steels, the parameters of the thermal regime at the deposition of molten metal droplets on the already hardened surface layers, we studied using the analysis of the microhardness measurements [2]. To evaluate the reheating, the experiment was carried out to simulate the heating conditions in the spraying process. Specimens of steel chromium steel wires like X20Cr13 and X40Cr13 with length of 60 mm (four specimens of each grade) were heated in a reducing atmosphere to 1000 °C and after 15 min exposure they were oil-hardened. Then, hardened specimens were tempered at 200, 300, and 400 °C, exposured for 5 min, cooled in air. The specimens were cut at the place of heating, poured into a plastic "Etacryl", some of which were boiling in water for 1 h. The boiling point of water, 100 °C, should be considered as the lowest tempering temperature. Microhardness was measured at thin sections. The obtained values of microhardness with a certain degree of approximation show the variation of this parameter for hardened steels like X20Cr13 and X40Cr13 when tempering. Figure 3.14 shows the

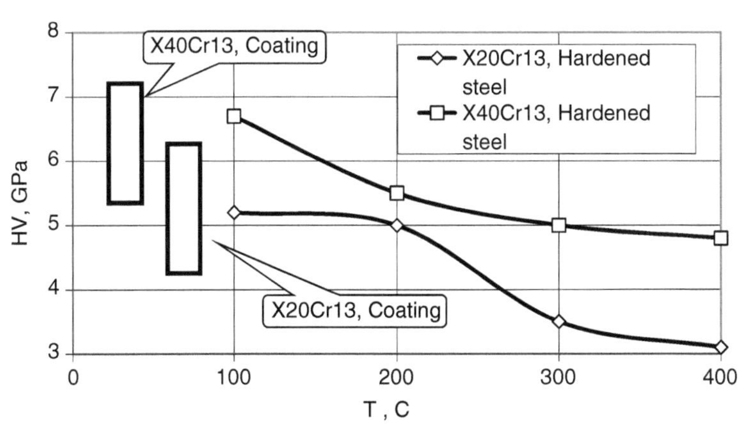

Fig. 3.14 Comparison of the steels microhardness in the sprayed and hardened state

results of wire hardness evaluations obtained compared with the range of values of the microhardness of steel coatings like X20Cr13 and X40Cr13.

The results show that tempering at temperatures above 200 °C leads to the lower hardness of steel than the average value in the sprayed layer. This allows us to conclude that the sprayed layers from steels like X20Cr13 and X40Cr13 are heated to a temperature not higher than 200 °C corresponding the microhardness of tempered (not above 200 °C) martensite.

3.2 Residual Stresses

Residual stresses (RS) are one of the main causes of coatings failure. To date, in a number of fundamental papers were published devoted to their study, and to the methods of determining the influence on the coatings characteristics [1, 3, 9–12, etc.].

Three kinds of internal stresses are distinguished [5]:

- Internal stresses of the first kind arise between the separate section zones and between different parts of the details;
- Internal stresses of the second kind occur within the grains or between adjacent grains. They arise between the various phases due to the difference in the coefficients of thermal expansion or due to the formation of new phases having linear sizes;
- Internal stresses of the third kind occur within the space of the order of several unit cells of the crystal structure. As an example of such stresses arising, we can mention the case where a foreign atom in solid solution creates around itself elastic stresses.

For solid materials the level of tensile strength is the following: the casting of high-strength steels—about 1700 MPa, at forging of high-strength steels—1200 MPa, high-strength low-alloy steels—550 MPa.

Unlike solid material, the bond between individual particles is considerably lower than the strength of the coating constituents. In TS coating, the tensile strength is 27.5–55 MPa, with a gradual increase up to 150 MPa at a different processing. From strength comparison of solid materials and coatings, it can be seen that the strength of the coating is largely determined not by the material strength of individual particles but by cohesive forces between the particles.

In the spraying process, the molten particles impact at high velocity on the substrate surface, then deform and fix to it by adhesion forces. During the crystallization of a single particle, RS of the second and third kind occur which are balanced in the volume of a single particle. For the study of RS in the coatings, Virnik et al. adopted a model in which instead of single particles a continuous layer is considered [13]. A series of subsequent studies [3, 9, 10] have confirmed the possibility of using this model. This allows performing calculations using the theory of continuum physics (strength of materials, theory of elasticity and plasticity, theory of heat and mass transfer, and others). On the basis of these calculations, the stresses of the first kind arising in the macro-body are estimated.

Currently, the following physical mechanism of their formation is adopted. During the coating deposition while coating thickness, increasing the surface of crystallization is moved along the normal to the substrate. The change in temperature leads to a heat expansion of the system. However, the thermal expansion occurs constrainedly because of stresses occur due to fixing of the coating on the substrate, continuous temperature increase and potential arise of its gradient along the cross-section. Stresses occur also due to impurities, inclusions, block boundaries, the difference in coefficients of thermal expansion and the lattice parameters, phase, and structural transformations.

Thus, with the growth of a coating the crystallization stresses are formed on its surface. During the transition of the superficial element into the coating volume, it is imposed by stresses associated with the process of continuous growth and gradual changes in temperatures, loads, and crystallization stresses influence of other layers. After the coating is formed, it is cooled to ambient temperature, and the substrate is released from the fixture. Therefore, stresses existing in a certain volume are superimposed by thermal stresses associated with cooling system and a fixing device. Their algebraic sum gives RS. If they exceed the limit of plasticity, it is necessary to conduct further calculations considering this aspect.

In the simulation, the system "Substrate-Coating" is replaced by homogeneous plate, which is coated with material having other thermal and elastic characteristics (Fig. 3.15). First, the calculation of stresses in the coating was carried out [13] with the following assumptions:

- the spot of two-phase flow is a normally distributed heat source;
- plane stress is implemented in the plate (the plane is parallel to the surface of the sample)
- thickness of the applied coating increases continuously during Up to final thickness δ spraying;
- the coating is applied simultaneously over the entire area of the plate.

For this case, the stresses in the sprayed layer, which lies at a depth of (y) are as follows:

$$\sigma_y = \frac{-\alpha_1 T_y + \left[-\frac{P_k}{b} + \alpha_2 E_2 h_2 T_{c2} - \alpha_1 T_{c1}(y - h_2) + \int_y^h \sigma(\xi)\mathrm{d}\xi\right]}{E_2 h_2 + E_1(y - h_2)} \cdot E_1 \qquad (3.16)$$

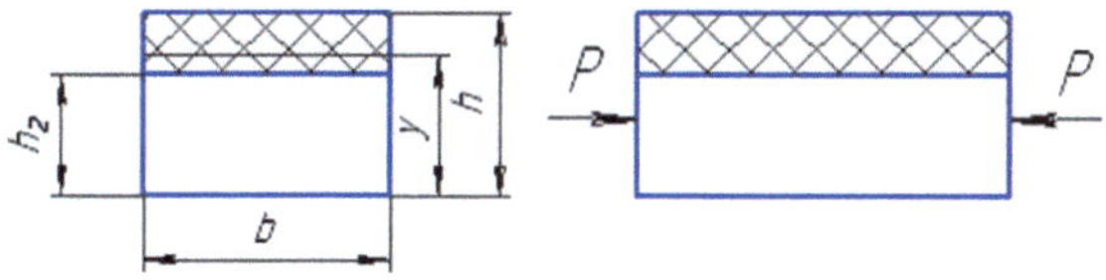

Fig. 3.15 Scheme of the coated specimen to calculate RS

where σ_y is the value of RS in the coating layer; α is a coefficient of thermal expansion; T_y is temperature field corresponding to the moment of formation of the sprayed layer at a depth y (can be determined analytically or experimentally); E is modulus of elasticity in tension; h, h_2 are the thicknesses of the coated sample and without it; T_c is the average temperature at the time of formation of the layer; ξ is current coordinate; P is force corresponding the reaction at fixed ends of the specimen, during the free expansion $P = 0$; indices "1", "2" refer to the coating and the base metal, respectively.

In subsequent studies, the Eq. (3.16) was clarified with respect to the existing methods of experimental studies and to the used mathematical apparatus [9–11].

Barvinok [9] developed models, proposed algorithms and implemented programs of the RS calculation in the system "substrate-coating" of the following cases:

- model homogeneous plate with thickness h, which is coated by material with other thermal and elastic properties not depending on the temperature; layering process of building layers is taken into account;
- n-layered ring of materials with different elastic modules, but close Poisson's ratios;
- narrow n-layer plate with different elastic modular but close Poisson's ratios.

For the model plate, purely analytical method of calculation is used. For multi-ring and plate, the analytical calculation is performed on the basis of experimental data on the deformation after sequential removal of the coating layers.

Relatively simple experimental–analytical method for the RS determination was offered by Puzryakov, this method was based on strain measurement after consecutive removal of layers by electrochemical etching (Fig. 3.16) [11].

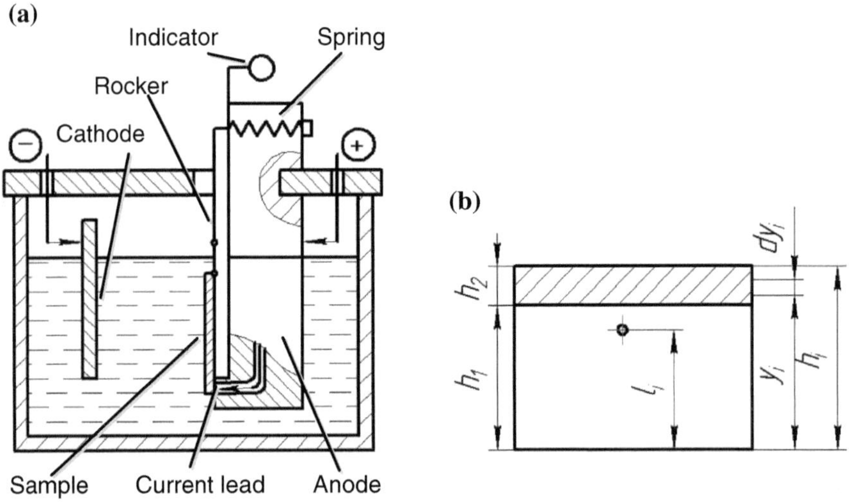

Fig. 3.16 Determination of RS by electrochemical removal method: **a** the scheme; **b** cross-section of the sample

The RS value in the sprayed layer and in the substrate is experimentally determined according to the deformation of the specimen during the gradual removal of a layer. To study RS, the specimens spraying with size 140 × 10 × 2 mm are sprayed. The width b = 10 mm is selected from the condition of stresses constancy across the width of the sample (b < 15 mm). Electrochemical etching carries out removing the strained layers when the mechanical effects on the specimen and the amount of RS are minimal. Electrolyte composition depends on the coated material.

The deformation of prismatic rectangular samples in process of the etching is determined according to the scheme in Fig. 3.16, which is recommended for long specimens. At a point located on the top-of coating at a distance a_i, residual stresses $\sigma(a_i)$ is calculated according to the formula:

$$\sigma(a_i) = \frac{8E_1}{l^2}\left\{\begin{array}{c} f_1[h - l(a_i) - a_i] + N(a_i)\dfrac{\mathrm{d}f(a_i)}{\mathrm{d}a} \\ -(h - a_i)f(a_i) - \displaystyle\int_0^{a_i} M(\varepsilon)\mathrm{d}\varepsilon \end{array}\right\} \tag{3.17}$$

where f_i is a bend formed by the unfastening of the sample; h is a height of the sprayed specimen; $l(a_i)$ is the distance from the center of gravity of the section to the surface opposite the removable layers; a_i is the thickness of the removed layers; $f(a_i)$ is the deflection due to the removal of the layer;

$$N(a_i) = \frac{\frac{E_1}{E_2}[h_1 - l(a_i)]^3 + [h - a_i - l(a_i)]^3 - [h_1 - l(a_i)]^3}{h - a_i - l(a_i)} \tag{3.18}$$

wherein E_1, h_1 are the modulus of elasticity and the thickness of the substrate; E_2 is the modulus of elasticity of the coating; l is the sagging between the support points of the specimen when measuring.

$$M(\varepsilon) = \frac{N(\varepsilon)}{h_1(E_1/E_2 - 1) - (h - \varepsilon)} - l(\varepsilon)\frac{\mathrm{d}f}{\mathrm{d}\varepsilon}. \tag{3.19}$$

The parabolic approximation is applied to calculate the derivative of:

$$\begin{aligned}\frac{\mathrm{d}f(a_i)}{\mathrm{d}a} &= f_{i-1}\left(\frac{\Delta_{i+1}}{\Delta_i(\Delta_i + \Delta_{i+1})}\right) + f_i\left(\frac{\Delta_{i+1} - \Delta_i}{\Delta_i\Delta_{i+1}}\right) + f_{i+1}\left(\frac{\Delta_i}{\Delta_{i+1}(\Delta_i + \Delta_{i+1})}\right) \\ &= f_{i-1}K_i^{(1)} + f_iK_i^{(2)} + f_{i+1}K_i^{(3)}\end{aligned}$$

where $\Delta_i = a_i - a_{i-1}$ is the thickness of the removed layer.

This formula is valid for all values of a_i, except for the two extremes. For the initial layer $i = 0$, $a_0 = 0$, $f_0 = 0$, and

$$\frac{\mathrm{d}f(0)}{\mathrm{d}a} = f_1\left(\frac{\Delta_1 + \Delta_2}{\Delta_1\Delta_2}\right) + f_2\left(\frac{\Delta_1}{\Delta_2(\Delta_1 + \Delta_2)}\right) = f_1K_0^{(2)} + f_2K_0^{(3)}.$$

For the last removed layer $i = n$ and

$$\frac{\mathrm{d}f(a_n)}{\mathrm{d}a} = f_{n-2}\left(\frac{\Delta_n}{\Delta_{n-1}(\Delta_{n-1} + \Delta_n)}\right) + f_{n-1}\left(\frac{\Delta_{in-1} + \Delta_n}{\Delta_{n-1}\Delta_n}\right) + f_n\left(\frac{\Delta_n + (\Delta_{n-1} + \Delta_n)}{\Delta_n(\Delta_{n-1} + \Delta_n)}\right) = f_{n-2}K_n^{(1)} + f_{n-1}K_n^{(2)} + f_nK_n^{(3)}$$

To calculate the integral the trapezoid rule was used:

$$\int_0^{a_i} M(\varepsilon)\mathrm{d}\varepsilon = \sum_{i=1}^{n} \Delta_i \frac{1}{2}[M(\varepsilon_{i-1}) + M(\varepsilon)]$$

The value $l(a_i)$ is determined by the expression:

$$l(a_i) = \frac{0.5\left[E_1\left(h_i^2 - h_1^2\right) + E_2h_i^2\right]}{E_1h_1 + E_2h_2},$$

where h_2 is the thickness of the coating.

By integrating the values of RS in thickness, the total RS of the sprayed layer can be obtained. The calculation results show that the RS values can be obtained both positive and negative depending on the conditions of parts fastening, coating, and substrate materials. The calculations results showed that RS may vary from +60 to −200 MPa in the plasma spraying coatings in an atmosphere (coating thickness up to 100 μm, coating materials are titanium and zirconium nitrides, substrate materials are alloys like NiCr28Al, aluminum, and titanium alloys) [9].

Applied to the arc spraying, the RS measurement was carried out in coatings of cored wires on the basis of ferrochrome and ferrochromaluminum during the sequential mechanical removal of layers [14]. The measurements have shown that tensile stresses act in the outer layer of the coating of 0.6 mm thick and are reduced when approaching to the substrate while at the substrate the depth of 0.5 mm act as the compressive stresses (Table 3.3).

The destruction of the coating in many cases is connected with the detachment on the sharp edges or with the swelling. This is caused by the fact that normal stresses

Table 3.3 Residual stresses in the arc sprayed coatings

Coating material	Maximum stresses values, MPa	
	Compressive	Tensile
30X25	115	180
CrAl 20 6	100	150

are formed during the spraying of powder particles to the surface of the substrate around the perimeter of the particles. These normal stresses cause the appearance of a bending moment, which creates a force breaking or clamping the edge of the particles to the substrate (see Fig. 3.17). Fracture stresses cause the detachment around the perimeter of the particle (or coating), clamping stresses cause swelling.

The formation mechanism of the stress state which causes coatings breaking due to the edge effect, is proposed in [10] by considering the scheme of deformation of a single particle. The spherical particle falls onto the substrate, flattens and takes the shape of a cylinder. During the cooling, the top of the particles is deformed freely and the bottom is compressed as a result of adhesion to the substrate surface. If the thermal shrinkage of the coating is greater than at the substrate, the fracture stress is calculated as follows. Deformations of coating and substrate surfaces lead to deformation of the generating line (δ) of the cylinder which is the flattened particle (Fig. 3.18). Here relative deformation is the following:

$$\varepsilon = (\delta - h_n)/h_n, \tag{3.20}$$

where h_n is the coating thickness;

$$\delta^2 = h_n^2(1 - \alpha_n T_n)^2 + (d^2/4)(\alpha_n T_n - \alpha_M T_M + \sigma_M/E_M)^2, \tag{3.21}$$

$$\sigma_M = (\alpha_n T_n - \alpha_M T_M)E_M/[1 + E_n h_n/(E_M h_M)].$$

Since the fracture stress in the coating is equal to $\sigma_n = \varepsilon E_n$ (excluding Poisson's ratio), then from the Eqs. (3.20, 3.21) they can be calculated with some simplifications by the formula:

$$\sigma_n = E_n\left[c(\alpha_n T_n - \alpha_M T_M)^2 - \alpha_n T_n\right], \tag{3.22}$$

where $c = 0.28 \cdot [1/(1 + ab)]^2 (d/d_0)^6$; α_n, α_M, T_n, T_M are coefficients of thermal expansion (CTE) and temperature of the coating and substrate, respectively; *a*, *b*—the ratio of the thickness and the elastic moduli of the coating and substrate, respectively; *d*, d_0—particle diameters at the time of the contact with substrate and

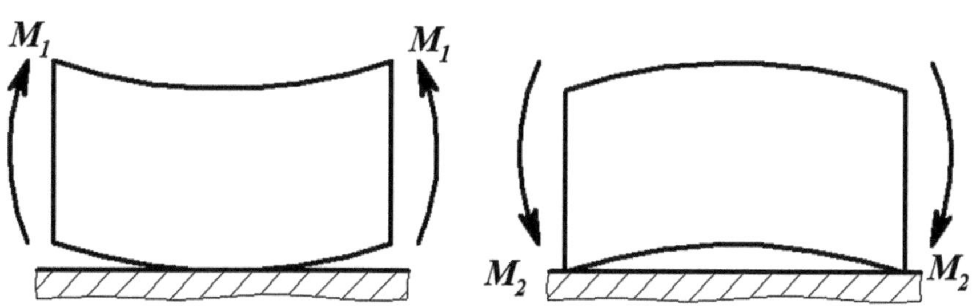

Fig. 3.17 Scheme of formation of fracture stresses

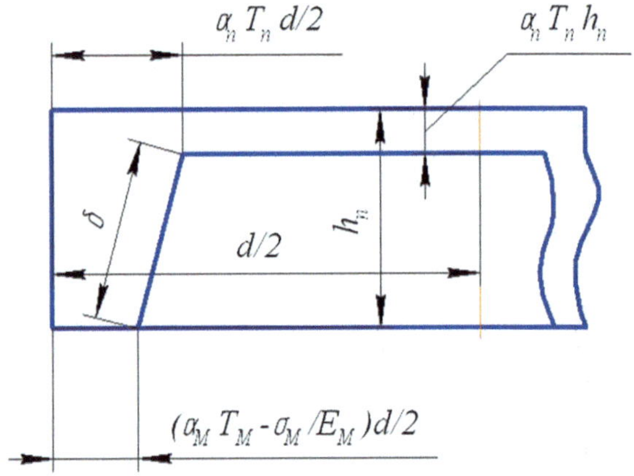

Fig. 3.18 The scheme for calculation of fracture stresses

after flattening, indexes "n", "M" refer to the coating and the substrate metal, respectively.

$$\sigma_n = E_n\left[c(\alpha_n T_n - \alpha_M T_M)^2 - \alpha_n T_n\right], \tag{3.23}$$

The maximum value of the fracture stresses is determined from the Eq. (3.23) due to its differentiation:

$$\sigma_{n\max} = k(\alpha_n T_n - \alpha_M T_M)^2, \tag{3.24}$$

where k depends on the deformation degree of the particle, which in turn depends on T_n.

The structure of Eq. (3.24) shows that the spraying by superheated particles (in spite of the high bonding strength) leads to the formation of high fracture strains because of the growth of fracture stresses. This confirms the experimental data (Chap. 1) of the increased danger of delamination of plasma coatings formed from the highly superheated particles in comparison with Arc spraying where overheat above the melting point is relatively much smaller.

The evaluation of fracture stress in the coatings with thickness of 1 mm [10] shows that the type of stress (compression or tension) and the gradient depend strongly on the material being sprayed (Fig. 3.19) and the preheat temperature (Fig. 3.20). Thus, the particles at temperature of 100 °C the tensile stresses are generated if the substrate is heated up to 100–300 °C. At lower heating temperatures, the breakaway may occur, at higher—the swelling.

The disadvantage of this approach is the lack of justification for the transfer of the calculations results for a single particle on the coating as a whole.

Based on the theory of elasticity Barvinok estimated the normal stresses in the zone of the edge effect, depending on the distribution of fracture stresses in the system [9].

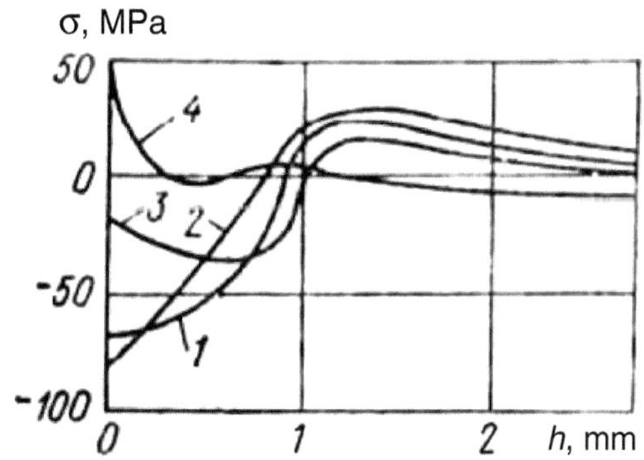

Fig. 3.19 Residual stresses in coatings of different materials: *1* ZrO_2; *2* ZrO_2+Ni-Al, the coating of variable composition; *3* ZrO_2+Ni-Al, a mixture; *4* iron oxide

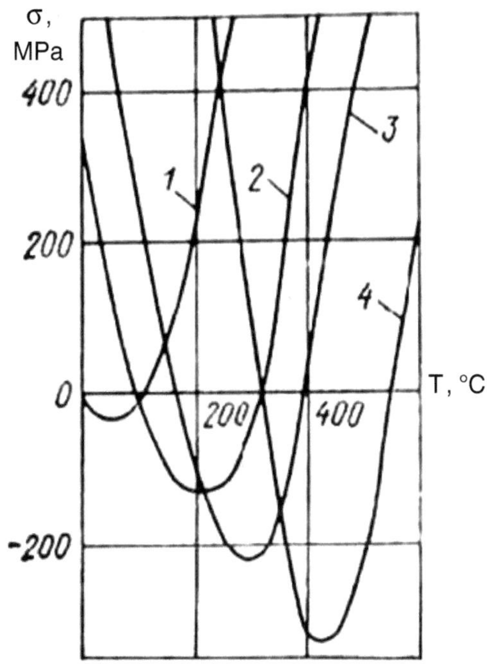

Fig. 3.20 Residual stresses in ferrous oxide coating at different temperatures of the substrate heating. Temperature of the particles in the coating formation: *1–4* 50, 100, 300, 500 °C

When coating on a flat surface with the parameter H/L ≪ 1 (*H*, *L*—thickness and length of the system) the state of stress is implemented (Fig. 3.21), described by the stress tensor [15]:

$$\sigma_x = \sigma_x(z);\ \sigma_y = \sigma_y(z);\ \tau_{xy} = \tau_{xy}(z);\ \sigma_z = \tau_{yz} = \tau_{xz} = 0. \qquad (3.25)$$

Coating on a plate or a single particle on a plane can form such system, and the stresses may be of different nature (elastic, thermal, residual). The stresses work on the almost entire part except the edge area, where there is a three-dimensional stress state,

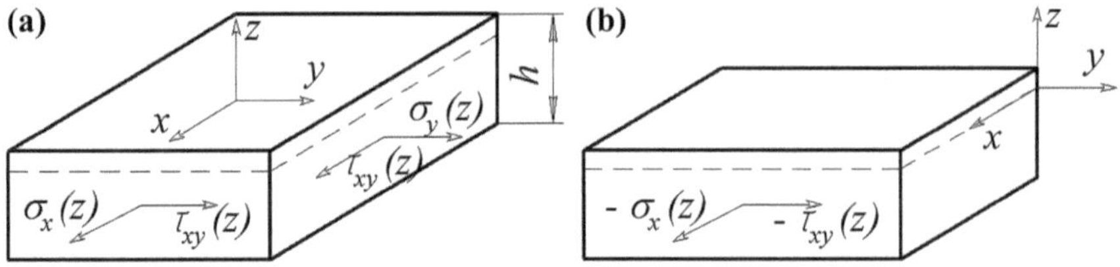

Fig. 3.21 Application scheme of contour load to the plate (**a**) and to the beam (**b**)

wherein $\sigma_z \neq 0$. Stresses (σ_z) are responsible for reducing of adhesive and cohesive strength of the coating, as well as for occurrence of cracks in the plate plane.

To solve the problem of the stressed state of the edge area the following method was used. We consider two states of stress. First, when in the whole detail including the edge area the main residual stresses (3.25) operate. Such a condition can be obtained if the plate at the ends is loaded with the stresses (3.25). As a second state of stress, the plate without residual stress is considered, also loaded with stresses (3.25) at the ends. The difference between these two states gives the stressed state of the edge area.

For the first case the plane stress state is realized, the normal component of stresses across the plate is zero. Consequently, all the stresses components are known for the first state.

For the second state of stress, it is taken into account that the applied load within the thickness of its system is self-balanced, as well as basic fracture stresses. In accordance with the principle of Saint-Venant the effect of such a load is distributed on the edge of the plate to a distance of the order of the thickness of the system [15]. Therefore, instead of treating the plate it is sufficient to consider a rectangular beam with the cross-section having the thickness of H and the width of, for example 1.5 H, with applied loads on one of the faces. Figure 3.21b shows such a beam for the edge perpendicular to the x-axis. Initial loads (3.25) are applied with the opposite sign that allows obtaining the desired σ_z immediately. It should be noted that the shear stress (3.25) causes only the beam torsion and makes no contribution to σ_z, and therefore the influence of contour τ_{xy} is not taken into account.

Thus, for a finding σ_z the plane elastic problem for a rectangular area with the normal stresses $\sigma_x^{(H)}(z)$, equal to the principal stresses (3.25), but with the opposite sign must be decided (Fig. 3.22). Moreover, the area consists of two strips of

Fig. 3.22 The scheme of loading of rectangular area of specimen

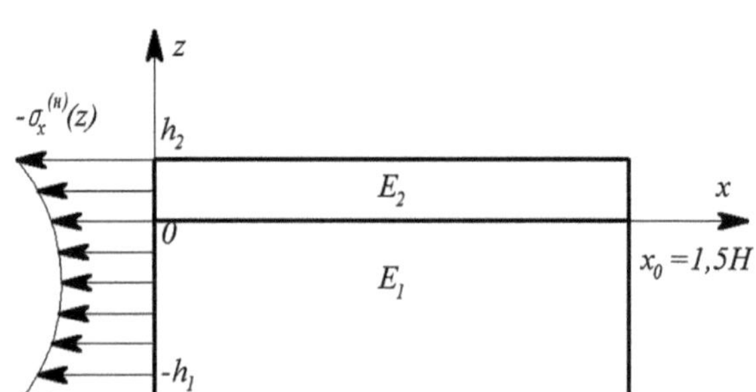

material (coating and substrate) with different elastic moduli (E), but the same Poisson's ratios (μ). The solution to this problem, made by Barvinok, showed that in the edge zone there are always peeling stresses σ_z, if the breakout stresses act in the main part of the component. The maximum value of both compression and tension stresses σ_z is always somewhat higher than the maximum values of the basic fracture stresses (Fig. 3.23).

There are tear off or downforce strains on cylindrical and spherical surface coatings depending on the ratio of the coefficients of thermal expansion (CTE) of the coating and the substrate, the location of the coating on the outer or inner surface. Puzryakov proposed analytical dependences for their determination [11].

In the case of spherical surfaces the radial stresses can be determined as follows:

$$\sigma_r = \frac{(\alpha_M - \alpha_n)\Delta T}{R^2\left[\frac{1-\mu_M}{E_M(2R+h_M)h_M} \cdot \frac{1-\mu_n}{E_n(2R-h_n)h_n}\right]}, \tag{3.26}$$

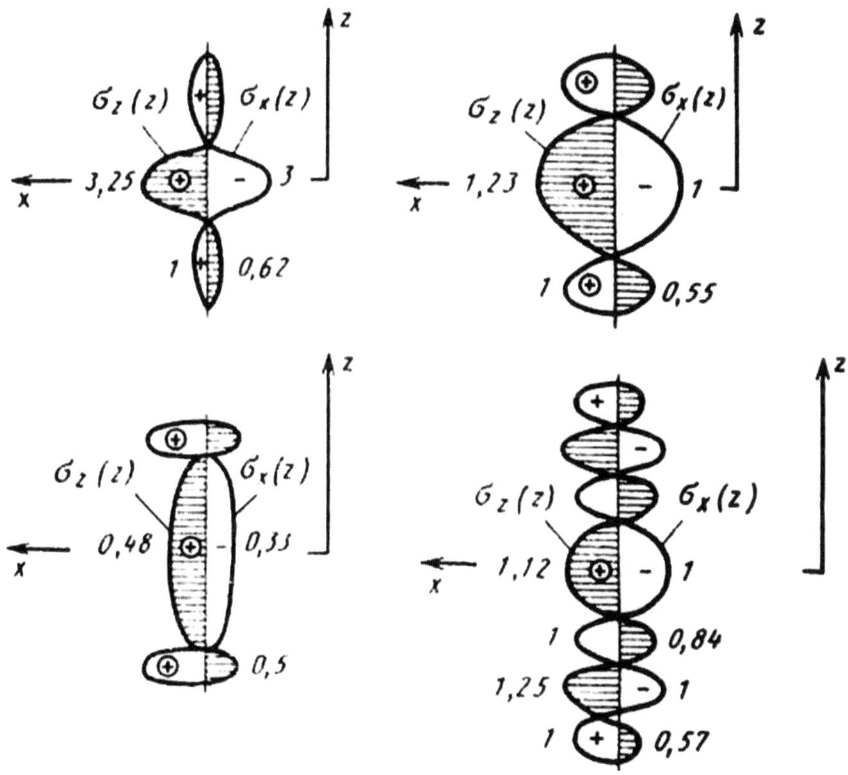

Fig. 3.23 Distribution of peeling stresses σ_z depending on the residual stresses $\sigma_x^{(H)}$ in the various cases of their formation of the coating

where R—inner radius, h—thickness, E—elastic module, indices "n", "M" refer to the coating and the substrate metal, respectively.

There are radial (normal) σ_r, shear σ_τ, and longitudinal (axial) σ_Z stresses on cylindrical parts.

Stresses for the inner layer can be expressed by the following relationships:

$$\sigma_r = \frac{K+N}{KL-2N^2}\left(1-(R_3/R_2)^2\right)(\alpha_M-\alpha_n)\Delta T \tag{3.27}$$

$$\sigma_r = \frac{K+N}{KL-2N^2}(\alpha_M-\alpha_n)\Delta T \tag{3.28}$$

$$\sigma_r = \frac{M+2N}{KL-2N^2}(\alpha_M-\alpha_n)\Delta T \tag{3.29}$$

where

$$\Delta T = T_n - T_M;\ K = q\frac{1}{E_n}-\frac{1}{E_M};\ N = q\frac{\mu_n}{E_n}-\frac{\mu_M}{E_M};\ q = \frac{R_2^2-R_3^2}{R_2^2-R_1^2};$$

$$L = q\left(1+(R_1+R_2)^2\right)\frac{1}{E_n}+\left(1-(R_3+R_2)^2\right)\cdot\left(\frac{\mu_n}{E_n}-\frac{\mu_M}{E_M}\right) - \left(1-(R_3+R_2)^2\right)\frac{1}{E_n}$$

R_1, R_2—the inner radius of the pipe coated and uncoated, respectively, R_3—the outer radius of the pipe; T—temperature, indices "n", "M" refer to a coating and the substrate metal, respectively.

The expressions (3.27)–(3.29) show that the highest value have stresses directed along the pipe axis. For exterior coatings, the CTE of the coating and the substrate should be as close as possible.

For internal coatings, the compliance with this condition is not necessary. Stresses in coatings vary depending on the ratio of CTE of the coating (α_n) and the substrate material (α_M) at external and internal spraying. Table 3.4 shows the scheme of the internal stresses in the coating peeling from the substrate material or its downforce.

Since all dependencies (3.27)–(3.29) have modulus of elasticity (E) and Poisson's ratio (μ), then they should be calculated taking into account the porosity due to dependences:

$$E = E_0(1-aP) \tag{3.30}$$

$$\mu = \mu_0(1-aP)$$

where E_0, μ_0—modulus of elasticity and Poisson's ratio of solid material; a—experimental constant; P—volume porosity.

Table 3.4 Scheme of the internal stresses in the coating

CTE	Stresses			Strain scheme
	Longitudinal σ_z	Shear σ_τ	Radial σ_r	
In outer layer				
$\alpha_n < \alpha_M$	Compression	Compression	Peeling strain	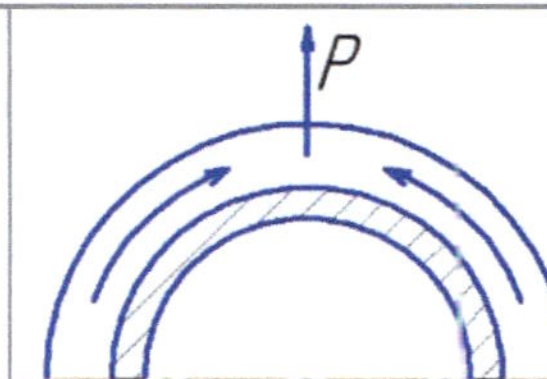
$\alpha_n > \alpha_M$	Tension	Tension	Downforce strain	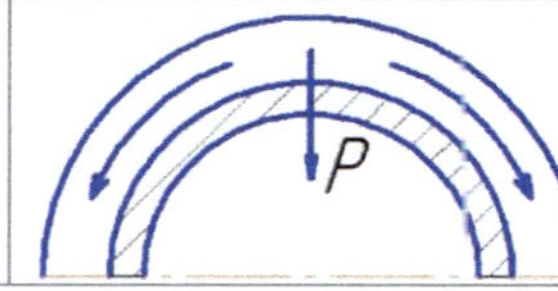
In internal later				
$\alpha_n < \alpha_M$	Compression	Compression	Downforce strain	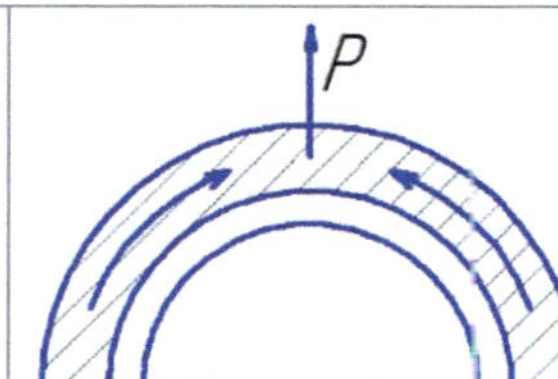
$\alpha_n > \alpha_M$	Tension	Tension	Peeling strain	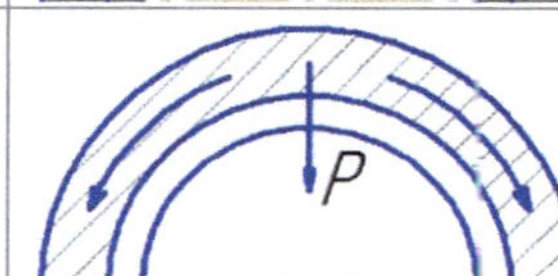

Based on these calculations, it is possible to point out some laws of formation of the coating stress state, characterized by temperature, crystallization, and other initial stresses. After the end of the spraying process in the coating-substrate system, the arising RS arise that are self-balanced in the volume of the system. Consequently, the material always has both the area of compressive and tensile stresses. The tensile stresses result in reduced strength of the material and can cause cracks perpendicular to the surface. After the occurrence of cracks, the RS level in the coating is reduced. Normal compressive stresses are formed in the edge effect zone in the layers where there are tensile stresses. In general, basic tensile RS negatively influence on the output characteristics of the coating, at the same time, the main compression RS have a positive effect. However, compressive stresses can lead to loss of stability of the layer in which they operate and in the edge effect zone, they give rise to peeling stresses. Therefore, the values of compressive stresses ought to be limited by a lowest value of strength in the tensile system, which usually corresponds to the strength of adhesion of the coating with a substrate.

3.3 Structure of Arc Sprayed Coatings

The properties of the coatings as a whole will depend upon the processes in two-phase stream and upon the forming a coating on a substrate. The main specific features that define the sprayed material will be:

- The properties of the contacts between the particles in the coating;
- Extremely fast crystallization and deformation of particles;
- Changing the composition of the material being deposited due to the interaction with the environment.

Let us consider how these features of a coating forming are reflected in its structure and determine the nature of the defects that can vary in size, density, morphology, and orientation.

The results are presented with respect to commonly used steel and babbitt coatings.

3.3.1 Steel Coatings

We present the results of metallographic research of thermal sprayed coatings.

Schastlivtsev et al. revealed the following [2]. On unetched metallographic polished sections with AS-coating of steel X20Cr13 the structure of the sprayed layer is clearly visible, Fig. 3.24. The layer consists of alternating very thin undulating white and gray-blue layers (on black-and-white photographs they are gray) actually representing, respectively, the solidified metal and oxides. There are also areas of mixed structure consisting of metal and oxides. The metal particles of

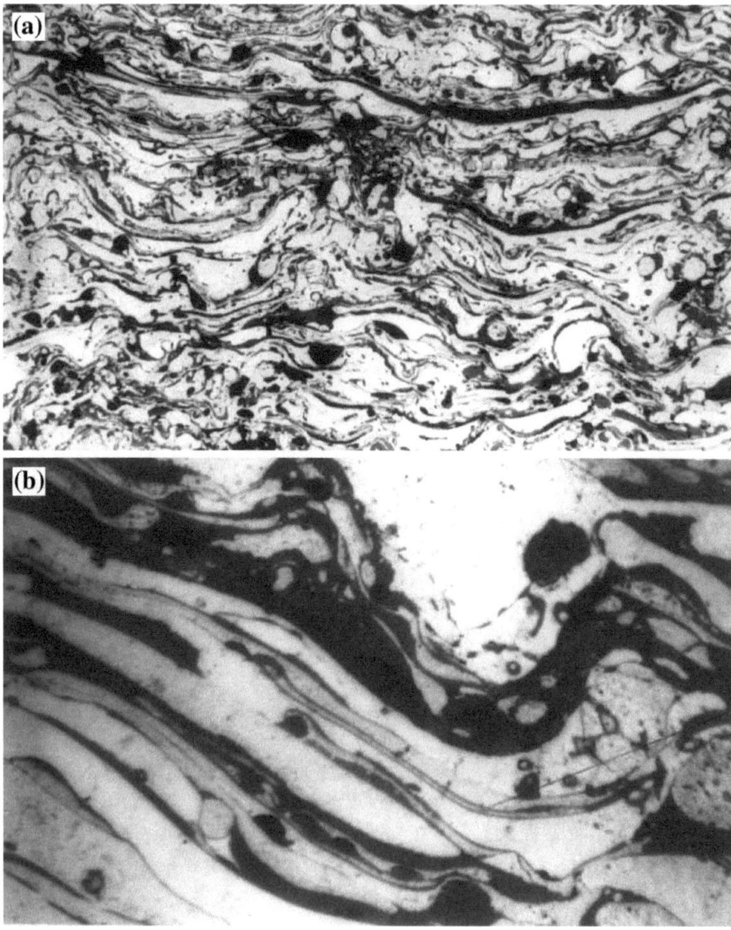

Fig. 3.24 Microstructure of thermal sprayed coatings without etching, steel like X20Cr13; **a** ×250; **b** ×625

more round form are visible, up to globular. Furthermore, there are globular and disk-shaped pores.

Proceeding from the conditions of deposition, it can be assumed that the disk-shaped pores can be formed between the layers due to insufficient layered droplets wetting. Globular pores are formed because of underfill of solidified layer.

Oxide films and, especially, large pores can contribute to metal delamination on the mechanism of internal friction, according to the terminology of Kragelsky [16], increasing the velocity of the mechanical wear of the metal coating.

During etching the specimens, it was noted that at first action of the etchant affects the thin layers of metal; globules, it is more difficult to etch large particles and thicker layers.

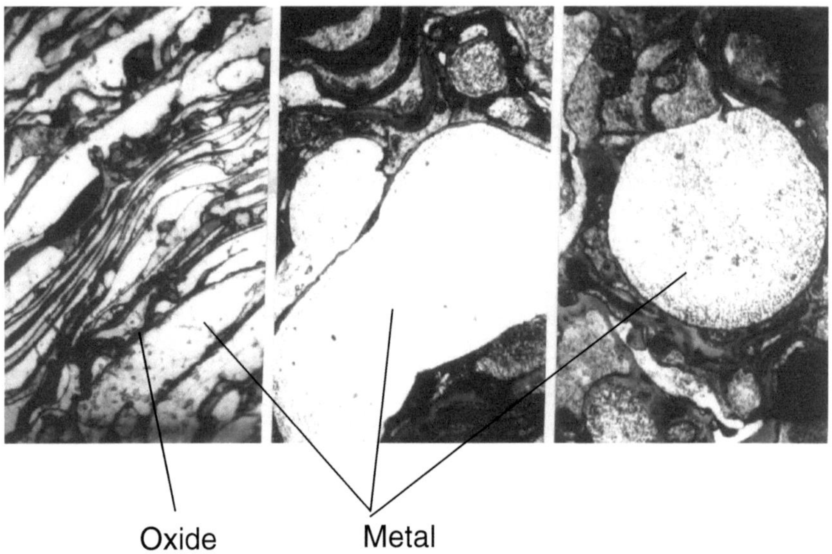

Fig. 3.25 Microstructure of arc sprayed coating after etching, steel X20Cr13, ×625

Poor etching of white components in comparison with other structural ones confirms the assumption that it is metal, wherein with high chromium content. In the globule and in thin layers of metal the structure of point character reveals, similar to the structure of tempered martensite (Fig. 3.25).

The combination of the small particle size and high velocity at the time of impact leads to high cooling rate. The analysis of the phase composition of sprayed coatings of steel like X20Cr13 and X40Cr13 showed that there are practically no carbides [4], although due the diagram of equilibrium state Fe–Cr–C for steels like X20Cr13 and X40Cr13 the structure consists of α—phase and carbides of type $Me_{23}C_6$ [17]. Probably, the high cooling rate inhibits the carbide precipitation from solution.

The study of the structure was completed with the microhardness measurements over the entire section, from nichrome sublayer to the surface. Since the dimensions of the areas of pure metal and some oxides are sufficient to measure, the microhardness and the measurements were performed both in areas of pure metal and oxides, and in areas of mixed structure consisting of metal and oxides.

Figure 3.26 shows the results of microhardness measurements of AS-coating of steel like X20Cr13 and X40Cr13. There were not found any dependence of the microhardness on the location of measuring points on the height of the layer. For steel like X20Cr13, the microhardness of sections of pure metal and mixed areas is practically the same; throughput the layer thickness it is very irregular, varies from 4.8 to 5.2 GPa. The oxides microhardness has a value from 5 to 6 GPa, i.e., a bit higher than the hardness of portions of the pure metal. The microhardness of

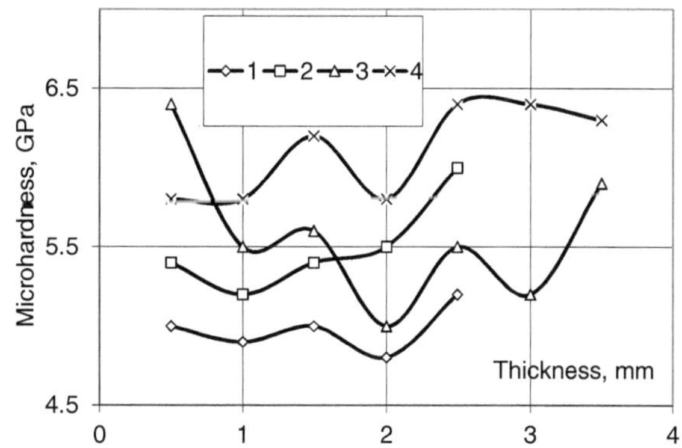

Fig. 3.26 Microhardness of AS sprayed coating. *1* X20Cr13, pure metal and the mixed structure portions; *2* X20Cr13, oxides, *3* X40Cr13, pure metal and mixed portions; *4* X40Cr13, oxides

spherical particles (globules) is in the range 4.5–5.9 GPa. Steel like X40Cr13 shows similar results, only the range of values is above 5–6.4 GPa.

The measurements of Abedini et al. showed that the microhardness of aluminum AS coatings is HV_{500} (51–52) ± 4 and practically does not depend on changes in velocity (from 109 to 143 m/s) and the preheat temperature of substrate in the range of 22–300 °C [7].

Layer-by-layer chemical analysis of the AS coatings showed that the element content is practically the same trough the depth of the sprayed layer but in the process of spraying the element's burnout takes place in comparison with the initial content (Table 3.5).

We can note that the amount of burnable carbon grows with increase of its content in the initial material. This complies with the laws of its behavior at steelmaking processes [18].

To assess the effect of the carrier gas atmosphere, the specimens were prepared by various arc spraying guns (Table 3.6). The commercially produced gun EM-14 was selected as a basis for comparison. The ADM-10 gun uses combustion products of propane-air mixture for spraying. The ADM-12 nozzle head is a closed system with the central nozzle.

The structure of the sprayed layer in all the specimens consists of alternating thin layers of the sprayed metal and oxides incorporating pores of different shapes and single round metal particles (Fig. 3.27). The latter, probably, hit the surface of the coating being solidified. The size of these rounded particles is less than 30 μm, which corresponds to our calculations for solidification of metal droplets at a working distance of spraying; the calculations are described in Chap. 2 (Figs. 2.36 and 2.37).

Table 3.5 Changes in the chemical composition of the sprayed material during spraying

Material	Elements content, weight%			
	Wire		Sprayed layer	
	C	Cr	C	Cr
X40Cr13	0.38	13.2	0.19	12.0
X20Cr13	0.18	12.1	0.13	11.7

Table 3.6 Parameters of the coating with a different composition of the carrier gas

Specimen number	Differences			Constant parameters
	Gun	$P_в$, MPa	$P_п$, MPa	
1	EM-14 [19]	0.6	0	Initial material—wire X40Cr13, ∅ 1.8 mm; I = 180 A; U = 28 V; $T_{melting}$ = 120 °C; L = 100 mm; $V_л$ = 1.8 m/s; δ = 2 mm
2	ADM-10 [20]	0.35	0.3	
3	ADM-12 [21, 22]	0.6	0	
4	ADM-12 [21, 22]	0.6	0.2	

Symbols $P_в$, $P_п$—pressure of air and propane; *I*—current; *U*—voltage; *L*—spraying distance; *Tпп*—temperature of samples preheating; *Vл*—linear velocity of the gun relatively to the sample; δ—coating thickness

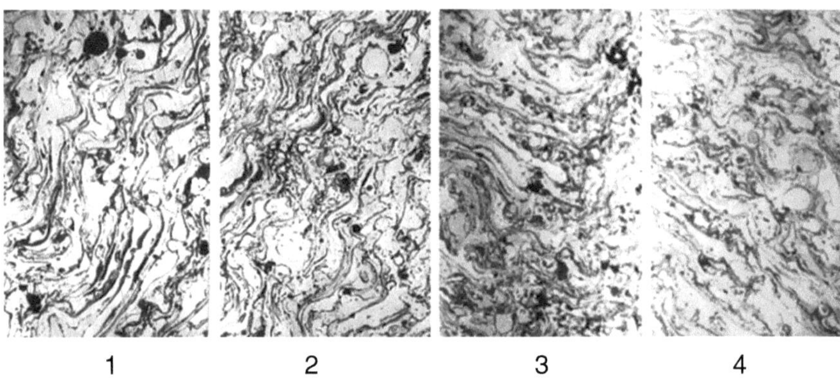

Fig. 3.27 The microstructure of the specimens without etching, ×250. Numbers correspond to Table 3.6

When comparison we see that AS coatings, specimen № 1, have the coarsest inhomogeneous structure with the presence of a large number of point and extended through the layers pores as well as single large particles. When using combustion products as the carrier gas, the specimen number 2, the structure is thinner, there are fewer voids and they are smaller, the number of extended pores is also significantly less. From a comparison of all specimens, it can be concluded that the structures of the steel layer are most preferable on the specimens made using the AAS guns with a combustion chamber and without it, with the injection of propane (specimens number 2 and 4). They differ in form and probably in oxides content. In the first case, there is relatively smaller number of oxides in the structure, in the second one-oxide films are thinner.

From the comparison of specimens number 3 and 4, it is seen that the coating on the specimen number 4, when propane was fed into the air stream, has more homogeneous and finer structure, Fig. 3.28.

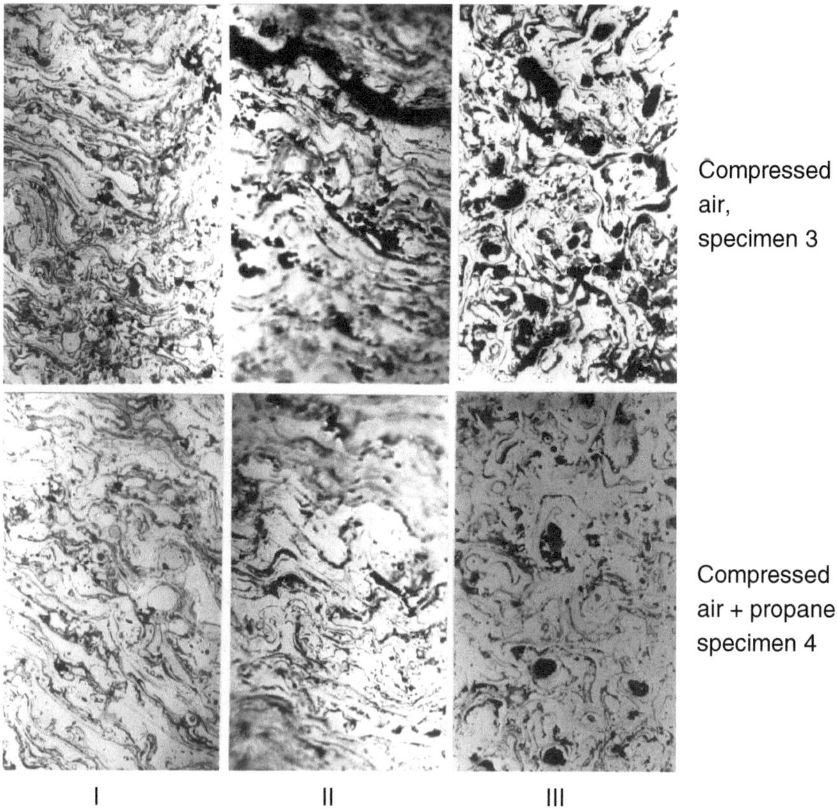

Fig. 3.28 The microstructure of the coatings with different composition of the carrier gas, ×250. *I* near the substrate; *II* a surface layer, thin section perpendicular to the surface; *III* surface layer thin section parallel to the surface

For this specimen in the sprayed layer, the minimum quantity of pores is observed and available pores are small, point. Very rarely, you can see not long pores extended on the structures layers. It does not show such extended pores between layers, as in the specimen № 3. This is clearly seen from a comparison of the structural state of the specimens number 3 and 4 in the surface layers (Fig. 3.28), especially when thin section was not made perpendicular to the specimen surface but parallel, with the removal of the metal layer of 0.5 mm. Sometimes you can see some not very large, round metal particles. The coating on this specimen has purer structure because of oxides.

The comparison of microhardness of coatings sprayed with a reducing carrier gas (AAS without combustion chamber) and compressed air (AS) showed an increase of microhardness for all components in the case of using the reducing atmosphere, Fig. 3.29.

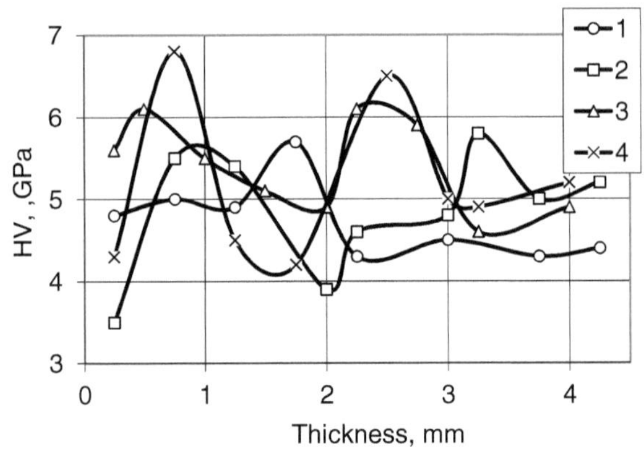

Fig. 3.29 Comparison of coating microhardness sprayed with propane injection (AAS) and compressed air (AS). *1* AS, pure metal and portions of the mixed structure; *2* AS oxides; *3* AAS, pure metal and portions of the mixed structure; *4* AAS, oxides

Table 3.7 Comparative properties of the coatings [63]

№	Coating method	Porosity, %	Hardness, HRc	Phase microhardness, MPa		Adhesion strength, MPa	Friction coefficient[a]
				Light	Dark		
1	Supersonic AS	2–3	35–37	1200	16,000	65	0.15
2	PS	6–8	28–30	1000	13,000	50	0.7
3	Typical AS	10–12	22–25	900	11,000	35	0.3

[a]The average value at a load of 5–10 MPa

The interval of hardness values is increasing by about 6–12 % for the various structural components.

Petrov gives similar data on the properties of sprayed coatings of stainless steel. These coatings were developed and made by the authors. A supersonic jet of combustion products as the carrier gas was used in the developed gun. It can be seen that the use of reducing mixtures leads to increased adhesion strength microhardness of strengthening phases, reduced porosity of coatings (Table 3.7).

Using a diffractometer DRON-3, the layer study of a phase structure was carried out by reading the diffraction patterns and debyeograms in thin sections, starting at a depth of 0.2 mm from the surface, and every 0.5 mm downward the layer to the nichrome (overall four measurements). This study was carried out for the coatings sprayed in a reducing atmosphere and by compressed air. Analysis of surveyed diffraction patterns and debyeograms showed that the phase composition of the coatings is the same. All specimens contain α-phase (martensite), iron oxides like Fe_3O_4, austenite (Figs. 3.30 and 3.31). There was no amorphous layer found in the structure.

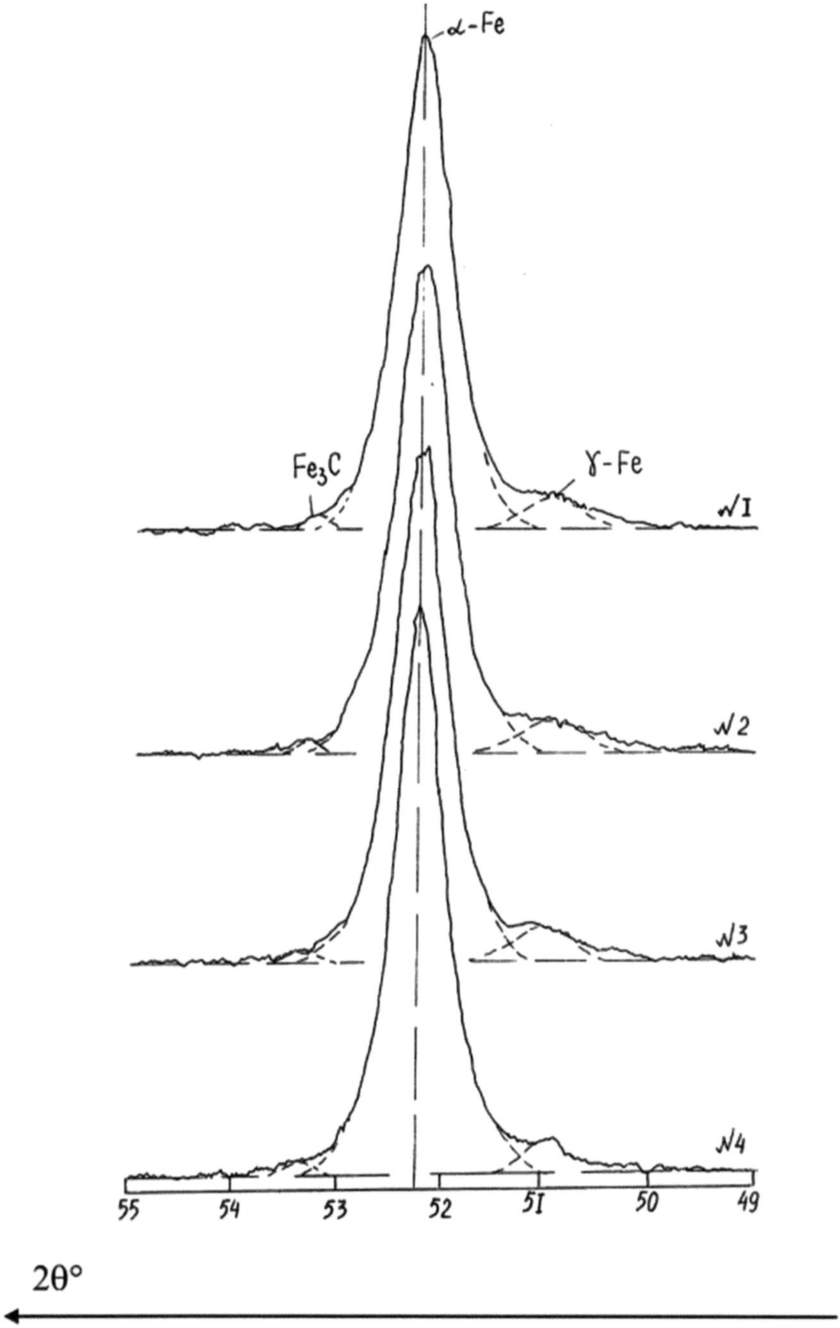

Fig. 3.30 Survey difractograms of the coating sprayed in a reducing atmosphere. *1, 2, 3, 4* numbers of layers from the surface to the base

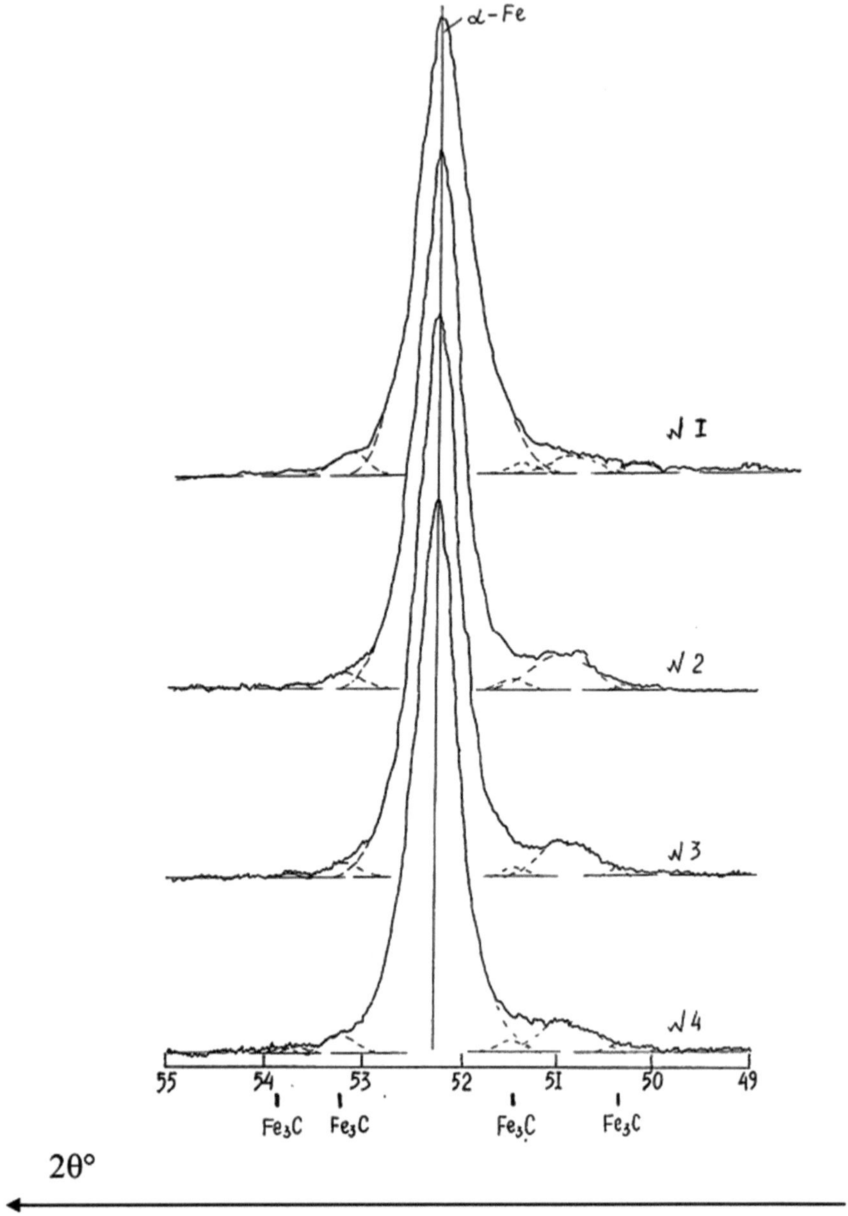

Fig. 3.31 Overview diffractograms of the coating sprayed with compressed air. *1, 2, 3, 4* numbers of layers from the surface to the base

Carbides amount is in the range of measurements sensitivity. The nature of the microstructure and high microhardness values let us suggest that α-phase represents slightly tempered structureless martensite and some oxides. Metal droplets falling on the surface of the parts are cooled at a high velocity causing a martensitic transformation. Subsequent layering of molten metal droplets on the already hardened surface layers leads to short-time heating the last, but probably the heating does not reach high temperature. This character of the layer formation corresponds to the low-temperature tempering of martensite.

The amount of residual austenite is not higher than 2 %. In the specimen sprayed with compressed air without propane, the amount of residual austenite varies according to the depth of the layer, increasing from the specimen surface. In the same specimen, the presence of the carbide phase is marked more clearly, Fig. 3.30. Obviously, this is due to the reduced cooling rate of the larger particles at AS.

The phase analysis of the specimens 3 and 4 showed the differences in structure depending on the presence of propane in the gas jet (Fig. 3.32). Specimen 3, which spraying was conducted without addition of propane, contains more oxides Fe_3O_4, their lines are clearly visible in the diffraction patterns, and the amount of retained

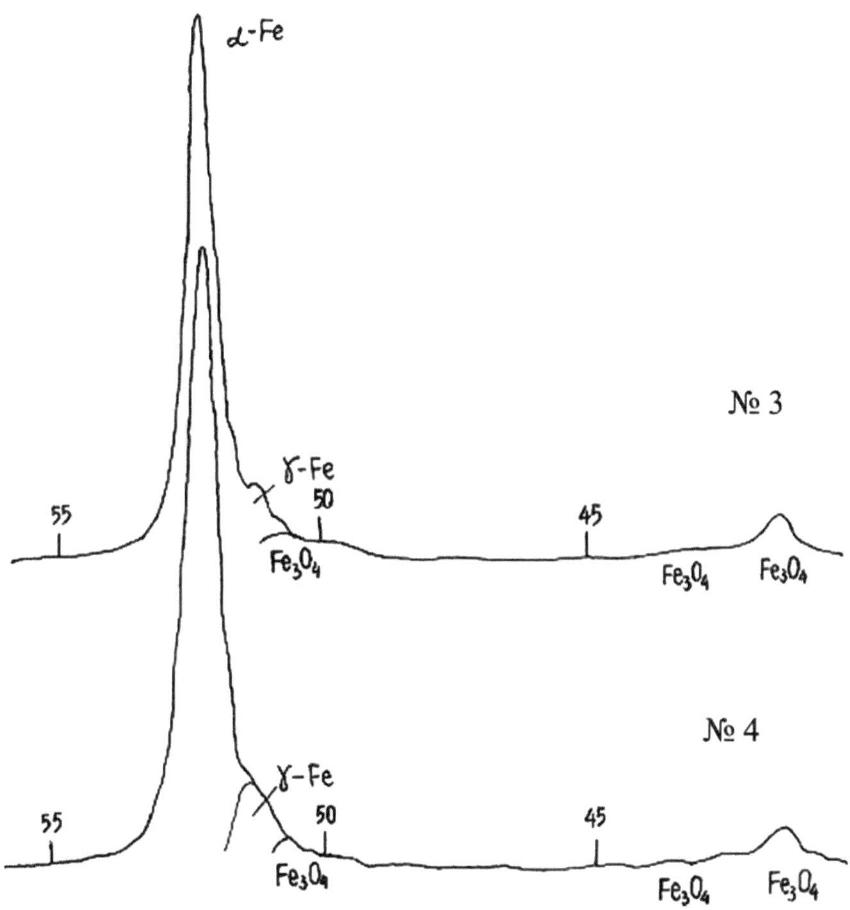

Fig. 3.32 Specimens 3,4. Difractograms, 2θ = (35–55)°

austenite is less than in the specimen 4. The amount of residual austenite in steel, fixed by hardness, depends on the position of the martensite point. The lower martensite point is the more the amount of residual austenite. Increasing of the carbon content decreases the martensitic point, thereby increasing the content of residual austenite. Such a picture observed in steels, is manifested in the sprayed layer. Addition of propane to the spray jet increases the amount of carbon in the coating (specimen 4), which leads to an increase for residual austenite.

On the basis of structural and phase analysis, Borisova et al. studied the properties of AS coating of solid wires, both from steel 1066 (66Mn4) separately and cospraying with aluminum.

At research of coatings from steel 1066 (66Mn4) [6] it is noticed that the spraying products are spherical particles coated with an oxide shell with dimensions of (1–600) μm. The particles sometimes have cavities and inclusions of iron oxides FeO and Fe_3O_4. During the spraying process, the carbon burnout to 0.28–0.30 % occurs. The microhardness of the coating layers is H_μ = 3700–5700 N/mm^2, there are bright crystals with hardness H_μ = 8500–11,700 N/mm^2, bordered by dark layers with hardness H_μ = 4500–5600 N/mm^2. The coating structure may represent a high-carbon martensite. In the layers adjacent to the substrate, the structure is a mixture of bainite, martensite, and its decomposition products.

Coatings researches of the system Fe–Al showed the following [23].

The porosity of the coating was 7 % when using a flux-core wire of Fe–Cr–C and 4 % in case of wire of Fe–Cr–C–Al. It has been established that the coating matrix phase is a solid solution of chrome or chrome with aluminum in α- or γ-iron. In the coating consisting of ferrochrome, the γ-phase prevails, and in the coating of ferrochromaluminum—α-phase due to a stabilizing effect of aluminum on ferritic structure. The microhardness of the matrix phase is not homogeneous, is in a range of 2200–5700 N/mm^2. In the structure of the coatings, the carbide phase is clearly distinguished, its microhardness varies within the limits of 10,000–13,000 N/mm^2. According to X-ray microanalysis, the presence of two types of carbides is most likely: Cr_7C_3 (harder) containing 10–12 % Fe, and $Cr_{23}C_6$ (softer) containing up to 20 % Fe. Besides, fine carbides are also arranged inside the grains having a eutectic structure. It has been shown that by increasing aluminum content above up to 10 %. Particles share of 20–30 % increases in size, and their shape varies from spherical to complex with spongy koralloidal and plate rims (Fig. 3.33).

The main structural components are the following:

- needle bainite–martensite structure of steel particles (H_μ = 3000–5000 N/mm^2);
- solid solutions based on iron (H_μ = 3500–7300 N/mm^2);
- eutectic Al + $FeAl_3$ (integral value H_μ = 550–1000 N/mm^2);
- intermetallics $FeAl_2$, Fe_2Al_5, $FeAl_3$, and phases with high (up to 82 % atm.) aluminum content type Fe_2Al_7, $FeAl_6$, (H_μ = 7500–10,500 N/mm^2);
- iron and aluminum oxides.

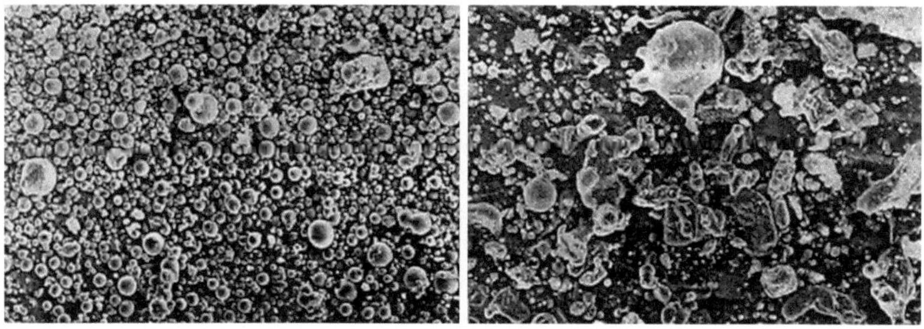

Fig. 3.33 Type of spraying products, ×100: **a** Steel 1066; **b** Steel 1066 + 30 % Al

Table 3.8 Chemical composition of flux-core wires [14]

Batch	Mass fraction of main components, %								
	Fe	Cr	Al	C	Si	O	Mn	S	P
Fe–Cr	71.11	25.58	–	2.96	≤0.02	0.1	0.24	≤0.03	≤0.03
Fe–Cr–Al	69.90	20.75	6.58	2.41	≤0.04	0.1	0.24	≤0.02	≤0.03

Shell—cold-rolled strip of steel 08кп, GOST 503-71 (like steel 1008)

With increasing of aluminum content, the microhardness of interaction products increases.

The great amount of investigations concerning the effect of Al, B on the structure and properties of sprayed coating of flux-core wires in the system Fe–Cr–C is made by Pokhmursky et al. [8, 14, 24–27].

Ferrochrome and ferrochromaluminum coatings (Table 3.8) are obtained using arc spraying of cored wires by the gun [28]. The porosity of the coating is 7 % when using cored wire of Fe–Cr–C and 4 % when using cored wire of Fe–Cr–C–Al. There are iron oxides FeO, F_3O_4, and aluminum oxides Al_2O_3 (in the case of the wire of Fe–Cr–C–Al) in the coating. With the addition of aluminum, the ratio of iron oxide phases changes toward F_3O_4, and the total oxygen content in the coating decreases by more than 1.5 times, 3.3–1.9 %.

It is established that the matrix phase of the coating is a solid solution of chromium or chromium with aluminum in α- or γ-iron. In the ferrochrome coating, the γ-phase prevails, and in the coating of ferrochromaluminum—α-phase (there is the stabilizing effect of aluminum on ferritic structure). The microhardness of the matrix phase is not homogeneous, it is in a range of 2200–5700 N/mm^2. The carbide phase is clearly distinguished in the structure of the coatings, its microhardness varies within the limits of 10,000–13,000 N/mm^2. According to X-ray microanalysis, the presence of two types of carbides is most likely: Cr_7C_3 (harder) containing 10–12 % Fe, and $Cr_{23}C_6$ (softer) containing up to 20 % Fe. Besides, fine carbides are also located inside the grains having a eutectic structure.

3.3.2 Babbit Coatings

As an anti-friction coating of inserts of high velocity heavy-duty plain bearings, the use of Babbitt is widespread as an anti-friction layer in the shell of high velocity heavy-duty plain bearings due to it is characterized by a low coefficient of friction and high wear resistance. It is prospectively to use the Babbitt coating by spraying, in particular, by the activated arc spraying (AS). This method of Babbitt deposition is characterized by low unit cost compared to alternatives.

Studying set of tribological and structural characteristics, the authors compared the quality of Babbitt coatings obtained by AS and by alternative ways: by turbulent shell lining (SL), propane–oxygen gas surfacing (GS), plasma spraying (PS) [29, 30]. The methods selected as alternatives have the highest performance of obtained Babbitt coatings in their groups [31].

Specimens were made in the form of steel prisms 6 × 6 × 12 mm coated with 1.5 mm thick on the end of 6 × 6 mm. The chemical composition of initial material of coatings is shown in Table 3.9.

Specimens 1 were sprayed by the gun ADM-10, with maximum currents and voltages for stable process: 1–26 V, 100 A. It corresponds to a feed rate 23.2 kg/h, the deposition efficiency was 70 %. Specimens 2 and 3 were cut from blanks made, respectively, by turbulent shell lining and gas surfacing according to [32, 33]. For spraying specimen 4, the following equipment was used: the plasma unit «Castolin Eutectic», current 170 A, voltage 32 V, plasma gas argon, the feed rate 3 kg/h.

For metallographic analysis, the micro thin sections were prepared in the following sequence: (a) grinding on abrasive paper with different grit, polishing on soft felt using a paste GOI number 2 TU 6-18-36-85 on grinding and polishing machine NERIS ZE881; (b) etching by 2 % solution of nitric acid (HNO_3) in alcohol. Etched and dried microsections were examined in reflected light using metallographic microscopes POLAM P-312, MIM-7 and Neophot 32.

The results of metallographic microstructure analysis (Fig. 3.34) showed that the main phases of tin babbits structure besides α-solid solution on tin base are SnSb (β-phase) and needle-shape crystals of u_3Sn (ε-phase). For PS and AS coating the dispersion structure is characteristic with the size of β-phase (SnSb) 10–30 m km, while the SL and GS provide babbitt with intermetallic inclusions SnSb of 120–290 m km and similar globular form.

Table 3.9 Description of alternative coatings of Babbitt B83

Specimen number	Method of coating	Chemical composition, mass %			Initial material
		Sn	Sb	Cu	
1	AS	Base	10.8	6.0	Wire Ø 3.0 mm
2	SL	Base	11.0	5.8	Pig according to GOST 1320
3	GS	Base	11.0	5.8	Rod Ø 8.0 mm
4	PS	Base	11.0	5.8	Powder 40–100 μm

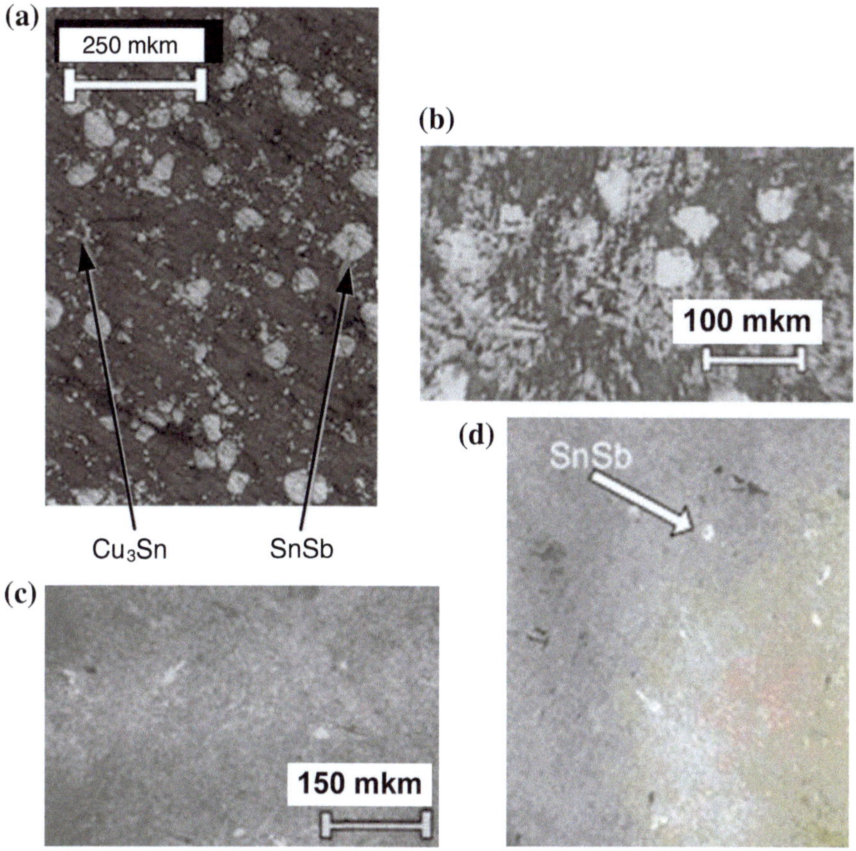

Fig. 3.34 Microstructure of Babbitt B83 obtained in different techniques: **a** SL; **b** GS; **c** AS (×100); **d** PS

For all methods, the particle size of the ε-phase is the same, with more roundness in sprayed (AS and PS) coatings. When spraying, the variation in size and intermetallic phases is also less marked in comparison with the SL and GS.

The difference in morphology of intermetallic compounds is due to the peculiarities of layers formation at different production methods. During the spraying (PS, AS), the cooling rate of the particles is (10^3–$10^{4\circ}$) with [34], which is 2–3 orders of magnitude higher than in the case of SL and GS. This is due to the small size of the droplets during spraying, 30–100 μm, and because of each droplet falls on the already cooled to solid state particles of previous layer [3]. As a result, when spraying the velocity of crystal nucleation is much higher than the velocity of growth in the solidifying alloy. The nucleation of crystallization centers is the predominated process in the crystallizing alloy, along with the slow growth of crystals. As a result, the structure of the sprayed Babbitt coating is on the order finer than at equilibrium crystallization, typical for SL.

Marked structural differences became evident in the tribological tests of the alternative coatings, see Sect. 4.1.3.

Conclusion. A method of forming, grain size, strength, and composition in TS-coating greatly differ from those of the materials obtained by rolling, casting, and forging. This makes a significant correction in the evaluation of coating properties. The properties can vary depending on the composition of the atmosphere, the presence of oxides at the substrate surface and its temperature.

There are three major types of defects in structures:

- discoidal pores between the layers, which are formed due to insufficient layered wetting of droplets, resulting in reduced cohesion;
- globular pores formed as a result of underfill of the solidified layer;
- vertical microcracks in the layers between the individual droplets.

These defects and the connected properties variation cause coarsening of the structure, the stresses in the coating, as well as changing its functional characteristics such as thermal and electrical conductivity. Adsorption of gases from the atmosphere, lodging of dust fractions, special behavior of particles at impact lead to the deterioration of the mechanical properties of interlayer coating borders.

3.4 Coating Strength

During the spraying, there is an interaction between the sprayed particles and the substrate surface (adhesion) and between the sprayed particles and the particles of already sprayed layer (cohesion).

Now, there is no single term evaluating the strength of coupling between the substrate and the coating per unit of their total surface. The most commonly terms used to evaluate the strength are "adhesion", "adhesion strength", "strength of adhesion of the coating to the substrate", "adhesive bond strength to the substrate". Such uncertainty in the terminology for one of the most important characteristics of the coating generates ambiguous data both in the special literature and in the technical documentation during the study of the properties of the coatings in the production conditions.

Adhesion (adhesion, sticking, seizure), as well as derivatives of this description do not always correctly reflect the various causes which lead to the bond of the base metal with coatings. The more correct term, assessing the specific force of this bond is "the strength of adhesion of the coating to the base metal" or adhesion strength [12].

Accordingly, the strength of the coating is proportional to its adhesive and cohesive strength. The destruction of the coating may occur if the external load or residual stresses exceed its strength. In case of exceeding, the adhesive strength the coating peeling, cohesive strength—cracking or peeling of the coating layers will occur. The results of studies [3, 9, 35–38] showed the following:

- The magnitude of adhesion strength depends on the mechanical, physical, and chemical interactions between the particles and the substrate;
- Cohesive strength depends on the following factors: the phase and chemical composition of the sprayed layer; the grain size of the sprayed layer; porosity of the coating.

There may be a large spread of data points of adhesion at various points in the coating. This is due to the difference in the mechanisms of coating damage caused by external influences. The classification of mechanisms of adhesive and cohesive failure was offered by Tushinsky et al. [12], Fig. 3.35.

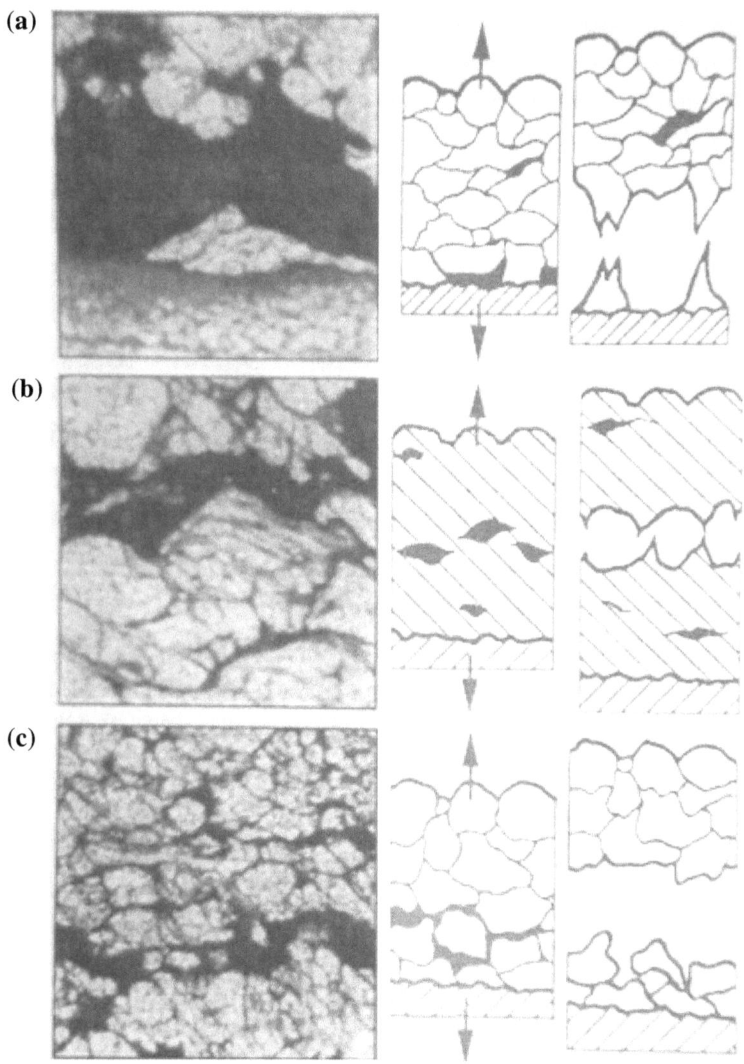

Fig. 3.35 The typical forms of cracks at the fracture of the coating, ×500

When the adhesive strength of the single particles exceeds the strength of the particles themselves, the fracture takes place. There can stay the fragments of the coating on the substrate elongated in the form of peaks and ridges, Fig. 3.35a (shown on the example of the aluminum coating, performed by cold spraying). It can be assumed that these firmly fixed particles are anchored, i.e., serve as centers of coatings formation. In this case, we have adhesive–cohesive nature of the fracture.

With good base preparation when the developed topography with a high propensity to chemical interaction is formed on the surface, there is a cohesive separation, Fig. 3.35b. The fracture of the coating occurs by the mechanism of cracks distribution in the material. The cracks appear at the edges of the flat pores, which are stress concentrators. The pores expand; herewith, blocking of neighboring pores takes place as well as fractures in weak contact bridges. When fracture of the strong bridges between increasing pores, the ridges bordering the pits occur. The surface topography shows the plastic deformation of the coating particles in the process of crack propagation. This can serve as a basis for predicting high mechanical properties of the coating.

With an increasing of the coating thickness the fraction mechanism changes. The cracking occurs at the contact surfaces of the particles, Fig. 3.35c. This contributes an increase in internal stress, increasing porosity, the appearance of interfaces.

The features of adhesive and cohesive components of coating strength and methods of its assessment are shown below.

3.4.1 *Adhesive Interaction*

The analysis of experimental study of the impact contact of droplets from various materials (Ag, Zn, Al, Cu, Ti, Zr, V, Fe, Ni, Mo) with different substrates (Ag, Al, Ni, Fe, Ti, Cu, Cr, Mo, W, CuO, NiO, Fe_2O_3, Cr_2O_3, SiO_2) showed the following [3, 12]:

- the strength of contact is determined by the mechanical interaction and by chemical bonds established during the spraying and by van der Waals forces;
- the preheating of the substrate strongly effects on the adhesion strength;
- the increase in the thickness of the coating is accompanied by the growth of residual stresses, reducing the strength of the coating to the substrate.

3.4.1.1 Mechanical and Chemical Interaction

Mechanical interactions are caused by the anchoring of deformed particles sprayed in the topography of the base metal. The development of mechanical interactions occurs due to increased surface roughness. To realize it, preparation of surface is made by threading, abrasive blasting, chemical etching, electrical spark machining, and by other methods.

Chemical interaction is stimulated by chemical preparation of the surface before spraying, including surface activation, removal of the oxide film, and cold work hardening.

High cohesive strength is observed while coating with materials whose melting temperature is higher than the one of base metal. In particular, this is based on the use of refractory metals, nichrome, molybdenum, as the material of the underlayer with high adhesion strength.

Moreover, a higher melting point than its oxide characterizes molybdenum. Therefore, at the molybdenum particle surface there are no oxides during the reaction with the substrate that improves the chemical interaction.

Depending on the spraying conditions and the ratio of the materials, the chemical reaction develops in the area of 40–70 % of the total contact surface. The latter is inversely proportional to a disk height obtained due to deformation of the particles upon impact (Fig. 3.3). The ratio of the disk height to its diameter before the impact is:

$$h/d = 0.045{-}0.111.$$

3.4.1.2 Substrate Preheating

For different material combinations of particles and substrate, there are ranges of heating the substrate (T_2) at which the adhesion strength ($\sigma(t)$) drastically increases, which is achieved at a given time (t) at the given heating the substrate. The maximum bond strength (σ_{max}), corresponding to the spreading on the setting centers of the entire contact surface (Fig. 3.36). In actual practice σ_{max} is not achieved, so the upper portion of strength curves is conditionally shown by dotted lines. The upper

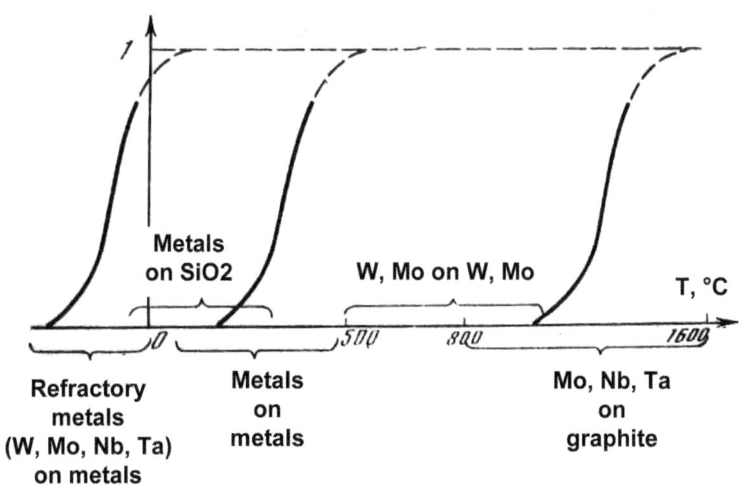

Fig. 3.36 Dependence of the relative adhesion strength on the substrate temperature (T_2)

limit of the substrate heating is limited by onset of intense growth of the oxide film on the surface. The low limit is caused by activation energy required for a chemical interaction.

The temperature dependence of adhesive strength is caused by the fact that chemical interaction varies proportionally to the reaction equilibrium constant. The temperature dependence of the reaction equilibrium constant K_c at constant concentrations of the reactants is expressed by the Arrhenius equation:

$$\mathrm{d}\ln K_c/\mathrm{d}T = E/\left(RT^2\right). \tag{3.31}$$

Hence, the curve of the relative adhesion strength change—preheating temperature of the substrate can be described by the equation

$$\sigma(t)/\sigma_{\max} = \mathrm{e}^{-E/(R(T+B))}, \tag{3.32}$$

where B is a coefficient depending on the type of metal particles and the substrate.

Depending on the nature of the metal, the activation energy (E) is in the range 5–50 kJ/mol. Its approximate value in case of overheating of 100–200 °C above the melting temperature T_m can be estimated by the empirical equation [39]:

$$E = 14.33\ Tm \tag{3.33}$$

Based on the above considerations, the methods of influence on the adhesion strength should be aimed at increasing the total contact area and decreasing the activation energy of interaction.

3.4.1.3 Increasing the Thickness of the Coating

Increasing the thickness of the coating is accompanied by growth of residual tensile stresses, reducing the bond strength of the coating. Figure 3.37 shows the dependence, connecting the thickness of the plasma coating with bond strength [3].

This dependence can be extended to other coating methods, taking into account their characteristics. Point (σ_x) corresponds to the bond strength at 100 % chemical interaction, i.e., the maximum bond strength, which is possible when spraying. Point ($\beta\sigma_x$) will characterize the strength of the connection of one layer of the particles in the coating; (β) is a factor equal to ratio of the area of places, where there was a chemical interaction (welding) to the total contact area between the coating and the base metal. The value of (β) depends on the spraying method and the nature of the sprayed metal. Thus, during plasma spraying of metals and refractory compositions $\beta \sim 0.1$; in detonation spraying under optimal conditions $\beta \to 1$. The thickness of a single layer is of 5–15 μm. The spraying of the next layer results in the rise of residual stresses, which will grow as the number of layers and the thickness of the coating (δ). Therefore, in practice, usually there is reduced bond strength equal to ($\beta\sigma_x - \sigma_0$). Increasing the thickness of the coating during

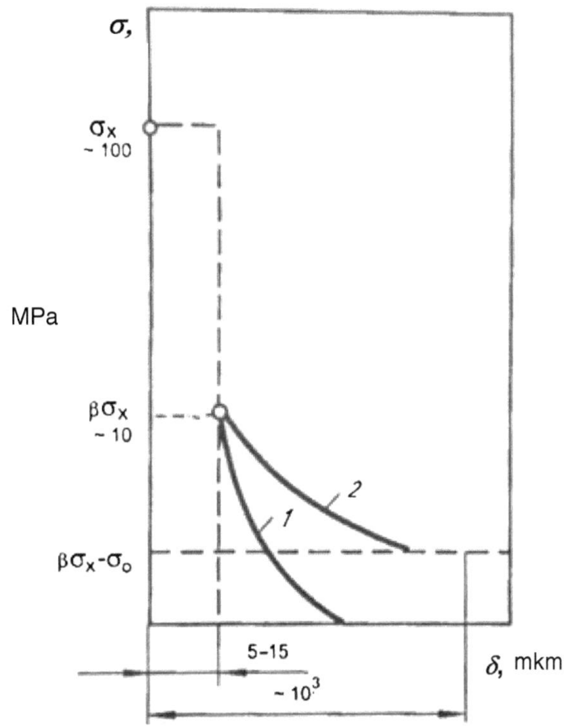

Fig. 3.37 Changes in coating adhesion to the base metal (σ) depending on the properties and thickness of the coating (δ) from brittle (*1*) and plastic alloys (*2*)

deposition of subsequent layers leads to (σ_0) increase. If brittle materials applied, when $\sigma_0 > \beta\sigma_x$ the coating spontaneously separates from the base metal (curve 1 intersects the horizontal $\beta\sigma_x - \sigma_0$). In the case of plastic coating, the residual stresses are partially relaxed, while remaining at a value less than ($\beta\sigma_x$), and there is a stabilization of coating–substrate strength.

3.4.2 Cohesion Interaction

Cohesive strength depends on the following factors: the phase and chemical composition of the sprayed layer; the grain size of the sprayed layer; porosity of the coating and its thickness.

Phase and chemical composition of the sprayed layer determines the strength of the particles material and the bonding strength between the particles in the layer. As studies [9] show, the latter factor prevails in increasing cohesive strength of the coating. Furthermore, the magnitude of the residual stresses of the second kind, which are balanced within particle's volume, depends on the structural characteristics.

Structure that is more homogeneous causes an increase in cohesive strength, as the inhomogeneity causes areas with reduced strength, which can cause the

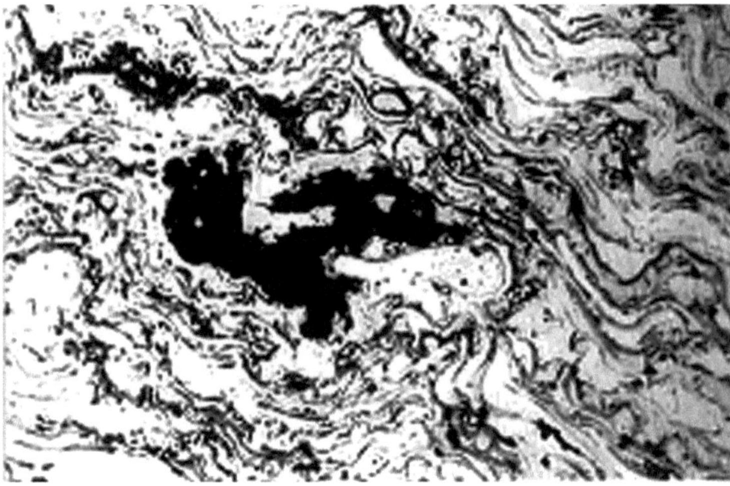

Fig. 3.38 The microstructure of the arc sprayed coating area with a local pores cluster, ×250

destruction of the coating. Calculations carried out in Chap. 2 showed that while thermal spraying the effect of electromagnetic forces in the zone of the arc prevents the forming a homogeneous flow of spraying particles. This creates, in turn, difficulties in obtaining a uniform coating structure.

In addition, the heterogeneity of the coating can be produced because of violation of the rules at preparation of materials and equipment: poorly treated wire, local distortions on the wire, contact tips wear. Characteristic defect caused by these reasons, is expressed in the local pores cluster (Fig. 3.38). Let us consider it on the example of the wire curvature. In case of accidental bending of one of them, the stability of arcing will be broken. It is resulted in tearing of unmelted pieces from the ends of the two wires and one interruption of arcing. If such an event occurs, then a portion of the metal is deposited on the surface of the part, which does not have the necessary spreading aerodynamic properties and fluidity. Under that portion of the metal forming of pores clusters is possible.

3.4.3 *Methods for the Evaluation of Adhesion*

Tests of adhesion between substrate and thermal sprayed coatings can be divided into methods for determining quantitative characteristics (for fracture on the adhesive and pin method, shear for cylindrical and prismatic specimens) and the methods for qualitative assessment (scratching, according to delamination, laser thermal shock, non-destructive methods).

3.4.3.1 Adhesive Methods

All variants of adhesive method can be combined into two schemes. Figure 3.39a reflects the principle of specially prepared test specimens in the form of a cylinder. The end surface of one specimen 1 after preparation is coated using coating 2. Specimen 4 having the same diameter is connected to the coating from another end using glue 3. Adhesive bond strength criterion 3 is a force P, at which the coating separation occurs at the boundary of the base metal 5 at tension, related to the surface area of the end. In international practice, the adhesive method of specimen tests in the form of a cylinder is taken as a base in determining the strength of adhesion of thermal coatings (ASTM C633, ISO 14916, JIS H8666-80, GOST 9.304).

Tests of coatings, which are sprayed on the product, are carried out according to the scheme shown in Fig. 3.39b. The coating 2 on a flat surface, is glued to cylindrical specimen 4. A tensile force (*P*) is applied to the latter until the separation of the coating from the base metal 5 on the border. These tests are regulated in accordance with ASTM D4541 and ISO 4624 (EN 24624).

When tested by the clue methods a number of features must be considered that affect the accuracy of measurements.

1. Tests are carried out on specimens having a fixed cylindrical shape of diameter 25 mm. Hasui showed that the adhesion strength is strongly dependent on the diameter of the specimen. When it falls from 55 to 25 mm the failure stress decreases almost a factor of 2 [40]. This is probably connected with the non-uniformity of the coating delamination. Coating fracture does not usually occur throughout the area of contact, even if it is on the border of "coating—base metal". The fracture extends first from the edge of the specimen, and then spreads to the center. Therefore, the bonding strength for this coating defined on a larger diameter specimen will be higher than for the specimen of smaller diameter due to the edge effect. With the diameter, increasing the edge effect is decreased.
2. To increase the strength of the clue bond, it is necessary to consider a number of factors.

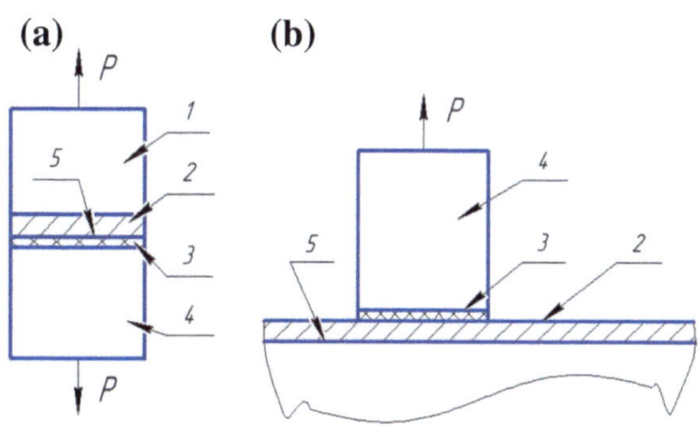

Fig. 3.39 Schemes determining bond strength of the coating to the substrate by clue methods. **a** Tensile strength testing, **b** pull-off adhesion test

The surface of the contraspecimen should be exposed to the grit blasting followed by degreasing with acetone.

To prevent the misalignment and to ensure the necessary contact pressure when bonding, it is desirable to use the centering device. One of the variants is shown in Fig. 3.40. Specimen 1 and contraspecimen 2 have been glued together by surface 8, are stacked in the bottom 3 and are screwed by the lid 4. Split liner 5 provides the coaxiality of specimen 1 and contraspecimen 2. For glue, curtaining in the split liner 5 the groove is provided. The screw 6 produces the compression of the specimen 1 and contraspecimen 2. Washers 7 protect the threaded holes in the specimens.

Glue curtaining from the cylindrical surface of the specimens is removed during its solidification, when it is in a plastic state, or after solidification.

3. The depth of penetration into the adhesive coating affects the nature of the fracture. Open porosity in the coating enables the glue penetration to the substrate that overestimates the test results. The likelihood of cohesive failure or separation of the coating by the clue increases with the lack of impregnation depth. To prevent penetration of the clue to the substrate, its surface is coated with a dividing layer of material, which is impermeable to the adhesive. However, this specimen preparation increases the likelihood of cohesive failure.
4. The magnitude of the stress in the fracture plane is affected by its location relatively to threaded hole. With regard to the representation of the geometry of the specimens, Fig. 3.41, stresses in the fracture plane were calculated using the finite element method and direct measurements of stresses [41].

The calculation showed that the proximity of the threaded hole to the surface of fracture, Fig. 3.41a, makes the specimen less rigid in the center than at the periphery of the cylindrical surface. This leads to stresses difference in the radial direction on the surface of fracture of about 50 % in the case of increasing the depth of the threaded hole up to 0.7 of specimen height. Increasing the length of the specimen and simultaneously reducing the ratio (depth of the threaded hole)/(length of specimen) to 0.4–0.5, Fig. 3.41b, the stress distribution on the fracture surface is aligned, Fig. 3.42.

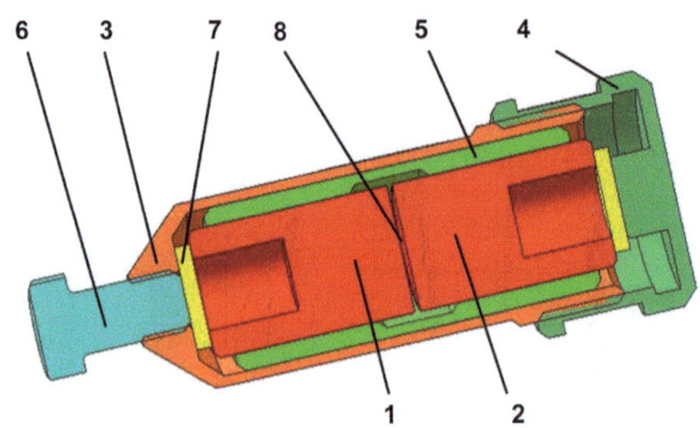

Fig. 3.40 Mandrell for specimens bonding

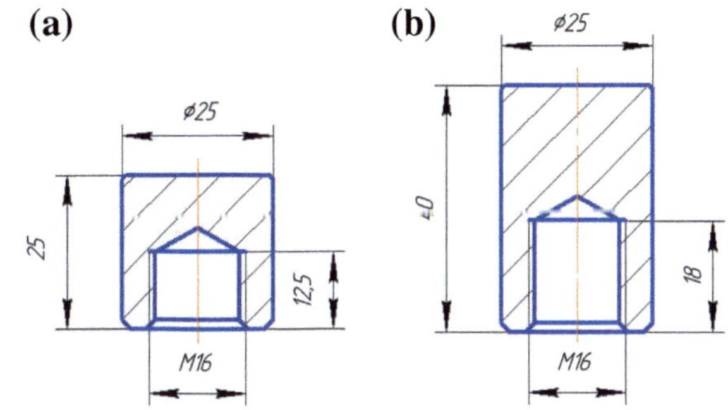

Fig. 3.41 Variants of test specimens

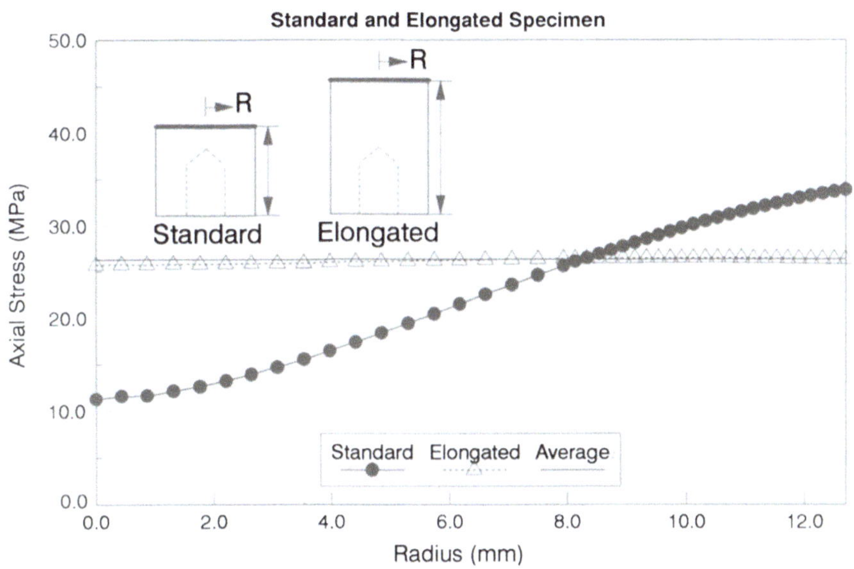

Fig. 3.42 Radial stress distribution on the fracture surface on the typical and extended specimen

5. To ensure coaxiality while load it is recommended to use the centering device based on Hooke's joint, Fig. 3.43. Its use will only allow creating compressive stresses and avoiding possible shear and bending stresses.
6. The strength of the coating adhesion in a single test and in glue is calculated by the formula

$$\sigma = F/S$$

where σ is adhesion strength, MPa; F—fracture force, H; S is cross-section area of specimens, m^2.

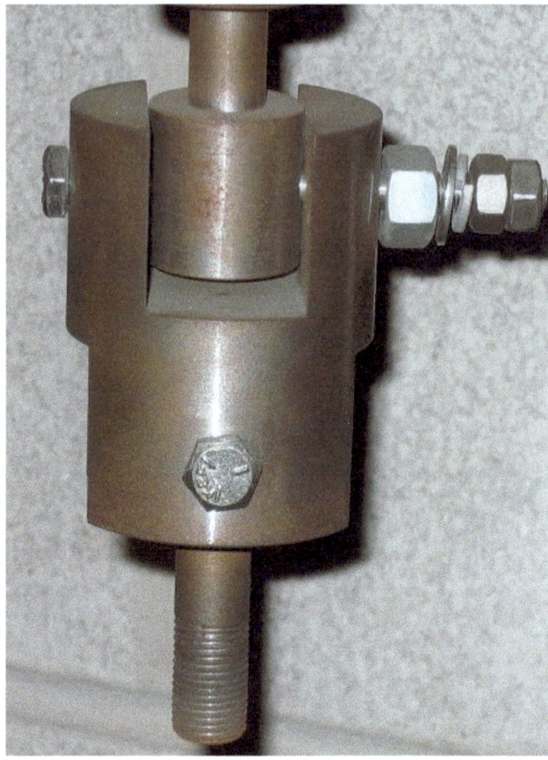

Fig. 3.43 Centering device for tensile tests

Adhesion strength of the test coating is calculated as the arithmetic mean of the five specimens tested.

7. When producing specimens, it is recommended to use mechanical or magnetic devices, which allow fastening simultaneously several specimens while coating. It provides the same conditions of spraying for each batch, and the manufacturing process of specimens is facilitated.

From the above description of the clue method can be seen that its main disadvantages are the following:

- High complexity of sample preparation;
- Possibilities are limited by strength of the clue, which connects fracturing surfaces;
- Glue penetration through the open pores to the substrate overestimates test results. At the same time, the lack of depth of infiltration contributes to cohesive failure of coatings. In each case, the depth of infiltration is chosen by trial and error.

However, when meeting the requirements for the specimens, it provides stable and reliable results.

3.4.3.2 Pin Method of Fracture Testing

In the pin method, the washer 1 serves as the basis, pin 2 is inserted into its hole in such a manner that its end surface is flush with the external plane of the washer. The total surface of the pin and the washer after preparation is coated with the coating 3. To avoid falling out during the coating, the pin is fixed with screw 4. The test consists of pulling the pin by applying a force (*P*), Fig. 3.44.

Ratio of the maximum force (*P*) to the square of the pin end is taken as the characteristic of adhesion of the coating. Tapered pin is used for the tests, as in the case of a cylindrical pin, the result is distorted due to the influence of friction between the mating surfaces of the pin and the washer.

The advantages of the pin method include:

- Reusability of specimens;
- The absence of preliminary stages;
- Efficiency;
- The ability to determine the strength of any coatings. However, the pin method has several disadvantages:
- Its use allows determining the strength of adhesion only if the coating thickness is more than 0.3–0.5 mm, with a smaller thickness the cohesive fracture mechanism is dominated;
- Size of the gap between the pin and the washer strongly affects the result. When there is a very small gap (details lapped), the van der Waals forces increase. When a gap is large, sprayed particles can get there and prop the mating surfaces. In both cases, the distortion results.
- The result is affected by the size of the specimen. Thus, for a detonation coating an increase in the diameter of the pin from 1 to 2.5 mm decreases the adhesion

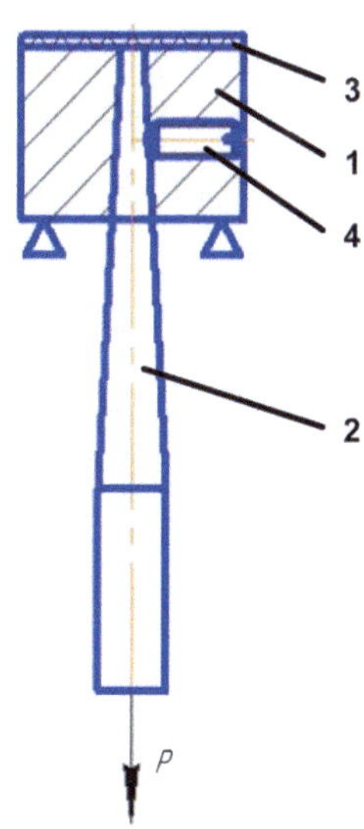

Fig. 3.44 Schematic pin technique

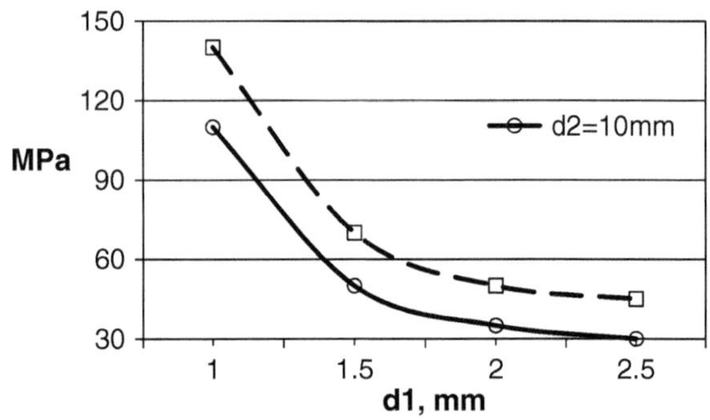

Fig. 3.45 Dependence of the adhesion strength between coating and substrate on the diameter of pin (*d1*) and the washer (*d2*)

strength, at further increasing remains essentially unchanged. Reducing the diameter of the washer reduces the strength of adhesion, Fig. 3.45. It is recommended to use the pin and the washer with diameters of 2 and 20 mm, respectively [12].

- When loading adhesive failure at the "coating–pin", interface is gone with the destruction of the coating on the perimeter of the pin. This is caused by the complex stress state, where besides normal stresses shear and bend stresses occur. The calculations made by Puzryakov [11] showed that the arising error of a result is negligible at small ratio of the pin radius to the thickness of the coating ($R/h \leq 2$). In other cases, this method can be used only for coatings that have much higher cohesive strength than adhesive strength of the coating with the base metal.

3.4.3.3 Shear Test Method

Shear test method is less labor intensive in the specimens' preparation; there is no restriction on the maximum adhesion strength. However, because of the difficulty in positioning the cutting plane, its precision is lower and its application is limited to the minimum coating thickness of 0.5 mm. Both prismatic and cylindrical specimens are used in shear tests.

Brief Description of the Shear Method, Prismatic Specimens

The method is to cut coating 2 deposited on the specimen end 1 with punch 5, Fig. 3.46. Punch 5 moves in the guide plates 3, 4 parallel to the end surface of the sprayed specimen at a loading rate of 50 m/s, Fig. 3.46. Tests are carried out on prismatic specimens.

The strength of coating adhesion in a single test is calculated by the formula

$$\sigma = P/S$$

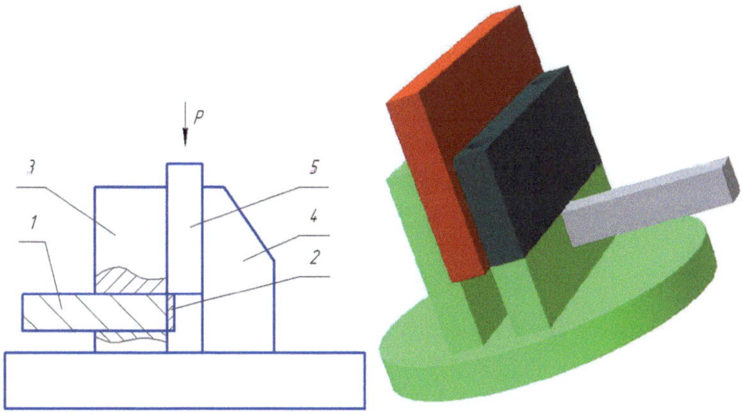

Fig. 3.46 Shear testing scheme to determine the adhesion strength of sprayed coatings to the substrate, *1* specimen, *2* coating, *3*, *4* the guide plates of hardened steel, *5* a punch made of hardened steel

where σ *is* adhesion strength, MPa; *P*—shear force H; *S*—cross-sectional area of the end surface of the deposited specimen, m^2.

Adhesion strength of the test coating is calculated as the arithmetic mean of the five specimens tested.

In world practice, shear test method is standardized according to EN 15340.

Brief Description of the Shear Method, Cylindrical Specimens

According to this scheme, Fig. 3.47, specimen 1 coated with coating 2 is set on the base 5; under the action of the force (*P*) coating 2 is cut with hardened punch 3

Fig. 3.47 Schematic shear test of cylindrical specimens. *1* specimen, *2* coating, *3* punch, *4* mandrel, *5* base, *6* top cover

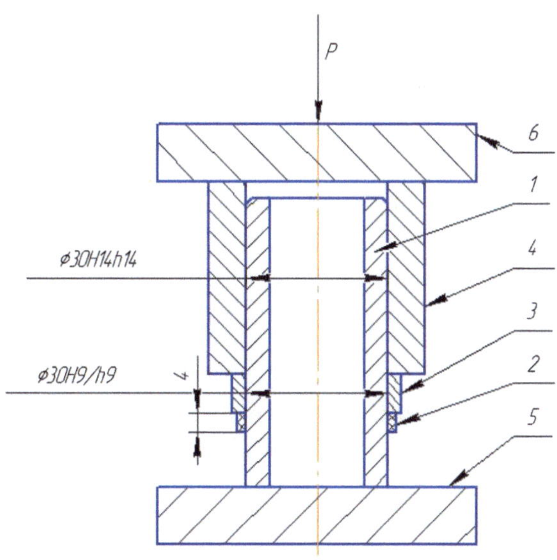

through the mandrel 4 and top cover 6. Slim punch 3 and wide mandrel 4 provide reducing the influence of friction on the surface of the specimen during loading. As can be seen, the dual size between punch 3 and specimen 1 was carried out by slide fit H9/h9 (accuracy ± 0.052 mm), used in particular in moving joint/connections in slow progressive motions and displacements. Transition mandrel 4 is matched with the specimen 1 at a clearance fit *H*14/*h*14 (accuracy ± 0.52 mm). The determination of adhesion is calculated by the formula:

$$\sigma = P/(\pi DB), \quad (1)$$

where *P* is shear force, *B* = 4 mm—coating width, *D* = 30 mm—specimen diameter.

3.4.3.4 Qualitative Methods

Evaluation of the Adhesion Strength by Scratching

Qualitative assessment of the adhesion strength of the coating to the base metal is determined by applying a scratch mesh according to GOST 9.304. For this purpose, the surface of controlled specimen is scratched with the tip of the cutting tool (the hardness of the tip material must be not less than 30 % higher than the coating hardness is). The scratch grid consists of squares with dimensions specified in Table 3.10. Scratches are cut until the base metal. Test is passed if no peeling of the coating is marked after scratching. After the control, the coating must be restored.

The method is used to evaluate the adhesion strength of zinc and aluminum corrosion resistant coatings.

Laser Shock Adhesion Test

When a high power laser (typically 40 J) with short pulse duration (a few nanoseconds) irradiates the surface of a material, the energy is deposited so quickly that a very thin layer of less than 1 μm of thickness of the material is vaporized. With the temperature increase, this vapor results in a hot plasma that absorbs the energy during the rest of the laser interaction. The release of the plasma generates a pressure loading. This pressure results in a shock wave that propagates in the material, the amplitude of which may reach several GPa typically. Consequently, laser shock generates purely mechanical effects within the tested material.

The shock wave propagation and reflections in the part lead to tensile stresses at the interface. Laser Doppler interferometer is applied to the rear surface to measure its velocity. The shape of the velocity profile is used to detect interface de bonding.

Table 3.10 Test parameters

Area with scratched grid	Distance between scratches, mm	Width of the coating under control, m km
15 × 15 mm	3	Up to 200
25 × 25 mm	5	More than 200

The study of the velocity profiles for various values of power laser flux allows the determination of adhesion thresholds [42].

Ultrasonic Evaluation of Coating Adhesion

Evaluation of Adhesive strength is possible by ultrasonic test. The NDE is based on spectral intensity at a certain frequency band of ultrasonic waveform.

Here ultrasonic waves pass through the boundary between the coating and base metal. After transforming the ultrasonic waveform into a frequency domain, adhesive strength is evaluated. Hatanaka with coauthors showed that spectral intensity at a certain frequency band increases with adhesive strength correspondingly [43].

3.5 External Influences on the Material During Spraying

The adhesion effectiveness and the subsequent forming of the coating are affected by the following:

No dirt or defects on the surface; a contact surface area; surface profile; contact temperature; time interaction; the cooling velocity; the kinetic energy of the particles.

There are several classifications of ways to influence the material during spraying: by stages of the process (pre-exposure to the surface coating, coating, and additional post-processing of the coating), by type of exposure (physical, mechanical, chemical, and thermal) [3, 9, 11, 44]. Based on the proposed logic of thermal spraying process description as a system model (see Chap. 1), these methods can be divided according to a place of influence: the substrate and the sprayed material.

3.5.1 Effects on the Substrate

Effects on the substrate are performed to activate and reduce the level of residual stresses. To do this, abrasive grit blasting, heating the substrate, removal of the oxide film by chemical, and physical effects are used.

(1) *Abrasive Grit Blasting (AGB)*

AGB cleans the surface and removes it from the state of thermodynamic equilibrium, releasing the interatomic bonds of the surface atoms, and also makes surface rough, increasing the area of contact. In a review paper on the impact of the AGB on the adhesion strength of thermal spray coatings, it is concluded that parameters of the microrelief of surface prepared by thermal spraying have the dominant influence on the strength of adhesion of the coating to the substrate [45].

AGB disadvantage is that the oxide film grows rapidly on the chemically activated surface. Therefore, the interval between the preparation and coating of the substrate is limited. Furthermore, the process is highly labor intensive and harmfulness. Despite its flaws, this is the main process of surface preparation. Mechanical thread cutting does not provide a surface activation; chemical etching was not widespread because it is non-technological.

Our assessment of the coatings quality with different methods of surface preparation has shown that such methods of details preparing as cleaning and degreasing are insufficient to provide stable adhesion strength. For example, for 0.2 mm zinc coating results are depicted in (Fig. 3.48). In addition, when a bad surface preparation insufficient heat dissipation to substrate takes place. It causes heating of the coating and reducing its mechanical properties. This can be seen in the difference of the bending angle. At using AGB, it is several times increased in comparison with other methods of surface preparation (Fig. 3.49).

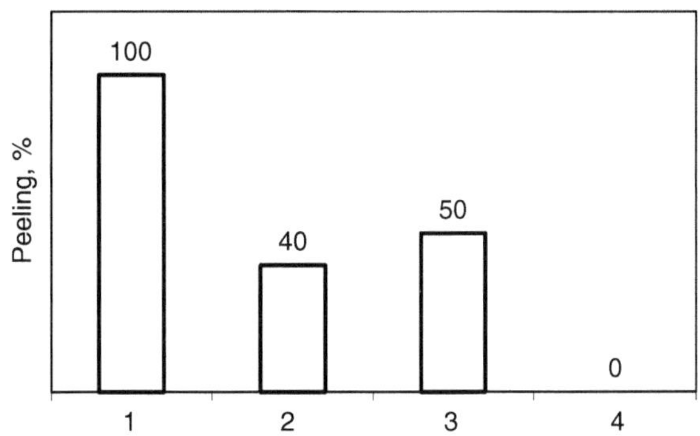

Fig. 3.48 Effect of surface preparation on the strength of adhesion (a comparison according to the proportion of peeled cut grid squares [46]): *1* degreasing; *2, 3* cleaning with abrasive wire discs; *4* AGB

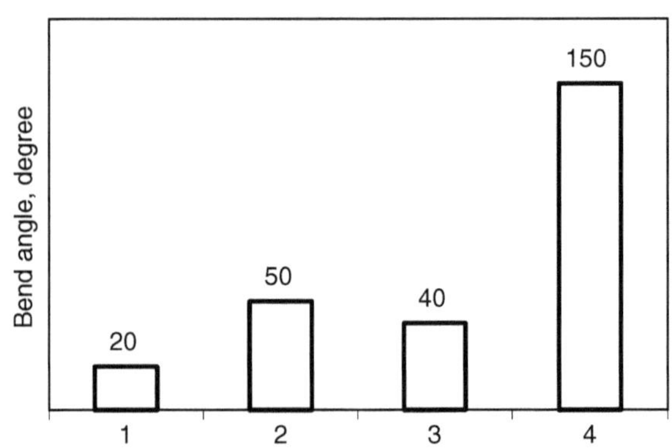

Fig. 3.49 Effect of surface preparation methods on the mechanical strength of the coating (the comparison according to bending angle of the coated plate): *1* degreasing; *2, 3* cleaning with abrasive wire discs; *4* AGB. The spraying material is steel 304, layer thickness is of 0.2 mm

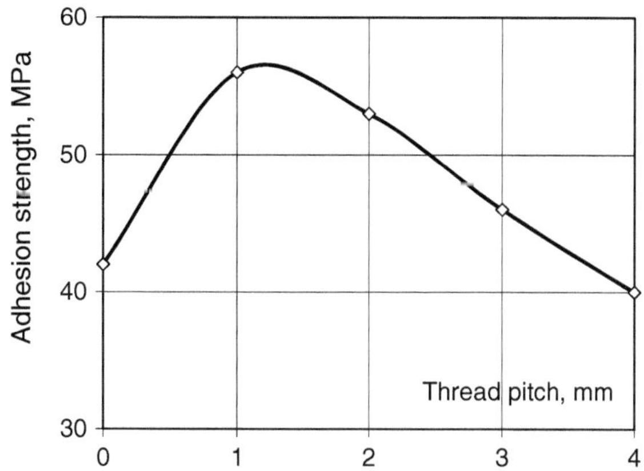

Fig. 3.50 Change of adhesive strength depending on thread, sprayed material is nichrome, the substrate is aluminum

Mechanical processing application in combination with grit blasting increases the adhesion of the coating due to increased contact area. Thus, when cutting threads on an end surface the adhesive strength is further increased by 1.3 times at optimum thread step of 1 mm, Fig. 3.50. Decrease in the adhesive strength at other thread steps is probably due to the worse conditions of drops spreading.

(2) *Substrate Heating*

For most combinations of coating and substrate materials temperature increase of the substrate causes a rise of deposition efficiency (DE), and the adhesive strength of the coating by reducing the level of residual stresses and improving the drops spreading, Fig. 3.51.

However, this method has several disadvantages:

- upper heating limit of the substrate is limited due to grow the oxide film, which divides interacting metals. This is because the activation energy of oxides is higher than the activation energy of most metals;
- the value of substrate heating is limited by the thermal deformation of the parts;
- possible growth of fatigue cracks, which adversely affects the parts operating under alternating and cyclic loads (e.g., crankshafts);
- high power capacity of process in the case of massive parts.

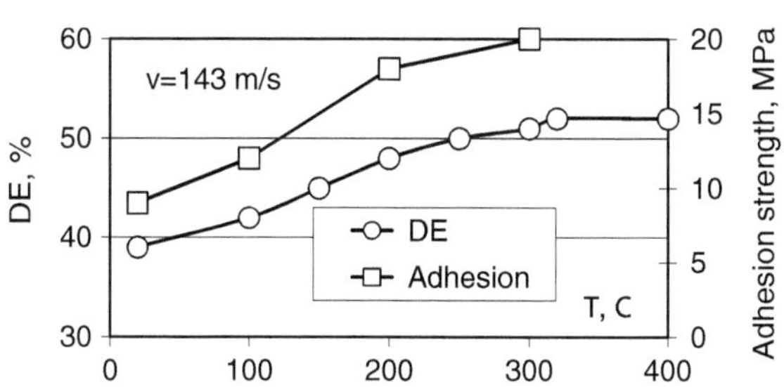

Fig. 3.51 Changes in DE and adhesive strength of the coating depending on substrate temperature, sprayed material is nichrome, the substrate is aluminum

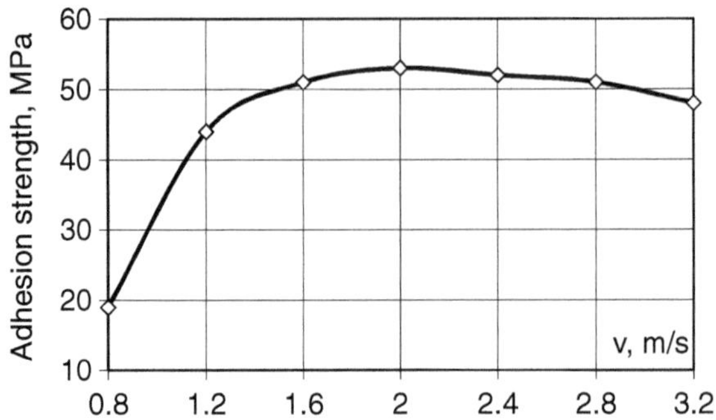

Fig. 3.52 The effect of the linear velocity of the sprayed surface movement with respect to the thermal spraying gun on the strength of the adhesion

As an example, we can mention the implementation of the substrate heating by changing the linear velocity of the sprayed surface with respect to thermal spraying gun, Fig. 3.52.

When too slow moving the overheating of sprayed coating occur, there are large residual stresses and adhesion strength deteriorates. Very rapid movement leads to a decrease in the value of the adhesive strength due to the growing share of the transverse component of the velocity of drops on impact. In addition, there may be gaps between the single sections of sprayed passes.

(3) *Removal of the Oxide Film by Chemical and Physical Effects*

These methods are applied to metals having high chemical affinity for oxygen. In related welding processes, the following is used to remove oxide film:

(1) reverse polarity welding, ensuring the destruction of the film due to the cathode sputtering [47];
(2) applying ultrasonic treatment to surface [48];
(3) surface tin-plating combined with an abrasive treatment [49].

In the practice of spraying, there are proposals to impose ultrasound on the substrate [50], a preliminary application of fusible material on the substrate [3], but they are not widely used, probably because of the technical complexity of the implementation.

To remove the oxide on aluminum before applying the thermal spraying it is effective to coat the substrate with fluorine-containing flux. This leads to improved wettability and the penetration of the fluorine-containing compound into the pores and other defects in the oxide film, which contributes to its mechanical destruction. Fusible fluorine compounds, replacing aluminum oxide, are destroyed when they are influenced by molten particles. At the boundary of "coating–substrate" is a gradual change in lattice parameter of the solid solution based on alumina at a depth

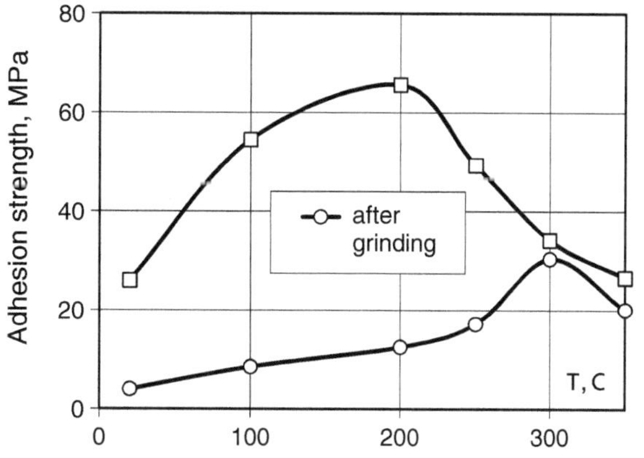

Fig. 3.53 Changes in adhesion depending on the preheating temperature and the type of surface preparation

of 0.4 mm [51]. This indicates that there is diffusion interaction. Using this method, results in a rise of adhesion strength by 2–3 times in comparison with the previously used method (Fig. 3.53).

3.5.2 Influence on the Sprayed Material

In this group, the following methods can be distinguished:

- the use of materials with high heat content;
- increasing the temperature of the particles;
- increasing the velocity of the particle

The Use of Materials With High Heat Content

As such, materials molybdenum and nichrome are the most common. Compared with steel they have higher melting point and electrical resistivity. This provides increased drops heat content of sprayed metal that reduces the activation energy of interaction.

A similar effect has the addition of aluminum into the spray material that can be implemented in the core wire. Pokhmursky et al. noted [25] that this effect is achieved by the heat of the exothermic reaction of aluminum and iron oxides of main components:

$3Fe + Al = Fe_3Al + 0.75$ kJ per gram of the charge
$2/3Fe_2O_3 + 2Al = 4/3Fe + Al_2O3 + 3.5$ kJ per gram of the charge,
$2/3Cr_2O_3 + 2Al = 4/3Cr + Al_2O3 + 2.7$ kJ per gram of the charge

Our results confirm these data. Comparative tests have shown that the addition of aluminum to the charge of core wire PPM-6/1 (Cr–C–Fe 8 3 bal.) [52] arises the coating adhesion strength by 10–15 % than that of nichrome coating, and three times higher in comparison with 08Г2C, GOST 2246 (steel like ER70S-6) coating (Fig. 3.54).

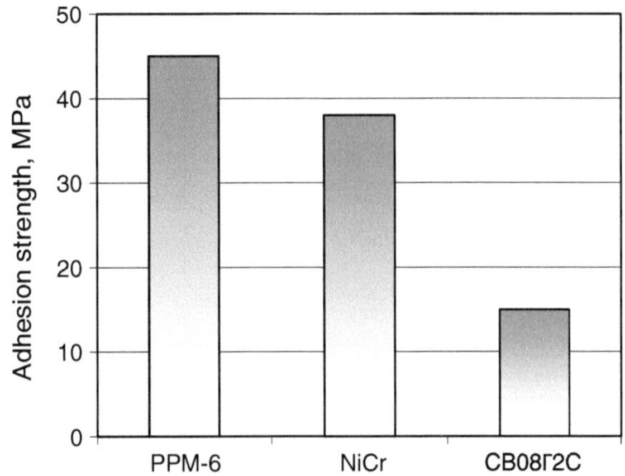

Fig. 3.54 Adhesion strength of coatings. Shear test method for AS coating thickness of 1 mm on the end face of the prism 10 × 10 mm [53]

Increasing the Temperature of the Particles

The overheating enables to improve significantly the adhesion strength in the methods of thermal spraying, providing excessive heat input to the sprayed material. For example, during plasma spraying of aluminum on aluminum, when the overheating is 900 °C above the melting point, the adhesive strength is 0.95 to a maximum, whereas when the overheating is 300 °C, the adhesive strength is only 0.2 to maximum [54].

However, the overheating is undesirable for several reasons:

- burnout of alloying elements and particles evaporation are possible; an exception is molybdenum, which has a higher melting point than the boiling point of its oxide [55]:

$$T_{boil}^{MoO_3} = 1563\,K;\, T_m^{Mo} = 2880\,K.$$

For this reason, there is no oxide film when atomization and in flight on the surface of the molybdenum droplets. It also shows that molybdenum is refractory metal. These features provide high adhesion strength, up to 60.8 MPa [34]. However, its use is limited due to the high cost and scarcity.

When the temperature of the metal considerably exceeds melting point, which is typical to arc burning zone at AS, the saturation of the liquid metal by atmospheric gases is drastically intensified due to its logarithmic dependence. As shown in Sect. 2.8, the solubility of oxygen in liquid iron is increased approximately 20 times, while increasing its temperature from the melting point to 2900 K. This leads to deterioration of the mechanical properties of the coating and the growth of the oxide film on the particles at flight due to oxygen precipitation from its liquid solution while lowering the temperature of the particles to the melting point;

- Overheating of the particles contributes to excessive heat input into the coating, which causes an increase in the level of residual stresses here. This is noticeably manifested when compared AS and plasma coatings. The first one have specific energy consumption per kg of the sprayed metal the order of magnitude lower (Table 1.2). As a consequence, AS coatings without cracks with thickness up to 2 mm are common practice, and in the plasma coatings the likelihood of cracking increases sharply with a coating thickness of more than 0.7 mm.

When AS, the particles overheating is realized by increasing the voltage and current.

Arc voltage at AS strongly influences the DE and adhesive strength. The voltage increase causes an increase of the arc length, which increases the enthalpy of the metal droplets. This, in turn, leads to greater progress of chemical interaction in contact "drop–substrate". At the same time, due to increasing the proportion of the vaporized metal the DE rate is reduced, which is measured by the mass ratio of the consumable wire and the coating. So, when nichrome spraying the steady arc burning is observed at 21 V, the DE is 0.8–0.9. Increasing the voltage to 30–32 V increases the adhesion strength by 25–30 % and reduces DF to 0.65–0.75. Further voltage rise slightly increasing the adhesion strength strongly decreases the thermal efficiency of the arc, Fig. 3.55. Likely, the increase of the adhesive strength is connected with the rise of the heat content of droplets, and the DE decrease is due to the increasing metal burnout at the arc length increase.

We can note that the likelihood of cracking at higher arc voltage increases in AS coatings due to the increase of the arc length, which is the main heat source.

The higher the current is the higher the processing rate and the heat input to the droplets. This improves the physical–chemical interaction "droplet–substrate" at the moment of impact, which leads to higher adhesive strength, Fig. 3.56.

However, after the current rises above a certain level the adhesion strength remains constant and even begins to decrease. Probably there is an excessive decreasing of the sizes of the droplets tearing off the end faces. As shown in Chap. 2, it increases the oxidation of droplets, which in turn impairs the "droplet–substrate"

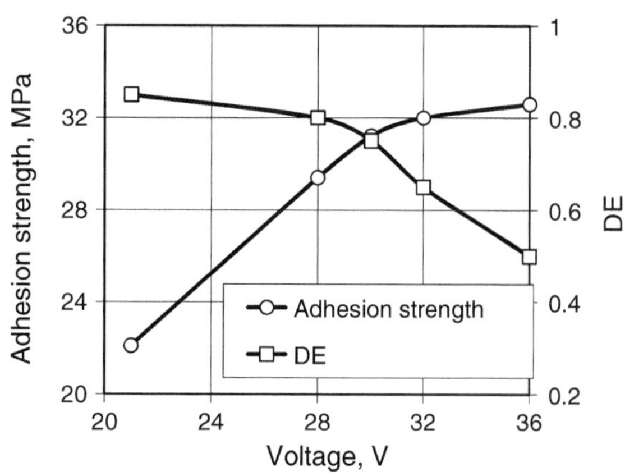

Fig. 3.55 Dependence of the adhesive strength and the deposition efficiency of the metal on the arc voltage

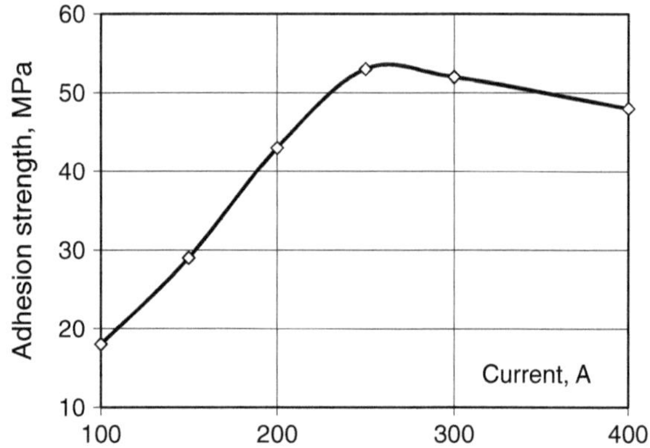

Fig. 3.56 Effect of arc current on the strength of adhesion

contact conditions. Furthermore, excessive heat input increases the likelihood of metal droplets splash on impact with the substrate, instead of spreading over it. Quantitatively, the maximum adhesion strength is achieved at a current of 250 A.

Droplets enthalpy increasing is possible also with increasing the electrodes extension, implemented, for example, by non-conducting inserts in the tips [56]. Additional Joule heat released in the wire when current leads to an increase in adhesive strength by 30–35 % while increasing electrodes extension from 10 to 40 mm, Fig. 3.57.

Increasing the Velocity of the Particle

The above-mentioned disadvantages of exceeding the melting point of the particles resulted in the modern trend of development of TS processes, and AS in particular, aimed at increasing the velocity of the particles while limiting their temperature (Fig. 1.1). The kinetic energy of the particles at impact provides their intensive deformation on the substrate.

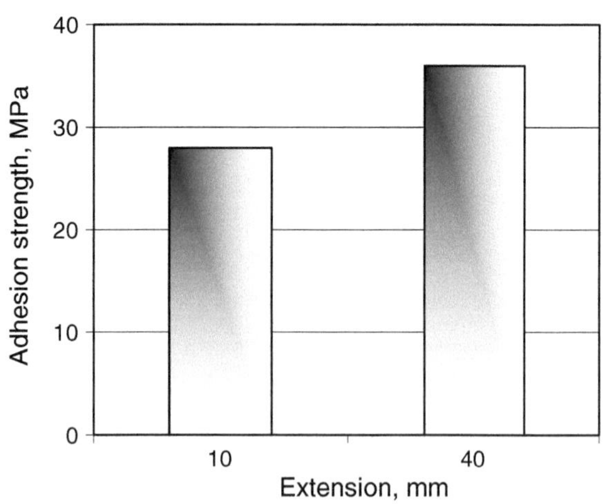

Fig. 3.57 Changing the adhesive strength depending on the electrode extension

For example, for aluminum particle the velocity growth from 109 to 143 m/s leads to an increase in values of adhesion strength and DE by 15–20 % [7].

For steel wire coatings from steel like 304 by increasing the average velocity of the particles more than three times (40–140 m/s), the adhesive strength increases 1.8–2.2 times [57, 58]. Deviation from the direct proportion adhesive strength—particle velocity can be explained as follows. In the latter case, the average particle size is reduced to 4–7 times. This causes (at the same temperature and the particles material) the decrease enthalpy of a single particle and the increase its degree of oxidation. These factors reduce the time of its crystallization time, the temperature and the completeness of chemical interaction in contact "particle-substrate".

Increasing the particle's velocity also leads to an increase in temperature in contact "particle-substrate" by transition its kinetic energy into thermal energy. Belotserkhovsky made assessing the influence of the particle's velocity on its heating at impact [44]. It is based on the solution of the problem of temperature change due to the collision of bodies [59]. The energy balance equation of the particle when it impacts the substrate can be written as:

$$E_{\mathrm{K}} + E_H = E_u + E_o + Q + \Delta E_n + E_y, \tag{3.34}$$

where E_K is the kinetic energy of the particle; $E_н$ is accumulated thermal energy; E_u—energy emitted by the particle; E_o—heat energy transferred to the substrate; Q—heat energy released by the deformation of the particle; ΔE_n—change of the surface energy of the particle while deformation; E_y—energy of elastic vibrations.

We shall estimate the value of thermal energy, which may be obtained by the particle at impact on the substrate. Let's assume that

$$E_K = Q + \Delta E_n + E_{y,} \tag{3.35}$$

$$\Delta E_n = \sigma \cdot \Delta S, \tag{3.36}$$

where σ is surface tension, J/m^2; ΔS—an area change at the deformation, m^2.

Taking into account the particle radius (r) before impact, and representing the particle after impact as a cylinder of height (h), the expression for the change of potential energy (3.37) can be written in the form

$$\Delta E_n = \sigma 4\pi r^2 \left(\frac{2r}{3h} + \sqrt{\frac{h}{3r}} - 1 \right). \tag{3.37}$$

In view of the short duration of the interaction, energy losses can be neglected. Assuming that the energy of elastic vibrations is converted to heat, we obtain:

$$\frac{mv^2}{2} = cm\Delta T = \sigma 4\pi r^2 \left(\frac{2r}{3h} + \sqrt{\frac{h}{3r}} - 1 \right), \tag{3.38}$$

wherein *m, v* are weight (kg), velocity (m/s) of the particle; *c*—specific heat of the sprayed material, J/(kg °C); ΔT—heating value due to deformation, K. Hence:

$$\Delta T = \frac{v^2}{2c} - \sigma\left(\frac{2r}{3h} + \sqrt{\frac{h}{3r}} - 1\right) \Big/ (c\rho r), \tag{3.39}$$

where ρ is density of sprayed material, kg/m^3.

The evaluation showed that the characteristic dimensions of the sprayed materials, $r = 20–80$ μm, $h = 4–10$ μm, the calculation error according to Eq. (3.39) will be less than 15 %, if you write it in the form:

$$\Delta T = v^2/(2c). \tag{3.40}$$

Hence velocity required for particles heating to the melting point:

$$v_1 = \sqrt{2c(T_{пл} - T_0)}, \tag{3.41}$$

Velocity needed for particles should be higher due to melting enthalpy:

$$v_2 = \sqrt{2c(T_{пл} - T_0) + h_{пл}} \tag{3.42}$$

where T_0 is temperature of the particles while approaching the substrate; $h_{пл}$—specific enthalpy of sprayed material melting, J/kg.

The results of the calculation (Table 3.11) show the values of the velocities, which are necessary to accelerate the particles of different materials while impact, they should be warmed up to the melting point or to the melting condition. If cold particles are accelerated using the methods of thermal spraying, such velocities can be achieved only with the use of detonation spraying (Fig. 1.1). If the particles when approaching are warmed to 200 K below the melting point, the Table 3.11 shows that the velocity of the particles necessary for their heating to the melting point at impact, is achievable using basic methods of thermal spraying. At thermal spraying, the use of this technique can help to develop both equipment and spraying

Table 3.11 Velocities required to heat cold particles to melt when they hit the substrate

Parameters		Steel	Nickel	Copper	Aluminum
c, J/(kg K) [60]		695	582	427	1030
h, J/kg [60]		1,340,000	1,148,000	346,000	1,062,000
$T_{пл}$, K		1809	1726	1356	933
v_1, m/s	$T_0 = 273$ K	1461	1300	962	1166
	$T_{пл} - T_0 = 200$ K	527	482	413	642
v_2, m/s	$T_0 = 273$ K	2194	1997	1272	1866
	$T_{пл} - T_0 = 200$ K	1720	1590	929	1592

technologies applying to core wires. In particular, for the latter case the thermal calculation, Chap. 2, showed the possibility of partial melting of the charge of wire.

Methods of increasing the velocity of the particles based on the equations of droplets motion in the two-phase flow and momentum (see Sect. 2.6). They are implemented by:

(1) increasing the mass flow rate of carrier gas, i.e., increase its velocity, pressure, density, cross-section of the nozzle;
(2) reducing the particle size to reduce its aerodynamic resistance.

The second method is realized at arc spraying by increasing pressure and velocity of the gas and lowering the arc voltage. Let us note that the latter is limited to a threshold, of stability disturbance at droplet separation. The droplet size may be reduced also by reducing the surface tension of the metal at the electrode end face, for example, by increasing the oxygen content in gas [61]. According to experimental data, the velocity of the particles increased for the proposed method by 20–50 % with a lowering of their diameter. The inclusion of surfactants into the composition of the wires also reduces the surface tension of the metal. Imposition of ultrasonic vibrations (IUV) on the current contact tips is prospectively [62]. It results in a narrowing of the particle interval. The maximum size of particles formed during wires spraying is reduced by reducing the surface tension of the molten layer, the minimum increases by reducing the temperature gradient in the liquid film on the end face of the electrode. According to experiments on the aluminum wires spraying, prior to application of IUV the diameters interval of sprayed particles was 3–120 μm, after—21–40 μm.

- Reduction of the oxide film on the surface of the particle. Oxide filming model, see Chap. 2, shows that for the implementation of this method of increasing the adhesion strength it is necessary:

  (1) creating a protective atmosphere, especially in the arc burning zone, where the droplets, according to calculations, get 85–95 % oxygen;
  (2) alloying of sprayed metal with deoxidizing elements;
  (3) ensuring the rational sizes of the sprayed particles through the gun design and technological conditions. The calculations have shown, for example, that the contribution of droplets of fractions 27–75 μm in the oxygen input is 1.5–2.5 times less than in case of the larger and smaller droplet size.

References

1. http://www.gordonengland.co.uktsc
2. Korobov YS, Polyakova AM, Yakovleva IL, Schastlivtsev VM, Pryadko AS (1997) Structure and properties of steel coatings sprayed by activated arc spraying. Weld Prod 1:4–6
3. Kudinov VV (1977) Plasma coatings. Nauka, Moscow
4. Carslow G, Jaeger D (1964) Heat conduction of solids. Nauka, Moscow

5. Gulyaev AP (1977) Material science. Metallurgiya, Moscow
6. Gubenko BG, Borisova AL, Kostin VA et al (1990) Structure heterogeneity of arc sprayed coatings of steel 65G. Avtom Weld 6:19–23
7. Abedini A, Pourmousa A, Chandra A, Mostaghimi J (2004) Splat formation in wire arc spraying. In: Thermal Spray Solutions—Advances in Technology and Application. Proceedings of ITSC 2004, Osaka, Japan
8. Pokhmursky VI, Student MM, Pikh VS (1986) Fundamentals of forming of protecting and restoring coatings by arc spraying from flux core wire. Phys Chem Mech Mater 6:11–16
9. Barvinok VA (1990) Tension state control and plasma coatings properties. Mechanical Engineering, Moscow
10. Kulik AY, Borisov YV, Mnukhin AS, Nikitin MD (1985) Thermal spraying of composite powders. Mechanical Engineering, Moscow
11. Puzryakov AF (2003) Theory of plasma spraying technology. MGTU, Moscow
12. Tushinsky LI, Plokhov AV, Tokarev AO (2004) Methods of material's research. Mir, Moscow
13. Virnik AM, Morozov AI, Podzey AV (1970) About the estimate of residual stresses in plasma sprayed coatings. Phys Chem Mater Process 4:53–56
14. Borisova AL, Mits IV, Kayda TV et al (1995) Structure and properties of coatings based on arc sprayed ferrochrome and ferrochromealuminium from flux core wire. Autom Weld 6:3–6
15. Timoshenko SP, Gudier G (1975) Theory of elasticity. Nauka, Moscow
16. Kragelsky IV (1968) Friction and wear. Mechanical Engineering, Moscow
17. Gudremon E (1959–1960) Special steels, Parts 1, 2. Metallurgizdat, Moscow
18. Linchevsky BV (1986) Thermodynamics and kinetics of gas interaction with liquid metals. Metallurgiya, Moscow
19. BAMZ (1982) Arc Spraying Gun EM-14: Specification. Barnaul, USSR
20. Korobov YS, Izoitko VM, Pryadko AS, Lukanin VL (2000) Part's restoration by activated arc spraying. Automob Ind 1:23–24
21. Korobov YS, Lukanin VL (1997) Spraying method. Ru Patent 2097144
22. Korobov YS, Lukanin VL, Novikov YN, Pryadko AS (2002) Arc spraying gun. Ru Patent 2186632
23. Borisova AL, Gubenko BG, Kostin VA et al (1990) Features of structure formation of steel—aluminum electric arc coating. Autom Weld 8:41–44
24. Pokhmursky VI, Student MM, Dovgunyk VM et al (2002) Flux core wires FeCrB+Al and FeCrB+Al+C for arc spraying. Autom Weld 3:32–35
25. Pokhmursky VI, Student MM, Sidorak II et al (2002) Phase composition of Fe-C-Cr-Al coatings, produced by arc spraying. Autom Weld 10:20–23
26. Pokhmursky VI, Student MM, Dovgunyk VM et al (2003) Structure and tribological properties of coatings obtained by arc spraying using cored wire. Autom Weld 8:13–17
27. Wielage B, Pokhmurska H, Student M, Gvozdeckii V, Stupnyckyj T, Pokhmursky V (2013) Arc-sprayed iron-based coatings from cored wires for applications at elevated temperatures. Surf Coat Techn 220:27–35
28. Borisov YS, Andreychuk VK, Dekhtyar LI et al (1987) Spray head to electric gun. USSR Certificate of authorship 1321476
29. Korobov YS, Nevezhin SV, Filippov MA, Ilyushin VV, Potekhin BA, Gogolev LV (2012) Comparison of the tribological characteristics of Babbitt coatings obtained by activated arc spraying and alternative methods. Weld Diagn 1:36–39
30. Korobov YS, Gogolev LV Potekhin BA, Ilushin VV, Nevezhin SV (2011) Comparative characteristics of coatings produced from Sn-Sb-Cu Babbitt by arc spraying, gas surfacing and pouring. In: Proceedings of international thermal spray conference and exposition, ITSC 2011. Hamburg
31. Potekhin BA, Gluschenko AN, Ilyushin VV (2006) Properties of Babbitt B83. Metal Technol 3:17–22
32. Potekhin BA, Gluschenko AN, Ilyushin VV (2004) Device for filling plain bearings. Ru Patent 38649

33. Ilyushin VV, Potekhin BA (2010) Influence of flame forming gases used in surfacing, on the structure and properties of Babbitt B83, B88. In: Proceedings of the 12th international scientific and practical conference "Resource-saving technologies of repair, restoration and strengthening of parts of machines, equipment, tools and tooling," Part 1. Publishing House of the Polytechnic University Press, St. Petersburg
34. Bobrov GV, Ilyin AA (2004) Application of inorganic coatings. Internet Engineering, Moscow
35. Bartenev SS, Fedko YP, Grigorov AI (1982) Detonation coatings in mechanical engineering. Mechanical Engineering, Leningrad
36. Kostikov VI, Shesterin YA (1978) Plasma coatings. Metallurgiya, Moscow
37. Maksimovich GG, Shatinsky VF, Kopylov VI (1983) Physical and chemical processes in plasma spraying and destruction of coated materials. Naukova Dumka, Kiev
38. Davis JR (ed) (2005) Handbook of thermal spray technology. ASM International, Materials Park
39. Rizhonkov DI (ed) (1989) Theory of metallurgical processes. Metallurgiya, Moscow
40. Hasui A, Morigaki O (1985) Surfacing and spraying. Mechanical Engineering, Moscow
41. Han W, Rybicki EE, Shadley JR (1993) An improved specimen geometry for ASTM C633-79 to estimate bond strengths of thermal spray coatings. J Therm Spray Technol 2:145–150
42. Ducos M et al (2004) Non-destructive adhesion testing of plasma-sprayed coatings using ultrasounds and laser shocks. In: Thermal Spray Solutions—Advances in Technology and Application. Proceedings of ITSC 2004, Osaka, Japan
43. Hatanaka H et al (2000) Ultrasonic examination of thermal sprayed coatings with frequency analysis. In: Proceedings of 15th world conference on non-destructive testing, Roma, Italy
44. Belotserkovsky MA (2004) Technologies of activated flame spraying of anti-friction coatings. Tehnoprint, Minsk
45. Kharlamov YA, Borisov YS (2001) Influence of surface microrelief on the bond strength of thermal coatings. Autom Weld 6:19–26
46. GOST 9.304-87 (1990) Thermal spray coatings. General requirements and methods of control. Publishing House of Standards, Moscow
47. Nikiforov GD (1972) Metallurgy of fusion welding of aluminum alloys. Mechanical Engineering, Moscow
48. Jenkins WB (1976) Fluxless soldering of aluminum heat exchanges. Weld J 1:28–35
49. Smirnov GN (1981) Progressive methods of aluminum brazing. Metallurgiya, Moscow
50. Vasiliev AA et al (1998) Method of surface preparing at plasma spraying. Ru Patent 1274328
51. Korobov YS, Polyakova AM, Yakovleva IL, Schastlivtsev VM (1996) Investigation of the structure and properties of the transition zone in the aluminum parts with steel sprayed coating. Weld Prod 12:20–22
52. Korobov YS, Shalimov MP, Shumyakov VI et al (2009) Cored wire for arc spraying. Ru Patent 80378
53. Korobov YS, Pryadko AS, Cherepko AE et al (2001) Development and application of core wires for activated arc spraying. In: Proceedings of the 8th IRTC, technology of repair, restoration and strengthening of machine parts, machinery, equipment, tools and tooling. St. Petersburg
54. Kudinov VV, Ivanov VM (1981) Refractory coatings by plasma. Mechanical Engineering, Moscow
55. Turgdogan ET (1985) Physical Chemistry of high-temperature processes. Metallurgiya, Moscow
56. Korobov YS, Boronenkov VN, Baranovsky VE (1995) Head of Arc spraying gun. Ru Patent 2,047,395
57. Pryadko AS et al (1999) Studies of the properties of steel coatings deposited by different methods of arc spraying. Zaschitnye pokrytiya

58. Korobov YS et al (2007) Application of activated arc spraying to protect the power engineering parts In: Proceedings of the 9-th international conference and exhibition, technology of repair, restoration and strengthening of machine parts, machinery, equipment, tools and tooling, in two parts. St. Petersburg
59. Kilchevsky NA (1969) Theory of solid's impacts. Naukova Dumka, Kiev
60. Elliot F, Glaser M, Ramakrishna V (1969) Thermal chemistry of steelmaking processes. Metallurgiya, Moscow
61. Korobov YS, Boronenkov VN (1997) Method of arc spraying. Ru Patent 2,072,699
62. Borisov YS, Ilienko AG, Gaidarenko AL et al (1989) Device for arc spraying. USSR Certificate of authorship 1777966
63. Petrov SV, Saakov AG (2000) Plasma of combustion products in surface engineering. Topas, Kiev

Chapter 4
Output Coatings Parameters

Effective development of equipment and technological processes is possible based on the modeling and coating properties studying. The surface properties of parts in many cases determine their performance. Application of thermal coatings enables to provide the desired output characteristics: wear and corrosion resistance, electrical conductivity, high-temperature strength, heat resistance, and others. Quantitative values of these parameters are due to the structural features of the coatings. Furthermore, these properties are interrelated and it is often necessary to provide a combination thereof. Problems of wear and corrosion resistance of parts will be described; metal sprayed coatings are used to improve these features.

4.1 Wear Resistance

4.1.1 The Concept of Wear[1]

Wear is defined as the change in size, weight, shape, or condition of a solid surface under the influence of the external environment. Wear may be caused by fretting of machine parts, by influence on the working medium surface of the flow of liquid, gas, or by contact of the surface with solid particles.

When fretting, the processes of elastic-plastic strain are localized in micro volume of surface layer adjacent to the contact spots and are exposed to the action of contact stresses. Depending on the nature of friction, the shear stress distribution varies in the depth of the surface layer, Fig. 4.1. In sliding friction, the maximum level of shear stresses is achieved directly on the sliding surface. At rolling friction, the maximum stress level is realized at some distance from the rolling surface, which depends on the contact geometry, normal load and deformation characteristics of the material (elastic modulus, Poisson's ratio).

Micro volumes of metal adjacent to the contact spot are plastically deformed under shear pressure. Each single micro roughness embedded in the counter face surface creates compressive and tensile stress zones when moving. Ahead of the

[1]Monograph [1] was used to write that chapter.

V. Boronenkov and Y. Korobov, *Fundamentals of Arc Spraying*,
DOI 10.1007/978-3-319-22306-3_4

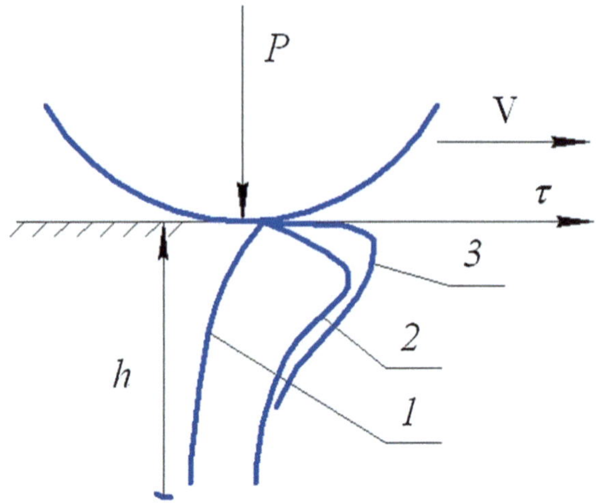

Fig. 4.1 Distribution of shear stresses (τ) on the depth of the surface layer (h) in the area of frictional contact of solids at different types of friction: *1*—slip; *2*—rolling; *3*—rolling with sliding [2]

front of micro roughness the roll of deformed metal is formed which applies compressive stresses. Behind micro roughness because of frictional forces, tensile stresses are generated in the material. Adhesive interaction occurs at the boundary zone between compressive stresses and roughness. As the micro asperity is moving within space, which is limited by strength of adhesive bonds on contact, the roller increases due to material forced out at movement. Behind the micro asperity tensile strain increases accordingly. Upon reaching a critical amount of roughness movement, molecular bindings on the contact are broken; wherein the asperity crushes the roller or jumps over it. Thus, the contact material in a single pass of micro asperity is exposed to consistent effect of compressive and tensile stresses. Wherein, a movement of as asperity and micro roughness is uneven, spasmodic, Fig. 4.2.

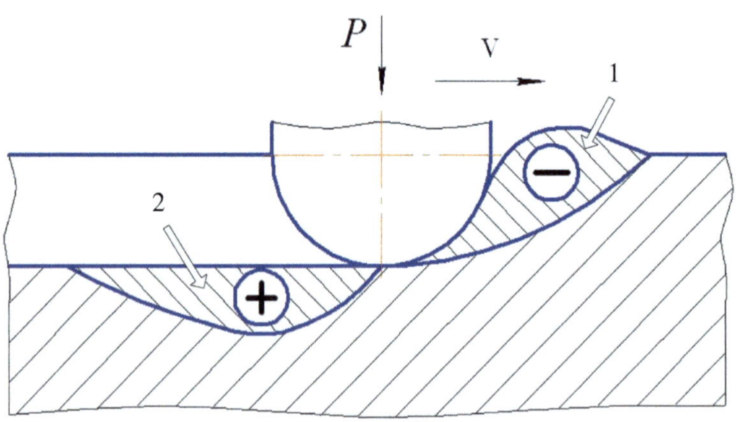

Fig. 4.2 Interaction scheme of a moving rigid cylindrical roughness and perfectly smooth metal surface: *1*—zone of compressive stress; *2*—zone of tensile stress [3]

The mechanism of failure of the working surface of the contacting bodies depends on the nature of the interacting bodies and friction conditions. There are the following basic types of failure [4]:

Scratching. The particle formed or appeared on the friction surface moves in the direction of sliding and crushes the material down, leaving a scratch. The latter ends off at the exit of the intruded element from the contact area, by such reasons as disintegration the particle, its embedding or carrying over outside the friction area. Repeated scratching in one place with the same intensity is rare, more often plastic deformation zone is covered with scratches. They are disposed almost parallel to the sliding direction, and between the scratches the material is located which has undergone to multiple plastic deformation. This material is peened that means it is exhausted the ability to deform plastically. When next loaded cracks can appear easily in this place which leads to the material separation from the substrate.

- *Microcutting*. When intruding into sufficient depth the solid particle of abrasive or wear product may produce material microcutting accompanied by forming chips. In terms of contact with the loose abrasive particles this kind of wear can become prevalent while crushing solids or sliding on the abrasive mass.
- *Delamination*. The material can be pressed from the friction surface at the plastic flow and after exhausting, the ability to further plastic deformation is delaminated. The material overlaps the oxide films while flowing and loses contact with the substrate. If at line or point contact of bodies, the stress in a certain depth from the surface became more than fatigue limit of the material, than cracks appear at processing which lead to the flaky separation of material. Delamination occurs on hardened or cemented parts, such as gears.
- *Pitting*. Because of material separation at fatigue wear pits are formed on the friction surface. This is a common failure mode of working surfaces in a rolling or sliding friction. Pitting is characterized by free form of pits with ragged edges. While the impact or contact fatigue the following components may be flaked: hard or soft inclusions or phases bordered by structural components with different hardness.
- *Tearing in depth*. Because of the bonding of the two contacting bodies under the influence of molecular forces, the local damage of the friction surface occurs in the form of broad and deep bands in the sliding direction. Areas adjacent to the tearing are plastically deformed to a greater or lesser extent.
- *Material transfer*. Because of the molecular forces, the material of one body is connected to the other one and, tearing from the first, remains on the surface of the second. The transfer is characteristic for most types of friction, can be detected, for example, at cutting, when the metal is transferred from the workpiece to the tool.

The main types of mechanical wear differ depending on the nature of the mechanical impact (GOST 27674-88):

- adhesive—tearing as a result of the setting, tearing in depth of the material, its transfer from one surface to another;

- fatigue—the result of repeated deformation of the contact surface areas of solids;
- abrasive—the result of cutting or scratching effects of free or fixed solids.
- erosion—the result of the impact of the flow of solids, liquids or gas moving at high velocities;
- cavitation—the result of numerous microhydraulic hammers from rupture of air bubbles or vapor liquid receiving surface of the solid. Hydraulic hammers create fluctuating stresses on the metal surface, resulting in its destruction. This type of corrosion leads to the formation of local deep ulcers on the blades of the propellers, mixing devices in chemical apparatus, etc.

The influence of corrosive environments significantly accelerates the mechanical wear of all these types.

In addition to these, there is also a fretting wear—the destruction of metal structures in places of slipping of the parts, which are tight, pressed together in the low-amplitude fluctuations, rotations vibrations. If mechanical impact is accompanied by an oxidizing action or electric current, the wear is becoming electrical mechanical or electrical erosive, respectively. This type of wear is widely spread in bolted and riveted joints.

Depending on the ratio of the internal properties of the material and external environments, various types of wear are implemented with different intensity, Fig. 4.3. For each kind, the number of specified properties and external influences can be ignored without great error in the measurements. The particular set of these characteristics will depend on the type of wear.

As the wear characteristics, absolute and relative characteristics can be used. In absolute terms, the wear is characterized by weight loss of items (ΔM), the change in its linear dimensions (ΔU) or volume (ΔV). Relative wear characteristics are the intensity and the rate of wear. For instance, with respect to a linear wear these characteristics can be defined as follows [3]:

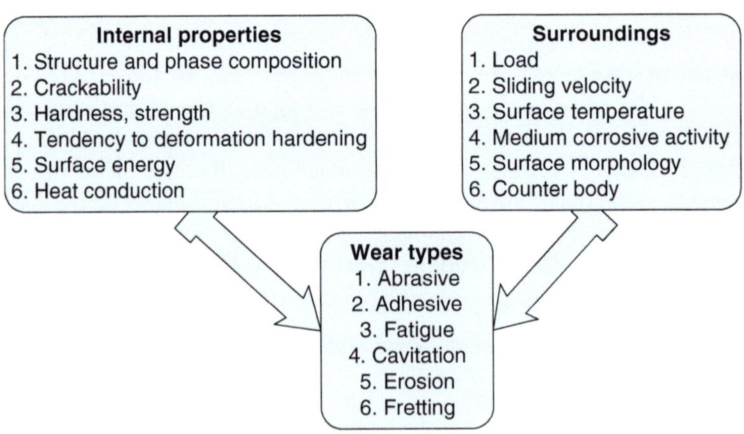

Fig. 4.3 The relationship of wear material type with its structure and terms of wear

- linear wear, ΔU, μm—resizing of surface when wear, measured perpendicular to it;
- wear rate $j = dU/d\tau$, μm/h—wear for the period of time during which it was formed;
- wear rate $J_h = dU/dS$—ratio of the rate of wear to the sliding path, where wear happened. Wear rate is determined more often using the ratio $J_h = Q/(\rho SP)$, where Q is weight material loss, g; ρ is material density, g/cm^3; S is friction path, cm; P is geometrical contact area, cm^2. In this case, J_h is dimensionless.

Allowable and critical kinds of wear should be distinguished. At allowable wear, a part's reliability is retained, despite the change in its shape and size as a result of wear. The critical wear is characterized by changes in the size and/or shape, when further operation results in a loss of efficiency of parts or the mechanism as a whole.

Let us study more carefully types of wear protection, which are characterized by the use of thermal sprayed coatings.

4.1.2 Abrasive Wear

When abrasive wear, the surface of the part is destroyed as a result of its interaction with the solid particles at the relative velocity. Abrasive wear can be caused by a single effect of the abrasive particles, which leads to the removal of very fine chips (microcutting) or multiple plastic or elastic deformation of microvolumes of metal, which causes fatigue failure and separation of the particles of the surface layer.

The process of abrasive wear can affected by the properties of the abrasive particles and the wear surfaces, corrosive environment, the nature of the interaction of particles and surfaces (sliding, rolling, impact), heating and other factors.

Types of abrasive wear are distinguished by the method of fixing the particles:

- *Wear by fixed abrasive particles*

The hardening (cold working) of the surface layer occurs due to mechanical impact of deformation. Khrushchev and Babichev established [5] that the relative wear resistance (E) of pure metals, annealed steels and polymers is directly proportional to their Brinell hardness (H):

$$E = b \times H, \tag{4.1}$$

where b is the coefficient of proportionality.

For heat-treated (quenching and tempering) structural and tool steels the following dependence is performed:

$$E = E_0 + b' \times (H - H_0), \tag{4.2}$$

where b' is proportionality factor depending on the chemical composition of steel; index "0" corresponds to the parameters of steel as annealed.

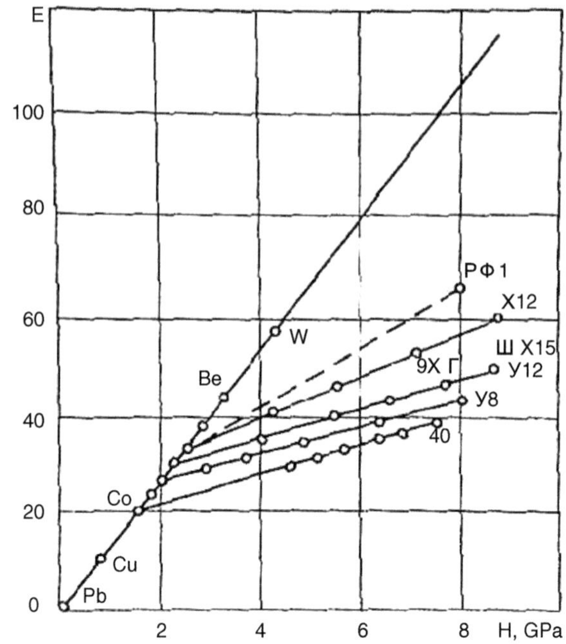

Fig. 4.4 Dependence of the relative wear resistance of pure metals and steels at abrasive wear (*E*) from their hardness (H) [5]

Dependence of the relative durability of their hardness for a number of commercially pure metals (Pb, Cu, Co, Be, W) and steels (РФ1, Х12, ШХ15, 9ХГ, У12, У8, 40) which correspond to steels like A11, D3, 52100, O2, W 112, W 108, 1040, is shown in Fig. 4.4.

• *Wear in the abrasive mass*

Wear resistance in this type of wear is directly proportional to the hardness of materials. Degree of filling of the mass with abrasive particles significantly influences on the value of wear. In each case, there is certain filling with the particles, when the wear material reaches a maximum. If we take the ability to wear out of clay soils for 1, then it will be 1.5 for sandy soils; for loamy soils 1.9; for sandy loam soils 2.3.

The presence of corrosive environments, such as wet soil in the presence of seawater or fresh water, accelerates the wear. As a result of redox reactions and tribochemical processes on the friction surface hydrogen is released, part of which diffuses into steel, intensifying the wear rate.

This type of wear is evident in the work tools of tillage, road and construction machines, excavator buckets, drags, etc.

• *Gas-abrasive and hydroabrasive wear* is caused by the abrasive solid particles entrained in a stream of gas or liquid. The latter provide loss of strength effect on the surface layer of the material. Gas-abrasive wear occurs in boilers, working on pulverized fuel, pneumatic transport parts; hydroabrasive—in details of pumps, hydraulic systems, valves, dampers, distributor valves, pipes, etc.

Wear by initially free abrasive particles is the result of microcutting of one surface by abrasive grains, which are embedded to another friction surface; or both surfaces are microcut by abrasive grains, which roll in the gap between them. Wear occurs due to fatigue failure under multi-cycle elastic or low-cycle plastic deformation by abrasive particles. Almost all machine components, which operate under the condition of hit by free abrasive particles the gaps between the mating parts, are subject to such kind of wear.

Evaluation of resistance to abrasive wear is produced at facilities that implement the various options of the Brinell—Hovart test scheme, Fig. 4.5.

However, this scheme does not adequately reflect the pattern of abrasive wear when impacting. The energy value of individual impact varies randomly within a certain range under realistic operating conditions. Furthermore, the mechanism of the loads action on the wear process may vary. For instance, a relatively small impact loads lead to hardening of the friction surface due to its hardening and embedding by abrasive particles.

High-level impacts cause the formation of cracks, chips, and other defects accelerating the process of abrasive wear. Complex effect, i.e., the alternation of hardening and softening processes can change the observed pattern. In tests of the coated specimens with the use of high-level impact can lead to premature coating failure that will distort the results. The coating can successfully operate under real

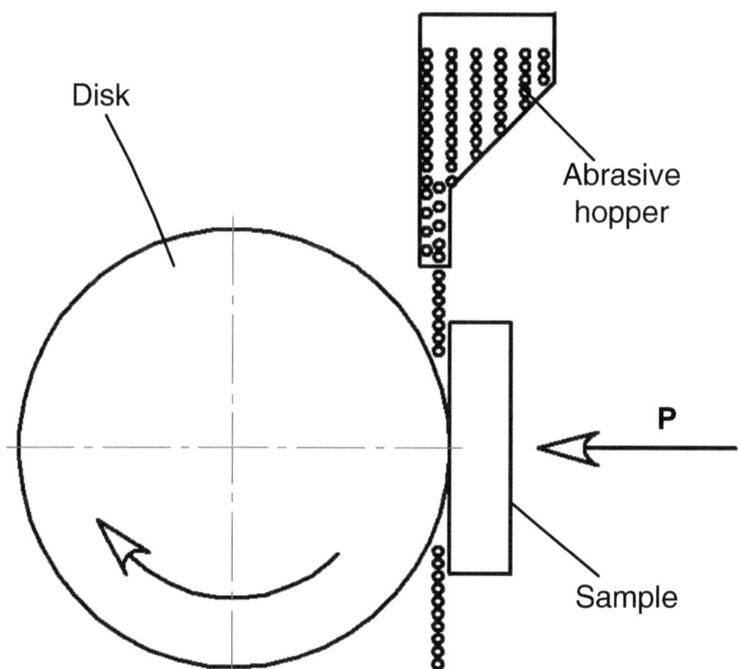

Fig. 4.5 Brinell—Hovart test scheme for abrasion testing

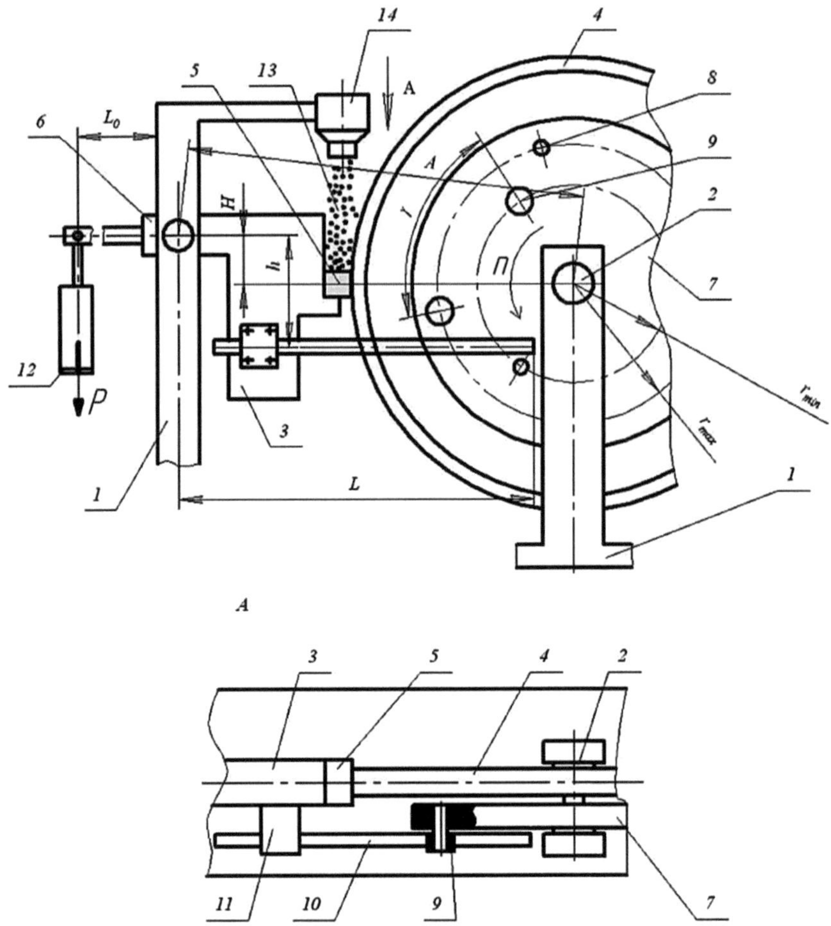

Fig. 4.6 Scheme of impact abrasive wear test [6]

load when the energy of individual impacts is variable, and it reaches the maximum value relatively rare.

To take into account, the features a scheme for testing the impact—abrasion was suggested (Fig. 4.6). The apparatus includes a frame 1, holders 2 and 3, according to contraspecimen 4 and specimen 5, double-arm lever 6 for their loading, rotation drive of contraspecimen (not shown). The device is provided with a disk template 7 with holes 8 installed coaxially the holder, where the fingers are fixed; sleeves 9, mounted on the fingers with the possibility of rotation; stop 10 attached to the node 6. Stop length is regulated by moving the platform 11. At one end of the double-arm lever 6 suspension with loads 12 is mounted, its other arm works as a holder 3. Abrasive 13 is fed to the friction zone of contraspecimen 4 and specimen 5 from the hopper 14.

Working length of the stop L is selected from the relation:

$$\sqrt{A^2 + r_{max}^2 - h^2} \geq L \geq \sqrt{A^2 - H^2} - \sqrt{r_{min}^2 - (H - h)^2}, \tag{4.3}$$

where r_{max}, r_{min} are the minimum and maximum radii of the circles where fingers through freely rotating sleeves are fixed; A is the distance between the axis of the holder 2 and the center of rotation of the lever 6; H is vertical dimension between the axis of the holder 2 and the center of rotation of the lever 6; h is vertical dimension between the center of rotation of the lever 6 and the stop 10.

The right part of this inequality distinguishes the minimum working length of the stop 10, which is provided by its contact with freely rotating sleeves 9, located at a radius r_{min}.

The left inequality is determined by the need to exit the stop 10 from contact with the sleeve 9 in such a position of the disk template that provides the maximum deviation of the stop 10 from the initial position. Violation of this condition will lead to the fact that the angle of deviation of the stop 10 will first increase to a maximum and then decrease by a certain amount and only after that a contact break occurs between the stop 10 and the sleeve 9. This will lead to a reduction in a given energy of individual impact and disruption of the test program.

Central angle (γ) between two consecutive fingers 8 located on the disk template is determined in accordance with the required value friction of path between every two blows with set impact energy. The minimum value of this angle must satisfy the condition:

$$\gamma_{min} > 2 \cdot \pi \cdot n \cdot \sqrt{2 \cdot L_0 \cdot \sin \beta / a} + \alpha, \tag{4.4}$$

where (α) is the central angle of the finger turn during its contact with the stop; (β) is the deflection angle of the right arm lever 6 from the horizontal during the contact between the stop 10 with a finger through the sleeve 9; (n) is rotary velocity of the disk template 7; (L_O) is the load lever arm of suspension with loads 12; a is the acceleration when returning the lever to its original position.

The first addendum on the right-hand side of condition (4.4) is the angle at the rotation of disk template 7 during the return of stop 10 to its original position after deflection from the horizontal by the previous finger at the angle (β). The initial position means the contact of contraspecimen 4 and specimen 5. The second addendum is an angle at which the disk template turns during the contact of the previous finger with the stop. Then their sum represents the angles of the disk template rotation during withdraw and return of the lever to its initial position after contacting it with the previous finger. Thus, if condition (4.4) is broken, then at sequent, the direction of rotation, fixing two fingers on the disk template it will not be impossible to ensure the return of the lever to its original position with the application of a shock force to the specimen. In this case, there will be premeeting of the next sleeve and the stop.

Using of independent rotary drive for contra specimen 4 and specimen 5 can extend the capabilities of the friction path control between single impacts. In this case, one rotation of the disk template, i.e., one cycle of impact stress change it would be possible to provide the required number of specimen cycles. We adopted the condition of no multiplicity, which provides constant displacement on the contraspecimen surface both the minimum and maximum impact loads, which leads to more uniform wear of specimen.

In evaluating the test results for the load value, the maximum specific load is taken [7]:

$$\sigma_{\max} = 0.542 \cdot \sqrt{P/(2 \cdot t \cdot \eta \cdot D)}, \quad (4.5)$$

where $\eta = (1 - \mu_1^2)/E_1 + (1 - \mu_2^2)/E_2$ is the elastic constant; μ_1, μ_2, E_1, E_2 are Poisson's ratio and modulus of elasticity for rubbing parts; t, D are thickness, diameter, m; P is pressure force of specimen, H.

By the proposed method, a flat specimen of the material under test is placed in contact with contraspecimen, is loaded, the abrasive is fed to the friction zone and the contraspecimen is getting rotational velocity. Single impact energy and the way of friction are changed cyclically according to a program; the duration of the cycle is not multiple to time of one rotation of the specimen. The preset sequence of the cyclic variation of the parameters allows estimating the value of the wear resistance of materials in the sequential action of the working surface of the impact load specimen with different levels of energy. This makes it possible to combine the application to the specimen as a hardening as destructive (cracked, chipped, pitting) loads. The difference between the cycle duration and rotation time of the specimen allows you to change the friction path for a program cycle within a wide range, which brings the performance test conditions closer to working conditions.

As contra specimen, the disk of steel 1020 with a diameter of 250 mm and a thickness of 5 mm was used. The diameter of disk template is 200 mm. Rotation velocity of specimen was 4 rev/min, of the disk template—0.7 rev/min. Fingers with free rotating sleeves were installed in the holes of disk template located on the concentric circles of radius 85, 65, and 48 mm. Central angles were chosen regarding the minimum angle (γ) according to Eq. (4.4).

According to the proposed scheme, the abrasion resistance (loss in weight) of various materials commonly coated in the spraying is compared [8]. Modes of coating spraying were the following: current 160 A; voltage 28 V; electrode diameter 1.6 mm, spraying distance 100 mm; compressed air pressure 0.3–0.4. MPa, pressure of propane 0.3–0.4 MPa (coefficient of oxidant excess in the carrier gas mixture α_{OK} = 0.8). Test modes are shown in the Table 4.1.

Table 4.1 Wear resistant testing modes

Regime number	Contact pressure (MPa)	Single impact energy (J)	Friction path (m)
1	233	0	30
2	129	72.8	30
3	233	133.6	30
4	233	133.6	122

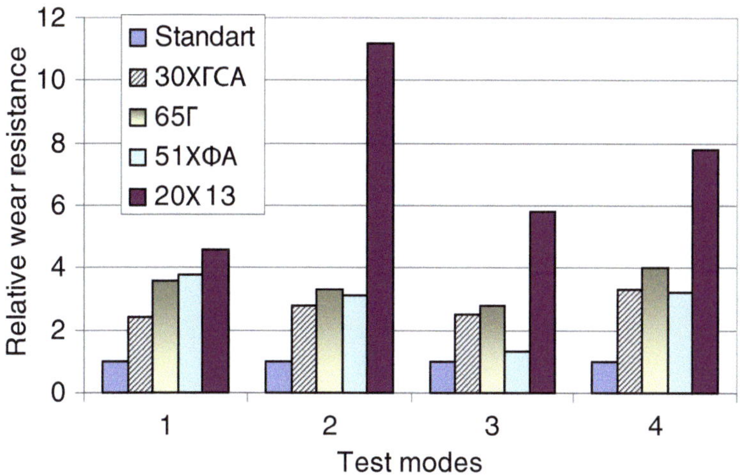

Fig. 4.7 Wear resistance of AAS coatings made of various materials with respect to the cast steel 20X (like steel 5120), load modes are in Table 4.1. Tested steels 30XΓCA, 65Γ, 51XΦA, 20X13 corresponds to steels like 30CrMnSi, 1566, 6150, 420

From the results, it is evident (Fig. 4.7) that the relative durability of the coatings tested is higher than that of the standard cast steel like steel 5120. When reducing the specific load (mode 2) and increasing the sliding path (mode 4) wear resistance of coating from high-chromium steel 420 becomes several times higher compared to other materials. Obviously, this is due to structural changes when impacting the substrate and cooling the coating material.

The difference in wear can be visually seen at comparison with standard specimens and steel coating of 20X13 after test completion (Fig. 4.8).

Fig. 4.8 Specimens after comparative shock-abrasive testing due to the scheme [6]. **a** The specimen with AAS-sprayed coating of 20X13; **b** specimen of standard steel 20X

The best results of the materials tested were obtained for the coating of steel 20X13. Its relative wear resistance is 2–5 times higher than that of other coating materials. This is ferritic-martensitic class steel; its hardness is higher than that of other materials considered. Furthermore, the coating of 20X13 represents tempered martensite, and its structural transformations will be accompanied by compressive stresses favorable for operational loads. This effect is obviously higher with the loads and friction path Increase.

The degree of oxidation of the metal drops can be changed by technological modes at AS. Oxide films forming during spraying play a dual role. On the one hand, it is one of the most solid constituents of the coating structure, its microhardness is 5–6 GPa (see Part 3). On the other hand, the oxides represent fragile part of the structure, which is characterized by low cohesion with other components. Therefore, for each type of coating there is interval of oxides content where these effects have optimal level.

An influence of degree of particle oxidation on the wear resistance was studied relating to coatings from steel 20X13. A specimen sprayed by AS gun EM-14 without the use of a reducing atmosphere, was adopted as the standard specimen. It was compared with specimens coated on various options by the gun ADM-10, which used combustion products of propane-air mixture as a carrier gas. It can be seen that the decrease in the oxidation degree of droplets, and as a result—decrease in coating oxidation degree lead to several times increase of the coating's wear resistance (Fig. 4.9). Additionally, spalling of individual particles took place in the coatings made with $\alpha_{OK} = 1.0–1.2$. While increasing the sliding path up to 122 m the spalling has become predominant in nature, which resulted in partial chipping of the coating for three specimens sprayed at $\alpha_{OK} = 1.2$ (in Fig. 4.9 not shown). Under the same spalling of individual particles of the coating sprayed when $\alpha_{OK} = 0.8$ was observed only on one specimen.

From a comparison of the data and the results of the structural studies on the influence of the carrier gas on the number and size of oxide's layers in the coating (see Chap. 3), it was concluded that there is a critical number of oxides in the

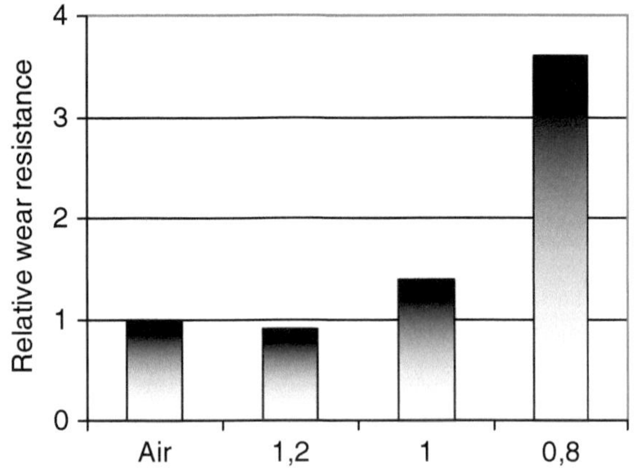

Fig. 4.9 Wear resistance of AS coatings of steel 20X13 with different ratios of air and propane in the carrier gas (average values for loads according to Table 4.1)

coating. It corresponds to the maximum of the wear curve for the specific material. Simulation of oxygen input to spraying particles (see Chap. 2) provides an estimating of process parameters interval, the implementation of which makes it possible to get the required amount of oxides.

At higher contact loads and impacts, the greatest danger is the wear of the internal friction mechanism. It is evidently for steel coatings, which we considered due to the emergence and development of microcracks in the oxide films, the amount of which increases with a decrease in reduction properties of the atomizing gas.

4.1.3 Adhesion and Fatigue Wear

There is the local connection of two solid bodies under the action of molecular forces when friction. This produces strong metallic bonds in the areas of direct contact surfaces.

Comparing the data and the results of metallographic researches, Chap. 3, it can be assumed that the increased wear resistance of coatings in comparison with cast metal is due to the various schemes of wear of cast steel and sprayed coating.

Under conditions of dry sliding friction of steel parts, the contact occurs on the oxide film. When it is destructing clean surfaces that are inclined to bonding, open on steel surfaces. In this case, the failure occurs at a weak spot in the depth of detail in the breakaway form (on the mechanism of internal friction). This greatly increases the wear as compared with the friction on the layer of boundary lubricant.

Due to the interaction of the metal drops with oxygen in flight and features of forming layer upon impact of droplets on the surface, the coatings have a lamellar structure: pure metal layer–oxide layer. After removal at wear the next lamella, the coating surface is opened, and there are alternate sections of pure metal and oxide films. In comparison with pure metal oxides, have lower friction coefficient and higher microhardness (Fig. 3.29, Table 3.6). At the same time, they have weak chemical bond with areas of pure metal of mating part. For these reasons, the wear according to the scheme of external friction caused by the interaction of the mated surface layers is most likely. Such wear mechanism is also favorable because oxides are arranged in relatively ductile matrix of pure metal layers. This contributes to more uniform load distribution. In [9] the use of oxides is considered as a solid lubricant in the case of totally ferrous oxide coatings applied to a pair of "piston–cylinder" in the car engine. In this case, oxides also act as a dry lubricant and reduce friction.

In tests on sliding friction (Scheme "pin-on-disk" with oil supply, contraspecimen bronze BpC-30, like bronze DIN CuPb30) friction coefficient and contact temperature reduce several times, wear resistance increases compared the arc sprayed coating with cast steel (Table 4.2). The possible causes are connected with the friction zone heating at wear: lubricant exudation from the pores [9]; formation of phases with high hardness and wear resistance in the surface layer [10].

Table 4.2 Tribological characteristics in contact with the sliding friction: friction coefficient (*f*), relative total wear (*Δ*), contact temperature ($T_к$) [11]

Specimen material	*f*	*Δ*	$T_к$ (°C)	Test conditions
Cored wire, the charge based on ferrochrome	0.018–0.023	1	25	Contact pressure 5 MPa, roughness R_a 0.63
Steel 1045, HRc 30–32	0.09	2.5	45	

The specific structure of the sprayed coatings and their porosity improves coatings performance capability in terms of sliding. Coatings impregnated with lubricant can operate for a long time without jamming after the oil supply termination. For instance, Babbitt plain bearing mated with sprayed steel coating was working after oil supply termination without jamming more than 190 h, while in hardened steel scores appeared after 2 h of working (Fig. 4.10).

For reverse pair of "Babbitt coating—steel solid substrate" our results [13, 14] show good tribological properties of AAS coatings in comparison with the alternatives—turbulent shell lining (SL), gas surfacing (GS), plasma spraying (PS). Methods of preparing the specimens are given above, in Sect. 3.5.

AAS Babbitt coatings have the friction coefficient at the workloads (pressure 1–6 MPa, the sliding speed 1.3–6.5 m/s) to 16–54 % lower and the rate of wear is 3.3 times lower than the SL and 32 % lower than PS (Fig. 4.11).

These differences in the tribological characteristics are due to the structural features previously described in Sect 3.5. The decrease in the friction coefficient at increasing pressure may be due to the changes in the ratio of hard intermetallic compounds and the substrate playing the role of a soft interlayer. This conclusion is confirmed by the results of researches on the compatibility of the rubbing surfaces [15]. The basis of a good adaptability of the rubbing surfaces to each other in the relative movement is favorable response to friction increase. In heterogeneous alloys, it is appeared by increase in the amount of metal transferred to the mating

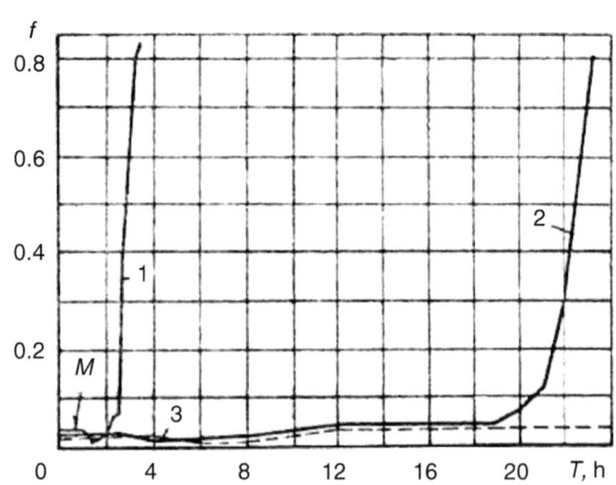

Fig. 4.10 Change in the coefficient of friction (*f*) in pair "Babbitt bearing—steel coating" after the oil supply termination [12]: *1*—hardened steel + oil И12А; *2, 3*—steel sprayed coating + oil И12А; 2 oil supply at the moment (*M*) is terminated

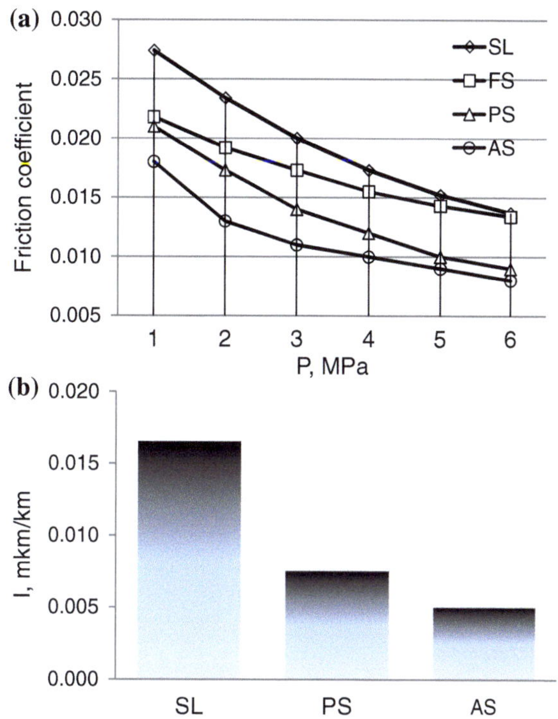

Fig. 4.11 Tribological characteristics of alternative Babbitt B83 layers, **a** friction coefficient (*f*); **b** wear rate (*I*)

surface from the soft phase due to deformation of the centers of light shift on the contact areas.

Crystals of soft α-phase with high antifriction properties serve in this case as a solid lubricant and prevent the accumulation of damage on the friction surface.

Lower friction coefficient and wear rate in sprayed coating compared with shell lining and gas surfacing are probably due to the smaller size and greater uniformity of intermetallic phases which are the most hard components of Babbitt coatings [16]. Sprayed coatings also have high oil ability of the porous acting layer which leads to less wear when insufficient lubrication [9]. Increased wear at PS compared to AAS is probably caused by less fatigue strength of PS coatings due to higher thermal stresses because of excessive heat input to the spray material in comparison with AAS.

Wear resistance of coating strongly depends on its mechanical properties: α—linear thermal expansion coefficient, *E*—Young's modulus; *G*—shear modulus; μ—Poisson's ratio; $\sigma_{cц}$—adhesion strength; σ_{-1}—endurance limit of coated steel; σ_{B}—temporary tensile strength, and others. Borisova et al. gives the experimental data on the mechanical properties of AS coatings of ferrochrome and ferrochrome—aluminum obtained by arc spraying from flux-cored wires, Table 4.3.

Shamko and Kovrigo give results of determining mechanical properties of sprayed coatings Tables 4.4 and 4.5.

Table 4.3 Mechanical properties of coatings [17]

Coating material	σ_{cu} (MPa)	*E* (GPa)	*G* (GPa)	μ	α ($\times 10^{-6}$, K^{-1})	Porosity (%)	σ_{-1} (MPa)
Fe–Cr	18.0 ± 1.6	72	29	0.23	10.8	7.0 ± 1.8	215
Fe–Cr–Al	20.4 ± 1.4	78	32	0.22	10.9	4.0 ± 0.7	220

Table 4.4 Friction coefficients and linear thermal expansion coefficient (α), 1/°C, $\times 10^{-6}$ for AS coating [18]

Coating material GOST/analog	The coefficients of friction[a] under a specific pressure (MPa)			Coefficient of friction at move starting[b]	α
	5	8	10		
08/ASM 1008	0.29	0.32	0.35	0.398	12.9
08Г2С/ER70S-6	–	–	–	0.337	–
30ХГСА/CSN 14331	0.24	0.27	0.31	0.369	11.5
40Х13/DIN X39Cr13	–	–	–	0.404	–
Х20Н80/Nichrome	–	–	–	0.287	–
АК5/ER 4043	–	–	–	0.289	–
Ц1/Zinc	–	–	–	0.258	–
12Х18Н9Т/ASTM 321	0.24	0.28	0.31	0.293	15.4
У8/ASTM 1080	0.30	0.32	0.35	–	11.25
20Х13/ASM 420	0.18	0.21	0.24	–	–
08Х18Н10/AISI 304	0.29	0.32	0.35	–	–
30ХГСА+08Х18Н10	0.30	0.31	0.35		–
БСМГ (25Г+М1)/(ASM 1026+Cu)	0.24	0.27	0.30	–	–
08+АК5	–	–	–	0.374	–
08+Х20Н80	–	–	–	0.385	–
08Г2С+АК5	–	–	–	0.314	–
Х20Н80+АК5	–	–	–	0.285	–
Х20Н80+Ц1	–	–	–	0.272	–

[a]Counterbody-surface asbestos cardboard on Bakelite bond, the relative sliding velocity of 1.6 m/min
[b]Counterbody the surface of hardened steel ШХ (ASTM 52100)

Petrov's data show that the friction coefficient of AS coatings sprayed by supersonic jet of combustion products as carrier gas is lower than that of plasma and typical AS coatings sprayed by air from similar initial core wires, Table 4.6.

This is probably connected with differences in the conditions of the plasma coating formation and the difference in the velocity and oxidation ability of the carrier gas in comparison with the standard AS.

Table 4.5 Mechanical characteristics of AS coating [18]

Coating material	$\sigma_{в}$ (MPa)		E (compression) (GPa), under load (MPa)		
	Fracture	Compression	20	40	80
У8/ASTM 1080	79.2	923	100	80	40
12X18H9T/ASTM 321	105.8	890.6	4.8	6.7	9.1
30XГCA/CSN 14331	101.0	865.0	11.8	20.0	26.7
08/ASM 1008	148.0	741.8	–	–	20.1

Coating material	*E* (tensile) (GPa), under load (MPa)			μ (under load) (MPa)		
	20	40	80	5	10	20
У8/ASTM 1080	80	80	20.8	–	–	–
12X18H9T/ASTM 321	4.0	3.8	3.4	0.02	0.05	0.14
30XГCA/CSN 14331	11.1	5.0	4.9	0.15	0.19	0.21
08/ASM 1008	–	84	–	0.09	0.12	–

Table 4.6 Anti-friction properties of the coatings from FeCrAl type core wires [19][a]

Coating method	Supersonic AS	Plasma spraying	Typical AS
Friction coefficient	0.03	0.05	0.15

[a]Roller—Block Test scheme, load 5–10 MPa

Described sliding conditions are close to reality for a wide range of parts. These are different journals of shaft and axles, bearing housings in body parts. This makes it possible to use these results and similar studies in relation to the operating conditions. AS allows you to obtain reliable coatings for the seating surfaces of rolling and plain bearings, flat sliding surfaces, bimetallic "steel–bronze", "steel–Babbitt" parts (Figs. 4.12, 4.13 and 4.14).

4.1.4 Erosive Wear

Under erosion a material surface damage is meant under the action of electric discharges or by the mechanical action of a high-velocity flow of liquid, gas, steam. Erosive effect in the latter case is composed of a continuous flow friction and its impacts on the surface. This kind of friction differs from sliding and rolling in the momentum of the applied force. In each case, there may be great contact loads, but their nature, distribution, and the variation over time are different, Fig. 4.15.

Specimens' tests may vary in the method of particle flow acceleration. The main options that implement the acceleration under effect of gravity or gas flow (Fig. 4.16) are based on the DIN 50332. In these tests, abrasive parameters may

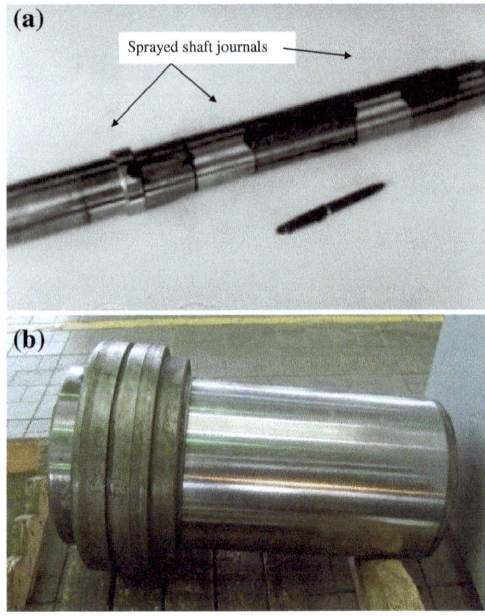

Fig. 4.12 Steel coating. **a** Transmission shaft of the nail machine drive, ball bearings are set on the sprayed support necks; **b** Piston plunger of press hydraulic cylinder, cylindrical sliding surface was restored to the length of 1090 mm

vary (flow velocity, angle of attack (α), fraction, hardness), and the temperature of the gas flow.

Comparison of erosion resistance of coatings made by methods of typical arc spraying (AS) and activated arc spraying (AAS) was held by Baranovsky et al. [20, 21].

Gun TAFA 8830 (Praxair-Tafa, USA) was used for AS, for AAS—gun HVAF-Arc (Unique Coat Technologies LLC, USA), where combustion products of propane-air mixture are utilized as the atomizing gas. As sprayed materials wires ∅1.58 mm from alloys (Fe-27Cr-3.5B-2C) and (Fe-29Cr-Ti-W-3.5BC) are used. The coating thickness on specimens is of 0.5 mm.

Tests for erosion resistance were performed by coatings blowing at elevated temperatures with pulverized coal from a boiler of a thermal power plant due to the blowing scheme by high-temperature gas stream [22]. As the pulverized coal abrasive particles based on silica with a small alumina content and sulfur and chlorine inclusions were used. The characteristics of the abrasive powder are the following: density 2700 kg/m^3, the average fraction of 0.3 mm, average powder microhardness of 780 HV_{50}. Test Modes: particle velocity—60 m/s; temperature—300 °C; angle of attack—30°; test time—5 h; load—375 g. Criterion of erosion resistance was the loss of the specimen's layer thickness. Test procedure of erosion resistance simulates the conditions of the inner surface of the distillation column for petroleum refining.

The microhardness of the coatings for HVAF-Arc and TAFA 8830 is similar, but macrohardness measured by the method of ball is higher at HVAF-Arc.

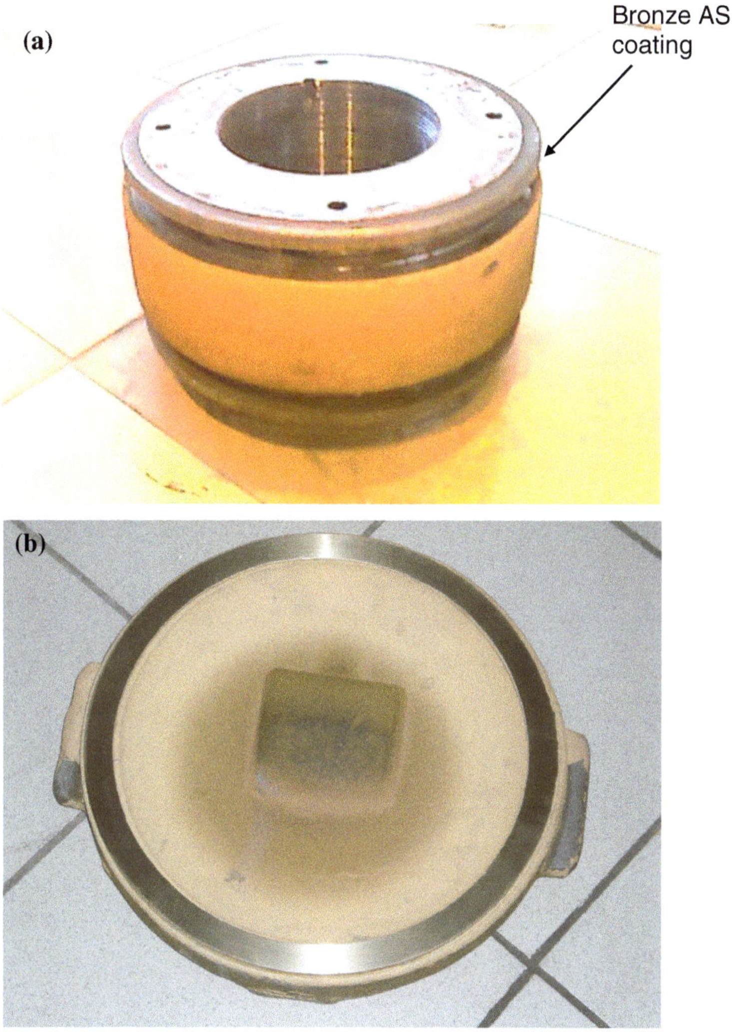

Fig. 4.13 Bronze coating. **a** Piston of hydraulic cylinder, bimetallic "steel—bronze", working pressure of 15 MPa, bronze coating on the cylindrical surface of sliding thickness of 5 mm; **b** Valve ∅300 mm for circular sliding surface, bronze coating with a thickness of 3 mm

Fig. 4.14 Babbitt coating of 5 mm thickness on a steel substrate, the processing allowance of 0.5 mm

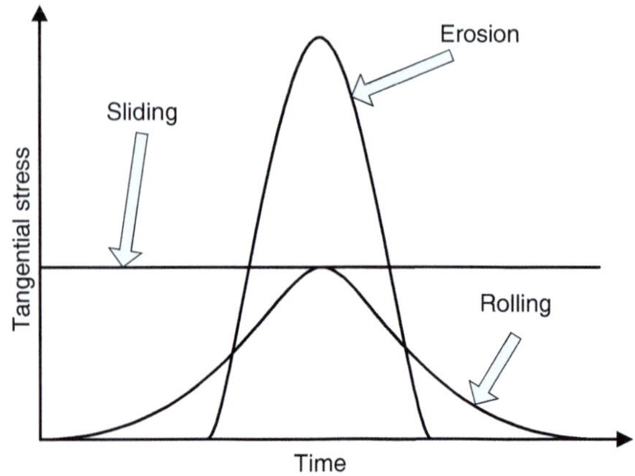

Fig. 4.15 Changes in tangential stresses over time for different types of friction

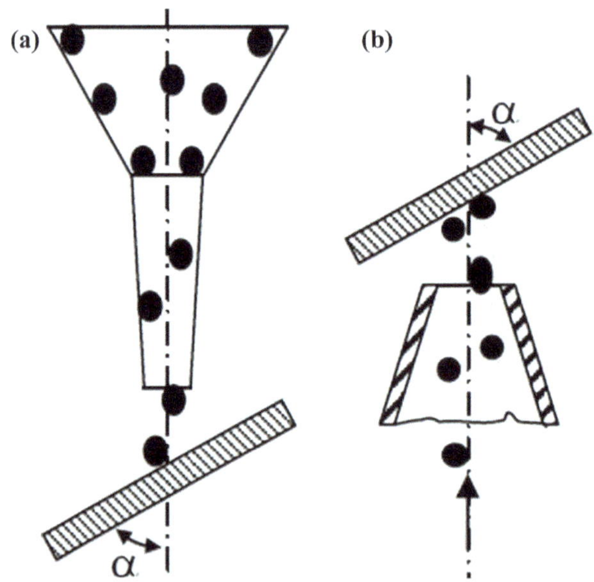

Fig. 4.16 Schemes of erosion resistance test in which the abrasive particles are accelerated. **a** Under the action of gravity; **b** By gas flow

For AAS, the adhesion strength is higher than 60 MPa, fine-grained structure of the coating has a porosity of less than 2 % and no visible microcracks, Fig. 4.17.

The tests have shown that the resistance of the AAS both coatings against high-temperature erosion is higher, respectively, 1.7 and three times (Table 4.7) compared to AS coatings. Thus, performance and processability of the processes are similar.

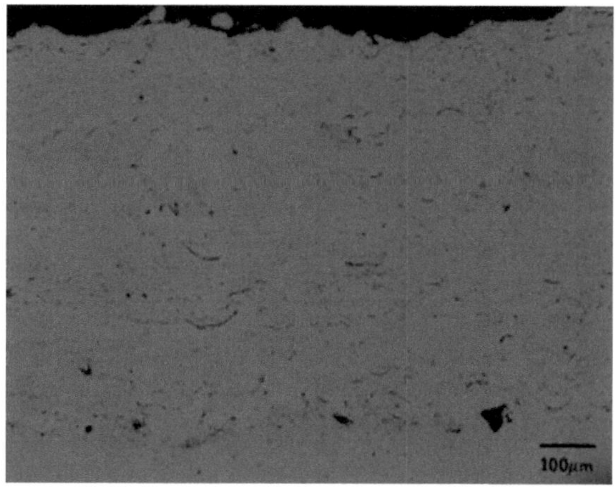

Fig. 4.17 Microstructure of AAS coating of wire Fe-27Cr-3.5B-2C

Table 4.7 Comparison hardness and erosion resistance of coatings

Feedback	Gun	Microhardness (HV_{300})	Macrohardness (HRc)	The loss of thickness from erosion (μm)
Fe-27Cr-3.5B-2C	TAFA 8830	740 ± 90	51.5 ± 1.5	116
	HVAF-Arc	750 ± 100	58.5 ± 2.0	70
Fe-29Cr-Ti-W-3.5B-C	TAFA 8830	755 ± 85	52.6 ± 2.2	161
	HVAF-Arc	785 ± 90	55.0 ± 1.8	53

4.2 Corrosion Resistance of Coatings

4.2.1 The Concept of Corrosion

The corrosion is determined as spontaneous destruction of metals and alloys because of chemical or electrochemical interaction with the environment.

Corrosion, depending on the nature of the metal, aggressiveness of the environment and other factors, leads to various kinds of damage. There are general, or continuous, and local corrosion (Fig. 4.18).

General corrosion covers the entire surface of the metal which is under the influence of aggressive environment, and can be uniform (Fig. 4.18a), i.e., corrosion of carbon steel in the sulfuric acid, and non-uniform (Fig. 4.18b), i.e., corrosion the same steel in the presence of chlorine ions.

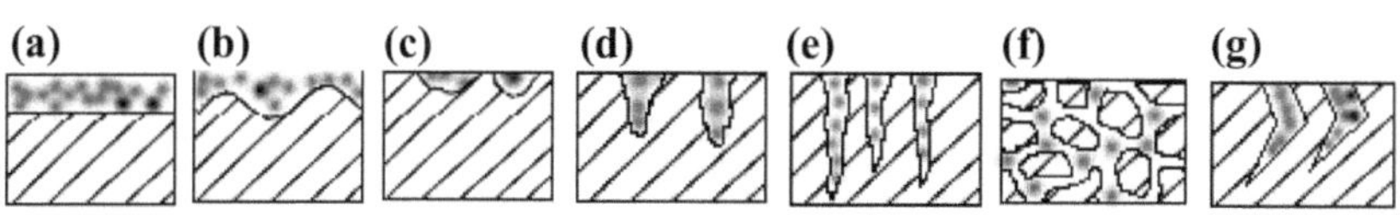

Fig. 4.18 The nature of corrosion damage [23]

Local corrosion covers individual portions of the metal surface. It is subdivided into following types of corrosion:

- spots (Fig. 4.18c), which occupy relatively large surface areas and are not distributed deeply;
- ulcers (Fig. 4.18d), which affect the metal to a greater depth and in a limited area;
- pointed on (Fig. 4.18e), which affects the metal at certain points to a greater depth, in some cases through;
- intergranular (Fig. 4.18f) is corrosion on the edges of the crystallites of metal; distributed over a large surface and at great depth;
- cracking (Fig. 4.18g)—metal corrosion simultaneously exposed to metal of corrosion environment and mechanical stresses, the rate of metal corrosion increases rapidly.

This classification is conventional because the real forms of corrosion damage may occur between characteristic types.

Let us consider the features of chemical and electrochemical corrosion which are important in the case of the use of protective TS coatings.

4.2.2 Chemical Corrosion

Chemical corrosion mechanism is associated with two coupled reactions on the phase interface—metal oxidation and oxidant reduction, and in space these processes are generally not separated. The accumulation of oxidation reaction products takes place also here. To carry out the reaction continuously metal atoms or ions on one hand and the atoms or ions of oxygen (or other oxidant) on the other are continuously diffused through the gradually thickening film of corrosion products. Depending on the properties of the resulting products there may be inhibition of the oxidation process.

Under the terms of the flow this type includes:

- High temperature or gas corrosion, i.e., oxidation of the metal in a gaseous medium at high temperature, typically above 500 °C, and the impossibilities of moisture condensation on the metal surface. Jet nozzles, combustion chambers, gas turbine blades, boilers, heating plants, incinerators are exposed to this type of corrosion. Also, these processes take place in fact at all stages of metallurgical processing, when heated metal is in contact with the atmosphere.
- Corrosion in non-electrolytes, i.e., liquids which do not conduct electric current, for instance, in petroleum products.

Let us consider the features of chemical corrosion on the example of widespread gas corrosion [24, 25]. TS coatings are successfully applied for the protection from gas corrosion. This type of corrosion has the following features.

Under the conditions of oxidative external environment, there is a high probability of metals oxidation. The most affordable oxidant is oxygen. The oxidation process refers to heterogeneous one with gaseous and solid phases. Even in the simplest case, when the metal forms only one oxide, the oxidation process comprises the following steps:

(1) oxygen diffusion to the surface of the oxide film;
(2) oxygen transfer through the interface gas—oxide (chemical adsorption act);
(3) metal diffusion in the oxide layer to the surface oxide—dioxide;
(4) oxygen diffusion in the oxide layer to the metal surface;
(5) crystal chemical interaction of the metal with oxygen and formation of an oxide on the inner and outer scale surface.

The initial oxidation conditions can be represented by the relation

$$\pi_O^g > \pi_O^s$$

where $\pi_O^g > \pi_O^s$ are oxygen potentials of gas phase and oxide.

The reaction product (oxide) is limited soluble in the metal and stands out in a separate phase dividing metal and gas. The formation of the oxide film changes the nature of the oxidation process due to the influence of the oxidant transfer through the oxide phase. Oxygen transfer to the metal surface can be carried out by diffusion through the oxide, or through pores and cracks therein. The thick film of scale without tearing can be formed only in the case when the oxidizing metal occupies a smaller volume than the forming scale. Otherwise cracks and fractures are formed as a result of low ductility of solid oxides. Evaluation of the continuity of the resulting film is produced by the ratio (Table 4.8):

$$V_{OK}/V_M = M_{OK} \cdot \rho_M/(m \cdot \rho_{OK} \cdot A_M) \quad (4.6)$$

where V_{OK}, V_M are molecular volumes of the oxide and metal spent on oxide formation; M_{OK} is the molecular weight of the oxide; A_M—atomic weight of the metal; ρ_M, ρ_{OK} are densities of the metal and oxide; m—number of metal atoms in the molecule of oxide.

The simplest analytical expression for the rate of metal oxidation is obtained by the following restrictions:

(1) isothermal process;
(2) the metal surface is flat and large enough;
(3) the concentration of the oxidant (oxygen) in the gas phase is constant;
(4) only oxygen particles transfer occurs on the oxide film,
(5) the reaction takes place at the interface metal-oxide.

If the rate of oxidation present as increase of the oxide film thickness per unit time per unit area, then at the kinetic response mode:

$$(d\Delta y/dt)_k = Kxp'_{O_2}. \quad (4.7)$$

Table 4.8 Ratio of metal and oxide molecular volumes

Metal	Li	Na	K	Mg	Ca	Sr
Oxide	Li_2O	Na_2O	K_2O	MgO	CaO	SrO
V_{OK}/V_M	0.57	0.59	0.48	0.79	0.63	0.66
Metal	Ba	Cu	Cu	Ag	Be	Zn
Oxide	BaO	Cu_2O	CuO	Ag_2O	BeO	ZnO
V_{OK}/V_M	0.73	1.67	1.74	1.58	1.67	1.58
Metal	Cd	U	U	Al	Ti	Zr
Oxide	CdO	UO_2	U_3O_8	Al_2O_3	TiO_2	ZrO_2
V_{OK}/V_M	1.27	1.96	3.12	1.31	1.76	1.60
Metal	Sn	Pb	Nb	Nb	Ta	Cr
Oxide	SnO_2	PbO	NbO	Nb_2O_5	Ta_2O_5	Cr_2O_3
V_{OK}/V_M	1.33	1.15	1.57	2.81	2.32	2.02
Metal	Mo	Mo	W	W	Fe	Fe
Oxide	MoO_2	MoO_3	WO_2	WO_3	FeO	Fe_3O_4
V_{OK}/V_M	2.18	3.45	1.86	3.36	1.77	2.09
Metal	Fe	Co	Co	Co	Ni	Pt
Oxide	Fe_2O_3	CoO	Co_3O_4	Co_2O_5	NiO	PtO
V_{OK}/V_M	2.14	1.75	2.00	2.42	1.52	1.56

And at the diffusion response mode:

$$(\mathrm{d}\Delta y/\mathrm{d}\tau)_d = D(p_{O_2} - p'_{O_2})/\Delta y, \tag{4.8}$$

where p'_{O_2}, p_{O_2} are partial pressure of oxygen on the surface metal-oxide and in the gaseous phase, respectively; Δy is thickness of the oxide film (Fig. 4.19); k is rate constant of the crystal chemical conversion; D is the diffusion coefficient of oxygen.

Considering that in the steady state

$$(\mathrm{d}\Delta y/\mathrm{d}\tau)_k = (\mathrm{d}\Delta y/\mathrm{d}\tau)_\partial, \tag{4.9}$$

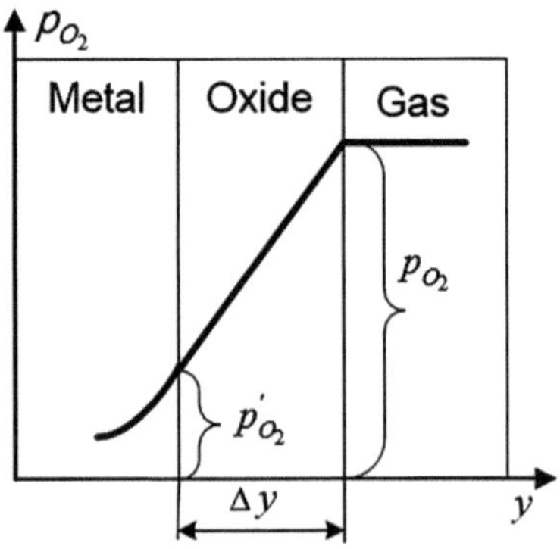

Fig. 4.19 Scheme of metal oxidation

from the relations (4.7) and (4.8) can be obtained

$$p'_{O_2} = p_{O_2} \cdot D/(k \cdot \Delta y + D). \tag{4.10}$$

Thus, the rate of oxidation is:

$$\mathrm{d}\Delta y/\mathrm{d}\tau = k \cdot p_{O_2} \cdot D\Big/(k \cdot \Delta y + D) \tag{4.11}$$

Integration with the boundary conditions Δy = 0 for τ = 0 gives:

$$\left(\Delta y/k + \Delta y^2/2D\right)/p_{O_2} = \tau \tag{4.12}$$

For most metals forming a complete and a thick oxide film ($\Delta y \rightarrow$ max) the initial period of the oxidation in the kinetic mode is very small, particularly at high temperatures, the role of the quadratic term of Eq. (4.12) can be decisive. In this case, the characteristic for the diffusion mode parabolic law of the oxide film thickness growth is true:

$$\Delta y^2 = 2D \cdot p_{O_2} \cdot \tau. \tag{4.13}$$

In the case of a thin oxide film ($\Delta y \rightarrow 0$, $\Delta y^2/2D \rightarrow 0$) the process rate is limited by the kinetic interaction (k-value is small). In this case, the film growth rate is subject to a linear law:

$$\Delta y = k \cdot p_{O_2} \cdot \tau \tag{4.14}$$

Such dependence is typical for an initial period of oxidation and for metals forming discontinuous films and having the relation (V_{OK}/V_M) < 1 (alkali and alkaline earth metals).

During the low-temperature oxidation (thin films, Δy = 10^{-9}–10^{-8} m), film growth process is mainly in the diffusion mode. However, there are significant deviations from the parabolic law (Eq. 4.13) since the diffusion phenomena in the oxide phase are complicated by the influence of the double electric layer at the metal-oxide film, tunnel effect and other features. At high temperatures (higher oxidation rate), there is also a deviation from the parabolic growth law of film thickness. In this case, due to the large internal stress the cracking of oxide layer occurs (abrupt decrease in the effective value Δy). It leads to a temporary acceleration of the process that affects the appearance of kinks in the curve of oxide layer growth, Fig. 4.20.

Quantitative calculation of the oxidation rates in the simplest cases is performed on the basis of the electron-ion theory of high-temperature oxidation proposed by K. Wagner.

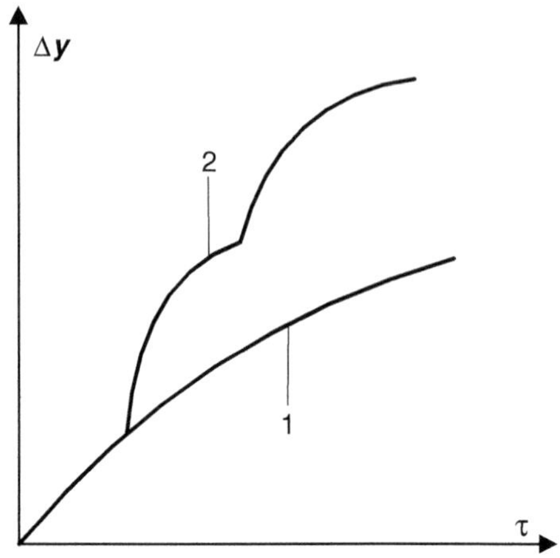

Fig. 4.20 Changes in the thickness of the oxide layer (Δy) over time (τ) in the diffusion regime of oxidation without cracks (*1*) and while cracking (*2*)

Oxidation of the metal Me in the diffusion mode can be represented by reaction

$$Me + 0.5O_2 = MeO \tag{4.15}$$

Ion crystal lattice MeO has a significant number of defects (vacant lattice sites, interstitial ions, etc.) at high temperatures. These defects facilitate the movement of particles and give rise to the electrical conductivity of ionic (χ_i) of solid oxides. In addition, they have ionic and electronic conductivity (χ_e). On the surfaces of oxide-metal and oxide-gas the processes of atoms oxidation Me to Me^{2+} ions and of recovery of O_2 molecules to O^{2-}

$$Me = Me^{2+} + 2e \tag{4.16}$$

$$0.5O_2 + 2e = O^{2-} \tag{4.17}$$

The transfer of electrons and ions through the oxide film may be identified with the flow of electric current in the galvanic cell of the chemical type. An oxide layer is here both the electrolyte with mobile ions Me^{2+} and O^{2-} and the external circuit with the electronic conductivity (Fig. 4.21).

Using Faraday's Law, the equation that describes the oxidation rate (v), i.e., the amount of metal oxidizing from the surface unit per time unit, moles/(sm^2 s) is derived from this scheme:

$$v = \frac{-\Delta G}{4 \cdot F^2 \cdot \Delta y} \cdot \frac{1}{\mu_{\Delta y} - \mu_0} \cdot \int\limits_{\mu_0}^{\mu_{\Delta y}} \chi \cdot t_i \cdot (1 - t_i) \mathrm{d}\mu \tag{4.18}$$

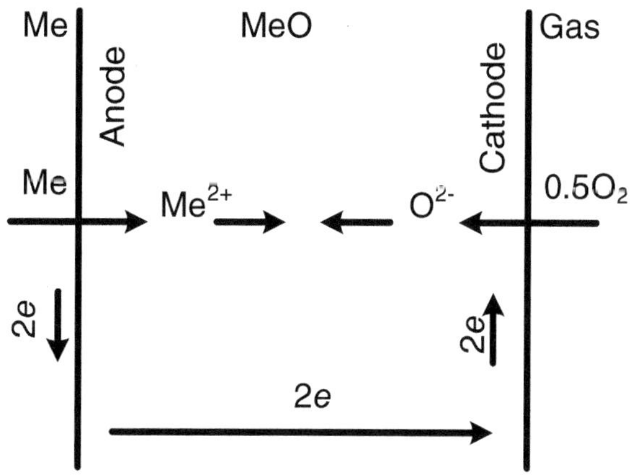

Fig. 4.21 Electrochemical scheme of metal oxidation by oxygen

where ΔG is Gibbs free energy: F is Faraday constant; μ is varying with the distance the chemical potential of oxygen in the oxide phase, the indices correspond to "*0*"—the surface, "Δy"—the withdrawal on the thickness of the film;

$\chi = (\chi_i + \chi_e)$ − total electrical conductivity; $t_i = \chi_i / \chi_e$ − the share of ion conductivity of the oxide layer.

By the nature of the dependence, v from Δy the expression (4.18) coincides with the previously derived Eq. (4.11) subject to the diffusion regime of oxidation ($k \gg D/\Delta y$), but they contain only those factors that can be calculated (ΔG) or found out from independent experiments (χ, t_i).

Equation (4.18) gives results in good agreement with the experimental data for simple systems, for instance, Cu|Cu_2O. They are also let to elucidate the influence of impurities on the process velocity.

Given oxidation laws are complicated by the structural features of the oxide films and the conditions of their formation. For the metals forming the cations of mixed valence stratified structure of the film is possible. Each layer may differ not only in the type of chemical compound but also in physical chemical properties.

Let us consider these features on the example of **iron oxidation**.

Interaction of solid iron with oxygen is a complex multistep process. At $T > 970$ K in a few minutes after the start of the process the parabolic law of changes the oxide film thickness is performed, i.e., diffusion regime of oxidation is realized. With time on the growth curve of the oxide layer thickness, there are fractures associated with the formation of cracks, (Fig. 4.20). In the early stages, with a small thickness of the oxide film and the possible crystal chemical transformations at interfaces kinetic inhibition cannot be excluded.

At lower temperatures, the variation of the oxide film thickness deviates from the parabolic (logarithmic, root, etc.). The reasons for deviation from the above mentioned theoretical positions are in multi-phase reaction products, in difference of physical and chemical properties of thin oxide films from the properties in the massive specimens, in the time variation of the porosity of the individual oxide layers, in scale cracking.

At low temperatures ($T < 845$ K), structural components of scale are the oxides Fe_2O_3 and Fe_3O_4. Thin primary film (~ 2 nm) occurring in air atmosphere at room temperature, consists of oxide γ-Fe_2O_3, having a kindred iron crystal lattice which facilitates the crystal chemical conversion of the metallic phase into the oxide. When heated to 420–520 K, it becomes thicker and is rebuilt from the outside to the lattice α-Fe_2O_3 (hematite). With increasing temperature up to 570–670 K at the iron boundary the magnetite layer Fe_3O_4 is formed. Further heating or long exposure lead to thickening of the scale, consisting of two layers (magnetite and hemtite), between them the finest intermediate zone γ-Fe_2O_3 facilitating heterogeneous process at the phase boundaries is saved.

Temperatures above 845 K are an area of stability FeO (wustite). If the partial pressure of the gaseous phase is sufficient to flow in the opposite direction of the reaction

$$6\ Fe_2O_3 = \ 4\ Fe_3O_4 + \ O_2, \Delta G^0 = 498.9 - 0.281T, \qquad (4.19)$$

So, the scale is composed of three layers: external-hematite, medium-magnetite, and the nearest to the iron-wustite (Table 4.9). Between hematite and magnetite, there may be intermediate zone γ-Fe_2O_3. In wustite, magnetite, and γ-Fe_2O_3 the diffusion of iron dominates and in magnetite—the diffusion of oxygen.

In particular concentration zone of the alloying component its oxide can form chemical compounds with the oxide of the base metal. In this case, the alloying is effective when the compound has better protective properties compared with the basic metal oxide. In the oxidation of iron protective properties of scale are greatly reduced when a wustite phase. Magnetite having a dense not much defective lattice of spinel type relatively well protects the metal from oxidation. Wustite, constituting most of the scale at high temperatures, has the weakest barrier properties due to its highly disordered structure. Therefore, when it appears the growth rate of the oxide film increases. Nature and concentration of alloying elements in steel have a significant impact on the stability of wustite phase in the scale in the range of 800–1200 K. According to the formulated by V.I. Arkharov principle of heat resistance, to reduce the tendency of steel to high-temperature oxidation the number and composition of the alloyed steel should be such as to prevent the emergence of wustite phase. Oxides of iron and alloying element should form a dense lattice of

Table 4.9 Content of iron oxide film due to the thickness, %

Temperature (K)	Oxide type[a]		
	FeO (wustite)	Fe_3O_4 (magnetite)	Fe_2O_3 (hematite)
873	50	50	–
1073	96	2	2
1273	95	4	1
1363	94	5	1

[a]Depending on the temperature and oxygen partial pressure the oxides deviate from the stoichiometric compounds

the spinel type with the lowest possible parameters. One such component is chromium. Its oxide forms with FeO a compound $FeCr_2O_4$, which lattice is little permeable to oxygen and iron. Along with this, the appearance of the chromium atoms in the lattice of hematite (outer layer scale) hardens it enhancing the protective properties of Fe_2O_3.

Heterogeneity and multi-phase scale in multicomponent systems do not allow, in general, to calculate the relative oxidation rates and to predict the composition of the resulting scale.

The question of influence of alloying component nature on the protective properties of the oxide layer is studied better. If the oxide of the metal introduced in a small amount is not released into a separate phase but forms a solution in the base metal oxide, its effect on the oxidation rate, it is usually possible to clarify using Wagner's theory of high-temperature oxidation.

For example, the oxidation rate of nickel to NiO at 1273 K is significantly increased with addition of small amounts of chromium. It is known that for NiO $\chi_e \gg \chi_i$, i.e. $t_i \ll 1$. Consequently, according to Eq. (4.18) the velocity is proportional to the ionic conductivity which is substantially increased by replacing of the part of the Ni^{2+} ions to Cr^{3+} in the cation sublattice of the oxide. The reason for this growth is the following. The crystal chemical formula NiO like wustite has the form $Ni^{2+}_{1-3y}Ni^{3+}_{2y}\square_y O^{2-}$. Since chromium ions have a higher valence, then replacing Ni^{2+} to Cr^{3+} electro neutrality of the phase is provided by an increase in the concentration of cation vacant lattice sites which causes an increase in the mobility of cations (growth of χ_i).

In some cases, the condensed phase oxidation is formed with a low electronic conductivity ($t_i \rightarrow 1$). Then, by Eq. (4.18) $v \sim \chi_e$ and the influence of small additions to the process velocity is explained by their influence on the electronic conductivity of the reaction product. If the alloying component of the alloy has a great chemical affinity for the oxidant, that when introduced the nature of the protective film can vary greatly. Thus, while oxidation of the alloy Cu–Be, containing <1 % Be, in an oxygen atmosphere, the protective film is formed consisting of almost pure oxide BeO. To make such way of metal protecting quite effective, high diffusion mobility of atoms, of the alloying element in the metal phase are required. Furthermore, its oxide should meet the condition of continuity and have low electrical conductivity (low mobility of ions and electrons of the conduction).

An example of protection against high-temperature corrosion is the application of AS coatings on the elements of the thermal power plants.

Such elements of the stations as boiler tubes, boiler covers are exposed to wear more intensively because of the impact of high-temperature corrosion. On a typical thermal plant, the surface area of these products (1000–3000) m^2 is subject to corrosion rate of 6 mm in thickness per year. It requires annual maintenance work at the station with cost about 50 thousand dollars. In addition, the daily loss of repair downtime up to 500 thousand dollars [26]. The way of solving this problem must satisfy the requirements of a large life protected elements, small-time repairs and reasonable price. Common methods of corrosion protection solve the problem

Table 4.10 Comparative effectiveness of various repair options of boilers elements [26]

Method of protection	Resistance (years)	Time of repair (days)	Relative costs
1. The use of flanged pipe connections instead of welded	3—4	20–30	3–4
2. Plasma spraying	3–4	15–20	3–4
3. Arc spraying	5–7	6–8	1
4. Chrome plating	3–12	30–40	5–6
5. The use of two-layer pipes (steel/stainless steel)	3–20	30–40	6–8

unsatisfactory. Widely used welding on of patches on the damaged areas leads to a sharp rise in the cost of repairs. The use of heat-resistant steel in large volumes is too expensive; the cost per ton of metal constructions is 15–20 times higher than in the manufacture of ferrous metal. Electroplating is non technological because of the large sizes of metal and the lack of stability in the environment of a working zone of thermal stations. As can be seen, among the ways to increase the resistance of the elements of boilers arc spraying is an effective alternative, Table 4.10.

Baranowski and coworkers [20, 21] carried out the relative resistance test of sprayed coatings for high-temperature corrosion. Spraying was performed using the following equipment:

1. The gun for activated arc spraying HVAF-Arc (Unique Coat Technologies LLC, USA), combustion products of propane-air mixture are used as atomizing gas.
2. Typical arc spraying gun TAFA 8830 (Praxair-Tafa, USA), compressed air is used as atomizing gas.
3. High-velocity oxygen fueled (HVOF) flame spraying gun Diamond Jet DJ1000 (Sulzer-Metco, USA).

Technological regimes are used on the recommendations of the manufacturers. Coating materials—alloys 625 (Ni-21Cr-9Mo-3Nb-Fe) and 671 (Ni-45Cr-1Ti); solid wire Ø1.58 mm for arc spraying; fraction powders + 16–45 μm for HVOF-process.

Specimens from low carbon steel 20 × 20 × 6 mm were coated. As the resistance criterion the weight loss in a gaseous medium composition (N_2-1 % H_2S-1 % HCl) at 400 °C, 1400 h and in air at 700 °C, 1000 h was selected. Specimens from low carbon steel 20 × 20 × 6 mm were coated. Before weighing the coating surface was cleaned with a brush. Such test conditions are close to conditions of a boiler using oil and coal dust with the afterburner nitric oxide as a fuel. In these boilers, the saturation with hydrogen sulfide in a reducing atmosphere is significantly limited by the presence of chlorine.

The resistance to high-temperature corrosion was evaluated by coatings gas permeability (see the methods of Maslovski [27]). Coatings of various thicknesses were applied on stainless steel filters with voids of 0.5 μm, specimens diameter of 25 mm, thickness of 3 mm. Sprayed specimens were installed in a glass with the

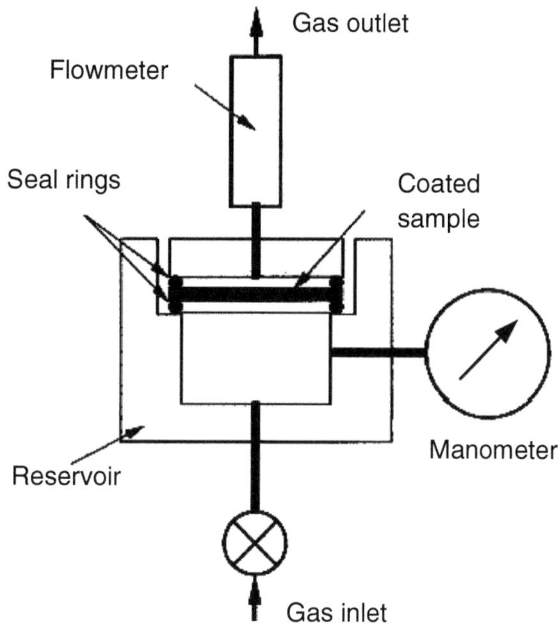

Fig. 4.22 Scheme of apparatus for measuring gas permeability

seal on the lateral surface, where the pressure drop of nitrogen ΔR under the specimen was measured by a pressure gauge (Fig. 4.22). The permeability was assessed by its flow rate (Q) at the output of the meter readings.

To evaluate the penetration of sulfur and chlorine into the coating the study of its structure and properties by using electron microscopy, X-ray analysis, microhardness and hardness (according to ball) was carried out. Plates having the parameters 75 × 25 × 6 mm were used as specimens.

Coatings made from alloys 625 and 671 performed by both arc spraying guns contain approximately equal amounts of oxides, Fig. 4.23. However, the HVAF-Arc coating structure is more fine-grained, the apparent porosity of the structure is of 1–2 %, which corresponds to the HVOF coating level.

According to the gas permeability HVAF-Arc coatings are denser than TAFA 8830 ones, Fig. 4.24.

With increasing of thickness, the permeability of the coatings is reduced, but for TAFA 8830 and HVOF coatings not linearly, as would be expected from their uniform porosity. Obviously, with increasing coating thickness above 700 μm the growth of residual stresses leads to the formation of new channels for gas flows due to the appearance of microcracks. However, this is not observed for HVAF-Arc coatings where at a thickness of more than 380 μm, the device does not show any permeability. There were no tests of HVOF-coating with thickness higher than 750 μm due to their delamination.

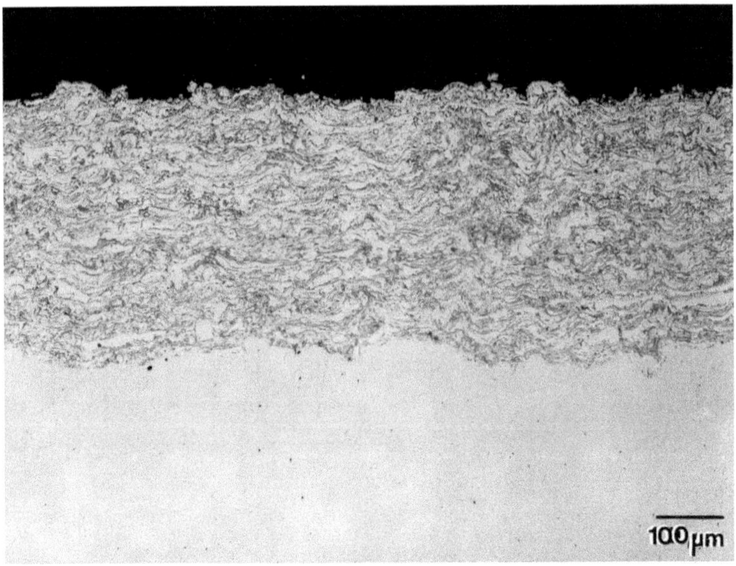

Fig. 4.23 Microstructure HVAF-Arc coating from alloy 625 (Ni-21Cr-9Mo-3Nb-Fe)

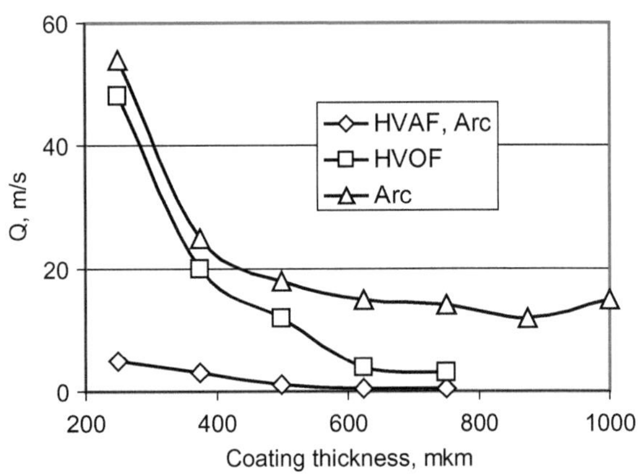

Fig. 4.24 Effect of coating thickness (δ) from alloy 625 by the amount of gas flow (Q) in tests on the gas permeability ΔR = 3.5 bar

From the corrosion resistance test results of coatings from alloy 671 (averaged data on five specimens), Table 4.11, it is clear that in these conditions the corrosive action of sulfur has less effect on the coating formed by HVAF-Arc gun.

In this case, it was not found any penetration of sulfur and chlorine to the substrate. HVOF-coating could not protect from chlorine and the TAFA 8830 coating proved to be unstable both to sulfur and chlorine.

Comparative evaluation of the oxidation resistance in air was carried out for the coating from alloys 671 and 625 in similar tests, Table 4.12.

Table 4.11 Coatings properties of Alloy 671 obtained by different methods

Guns		TAFA 8830	DJ1000	HVAF-Arc
Mass growth (mg/cm^2)		5.64	5.28	4.75
Results of X-ray analysis	Sulfur in the coating	Yes	Yes	Yes
	Sulfur on the surface of "coating—substrate"	Yes	No	No
	Chlorine in the coating	Yes	No	No
	Chlorine on the surface of "coating—substrate"	Yes	Yes	No

Holding in medium N_2-1 %H_2S-1 %HCl, 400 °C, 1400 h

Table 4.12 Properties of coatings of alloys 625, 671, obtained by different methods. Air exposure 1000 h at 700 °C

Guns	TAFA 8830	DJ1000	HVAF-Arc
Mass growth (mg/cm^2) Alloy 671 Alloy 625	3.23 3.49	1.89 2.29	1.68 2.91
Oxidation of the substrate surface under coating	Yes	Yes	Yes
The depth of intergranular oxidation on the substrate, micron (X-ray analysis)	150	50	50

From the presented data, it is evident that in an air atmosphere HVAF-Arc and HVOF coatings from alloys 671 and 625 have approximately the same degree of oxidation, which is significantly less than for TAFA 8830 one. For all the tested coatings the substrate is oxidized, but spreading of the process depthward to substrate for HVAF-Arc and HVOF coatings is three times less than for typical spraying.

To reduce the maintenance costs while protection against corrosion of the gas the use of refractory alloy coatings from alloys of the system Fe–Cr–Al in the form of core wires (CW), applied by arc spraying method (AS) is prospectively. CW application for AS, unlike solid wires (SW), allows to get the most heat-resistant coatings of the complex alloying system in the simplest way [28]. Varying the composition of the coating is necessary for its stability at various temperatures, composition of the gaseous medium, i.e., under different regimes of gas corrosion.

Chrome–aluminum alloys have high heat resistance due to the fact that on their surface upon heating an oxide film Al_2O_3 is formed, which has a high melting point, chemical, and thermal stability of [29], thereby providing a long service life of parts protected. Additional alloying with titanium and silicon prevents the formation of ferrous chromium carbides (Fe, Cr)$_7$C$_3$ and aluminum nitride (AlN), as well as the occurrence of local high-temperature corrosion (LHTC) [30]. LHTC mechanism is as follows. Oxidation of chromium carbides having in their composition a significant amount of iron causes the formation of porous ulcers filled with scale, which serve as channels for the diffusion of nitrogen into the metal.

Table 4.13 Calculated chemical composition of the coatings of the CW, mass%

Component	Chemical composition (mass%)		
	1	2	3
Aluminum	6.0	6.0	6.0
Chrome	28.0	28.0	28.0
Titanium	1.7	Less 0.5	Less 0.5
Silicon	Less 0.5	1.3	Less 0.5
Ferrum	Substrate		

Table 4.14 Combinations of CW used for spraying

CW due to Table 4.1	Symbol for a CW coating
3–3	a
3–2	b
3–1	c

Nitrogen binds aluminum to nitride, whereby the oxide films are enriched with iron and are reduced by carbon with its increased concentration on the surface boundary metal-oxide due to the selective oxidation of aluminum and chromium.

For the sprayed coatings of alloying system Fe–Cr–Al the input of titanium and silicon can prevent leakage of LHTC similar to solid chrome aluminum alloys. However, the conditions for the formation of AS coatings may affect additionally.

The authors investigated the effect of conditions of coatings formation of alloying system Fe–Cr–Al on characteristics that affect their heat resistance.

For comparison of the heat resistance CW were chosen with diameter of 2.0 mm, with a shell of low carbon steel, with different content of titanium and silicon. The calculated chemical composition of the coatings of the wires is shown in Table 4.13, the CW combinations for their spraying are shown in Table 4.14.

The criterion for choosing the initial composition of the wires was the content of chromium and aluminum, similar to the characteristic heat-resistant alloy grade X23IO5T GOST 109944 (CrAlTi 23-5). Compared to solid metal, the increased quantity of alloy elements was added into CW. It is caused by the conditions of the coatings formation at AS. In particular, the burn up of alloying elements and metal saturation with gases of the atmosphere takes place in the arc burning zone. The required content of the components in the wire was determined according to calculations by the model of sprayed particles oxidation at AS [31, 32].

Chromium and aluminum (28.0 and 6.0 wt%) content was accepted the same for all the CW to determine the effect of silicon and titanium on the heat resistance of the coatings. Additional chromium alloying up to 65 % and aluminum up to 10 % is irrational, because it does not increase the heat resistance of fechrals and promotes their brittleness after heating and cooling due to the emission of sigma phase [33].

The coating with thickness of 0.4 mm was deposited by activated arc spraying (AAS) [34] on the plates 30 × 20 × 1.5 mm of low carbon steel in the following modes: current—200 A, voltage—30 V, spraying distance—100 mm, propane pressure 0.4 MPa. Unlike typical AS at AAS, compressed air is not used as carrier gas but combustion products of propane-air mixture. It allows to achieve reducing atmosphere in the arc burning zone by lowering the partial pressure of oxygen from 0.21 to 0.035 MPa, i.e., six times which prevents the saturation of molten metal with oxygen [35]. The coating thickness is given by the condition to ensure their gas tightness. The thinner coatings have less gas tightness due to their porosity. Large coating thickness leads to an increase in residual stresses and the formation of new channels for gas flow due to the appearance of microcracks [32].

Heat resistance of coating specimens was determined according to GOST 9.312 as the value inverse to weight loss as a result of holding for 24 h at a temperature of 700 °C in air. The determination of heat resistance to weight gain is acceptable due to the short duration of the test.

The microstructure of the coatings was examined using an inverted metallographic microscope Nikon INSTECH Epiphot 200, the chemical composition of the coatings was determined by scanning with the electron microscope JEOL JSM 6490 with an adapter for energy dispersive and wave microanalysis Oxford Inca. Measurement of oxygen amount in the coating was performed by reductive melting in an inert gas current and vacuum on devices RO-116 and EAN-220.

The porosity of the coatings was determined by metallographic method using a software system SIAMS Photolab assessing the pore size distribution in number, area and volume. Adhesive strength of the coatings was investigated by the method of taper pin [36] using machine Instron 3382, the results were averaged for three specimens.

Research Results and Discussion

Study of the structure revealed that the lamells in the coating are separated by thin oxide films. There are no significant differences in the structure of different types of coatings, Fig. 4.25. The solid solution of chromium, aluminum, and other elements in iron forms the main coating phase in the coating structure iron oxide Fe_2O_3 is present predominantly, as well as inclusion of oxides of Al_2O_3 and Cr_2O_3.

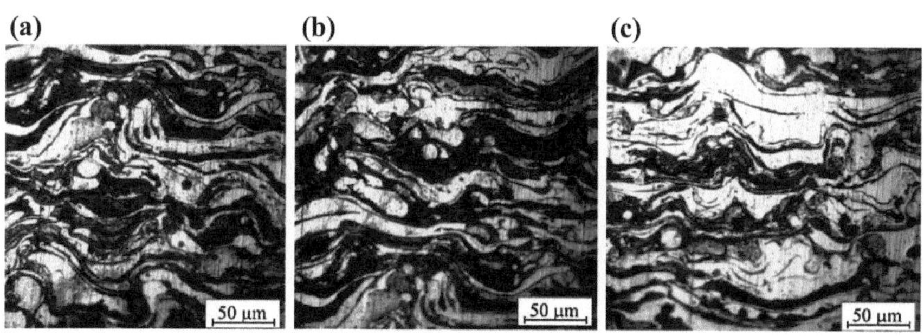

Fig. 4.25 Microstructure of the cross-section of coatings for CW combinations according to Table 4.14

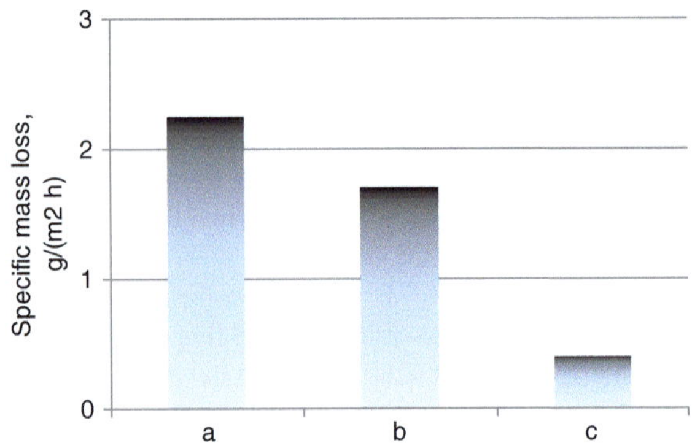

Fig. 4.26 The specific mass loss of specimens of heat-resistant coatings after holding for 24 h at a temperature of 700 °C for combinations of FCW according to Table 4.14

Table 4.15 Heat resistance of different steel grades

Steel grades	Steel class	Specific mass loss ($g/(m^2 h)$)
12X1MΦ	Perlitic	80
1X12B2MΦ	Martensite-ferrite	6
1X18H12T	Austenite	0.4
X23H18		0.1

Specific weight loss of coating specimens was 2.3–0.4 $g/(m^2 h)$, Fig. 4.26, what is significantly lower than similar values for pearlitic steels and martensitic–ferritic steels, such as 12X1MΦ GOST 20072 and 1X12B2MΦ GOST 5632 (like T11 и T122 ASME A213). Moreover, values of specific mass loss of the specimens can be compared to those for austenitic steels, such as 1X18H12T и X23H18 GOST 5632 (like Tempaloy A-1 и TP 310 ASME SA312), widely used in boiler manufacture, Table 4.15 [37].

After exposure there are preferably aluminum oxide Al_2O_3, minor inclusions of chromium and iron oxides, Cr_2O_3 and Fe_2O_3 (Fig. 4.27) on the coatings surface.

According to the test, the coating alloyed with titanium, has the best heat resistance, Fig. 4.26. The chemical composition of the coating before the heat resistance test determined by energy dispersive analysis revealed that the calculations error due to model [32] for the alloying elements was not more than 15 %, Fig. 4.28.

The chemical composition of the coating after the heat resistance test showed that the individual particle sizes (1–10) μm contain an increased amount of Ti (5–6 wt%), while the total content of Ti in the coating does not exceed 1.6 wt%. These particles are located at the grain boundaries in the coating and the oxide film Al_2O_3. Spectra corresponding to the localization zones of these particles also exhibit the several times increased content of aluminum and oxygen with several times reduced chromium and iron, compared with the average values for the

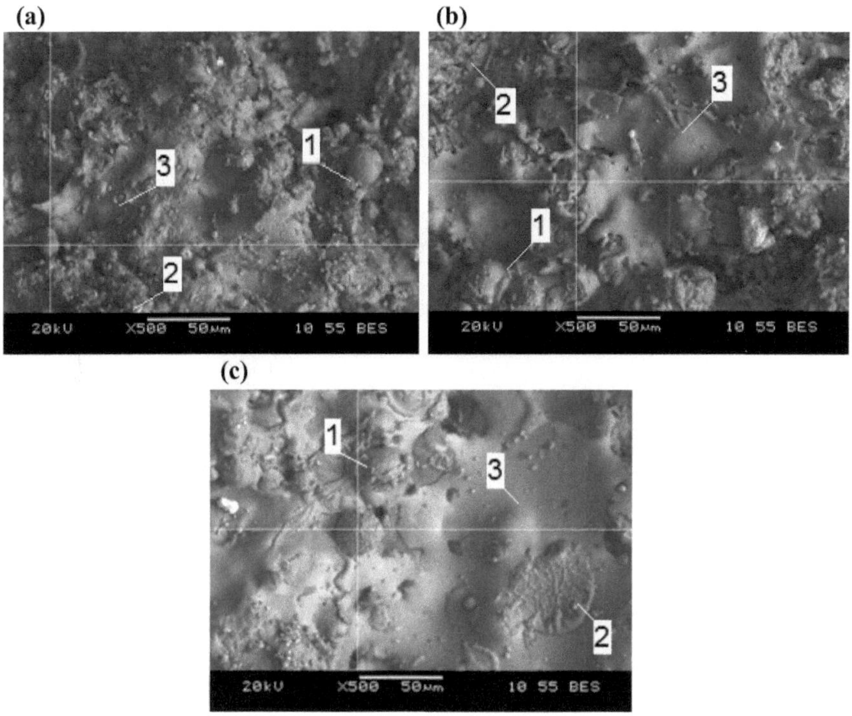

Fig. 4.27 Microstructure of coatings surface after exposure for 24 h at a temperature of 700 °C for CW combinations according to Table 4.14. *1* Fe_2O_3, *2* Cr_2O_3, *3* Al_2O_3

coating, Fig. 4.29. Presumably, these particles correspond to carbide TiC as thermodynamically more probable compound from possible in the coating [38].

For chrome–aluminum alloys, the formation of ferric carbides $(Fe, Cr)_7C_3$ and aluminum nitride AlN which initiate LHTC [30] is prevented due to the formation of thermodynamically stable and heat-resistant carbides TiC and underoxide diffusion barrier oxide layer. These data show that a similar effect is observed for the coatings. In the absence of these adverse structural components, the film of aluminum oxide Al_2O_3 is mainly formed on the surface when heating, Fig. 4.28, what ensures a high level of heat resistance.

The shown effect for coatings is achieved by the add an increased quantity of titanium and silicon into the charge of the base alloying system Fe–Cr–Al. Such quantities is 3.4 and 2.6 times higher than those for the standard chromium–aluminum alloy X23Ю5T GOST 10994 (Fe bal., Cr 22–24, Al 5–5.8, Ti 0.2–0.5, Si less than 0.5).

Adhesion coating strength of base alloying system Fe–Cr–Al, when titanium is introduced into the initial charge of wire (Table 4.14, variant c) is 44 MPa. This is comparable with the values for the CW coatings with similar purpose, Table 4.16.

(a)

(b)

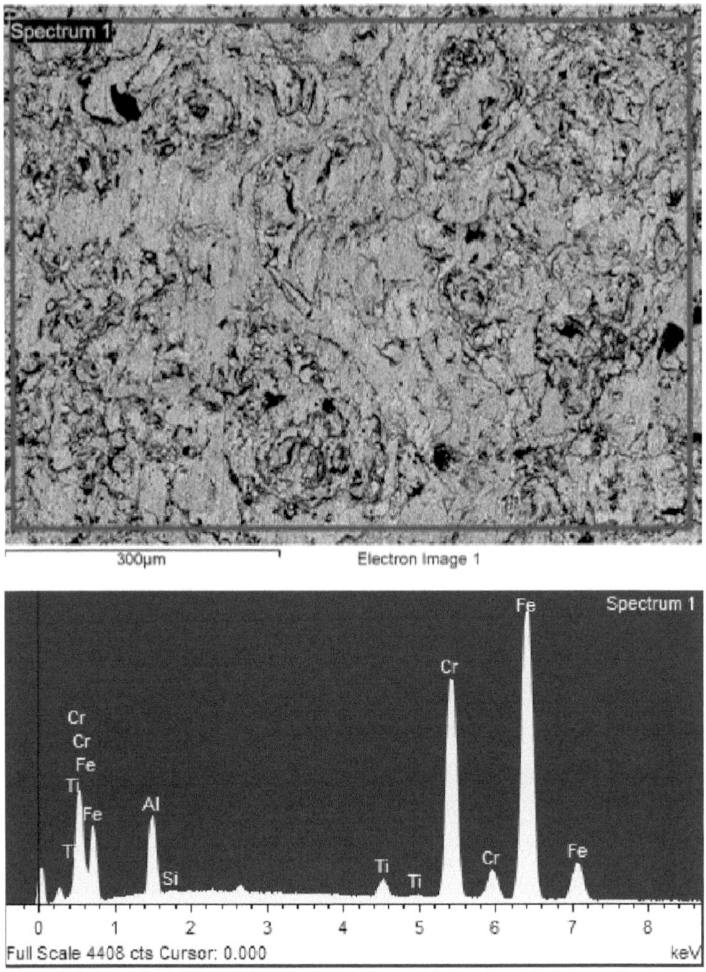

(c)

Element		Al	Si	Ti	Cr	Fe	Totals
Content	wt.%	6,16	0,26	1,60	29,30	62,68	100
	at. %	11,66	0,48	1,71	28,79	57,36	100

Fig. 4.28 Results of scanning electron microscopy for coating Fe–Cr–Al–Ti before the heat resistance test: **a** microstructure; **b** spectrum, **c** spectrum interpretation

The total porosity of the coating (including the open and closed porosity) is of 2.8 %, with a maximum pore size of 44.1 μm, a minimum of 0.5 μm. Share having a size up to 22 μm is estimated at 97.8 % and the major part accounts for pore size to 10 μm, Table 4.17. Since, according to the results of grain size analysis [31], the coating is formed from particles with a diameter up to 100 mm (weight content of 69.8 %), the data demonstrate the stability of the process of CW spraying.

(a)

(b)

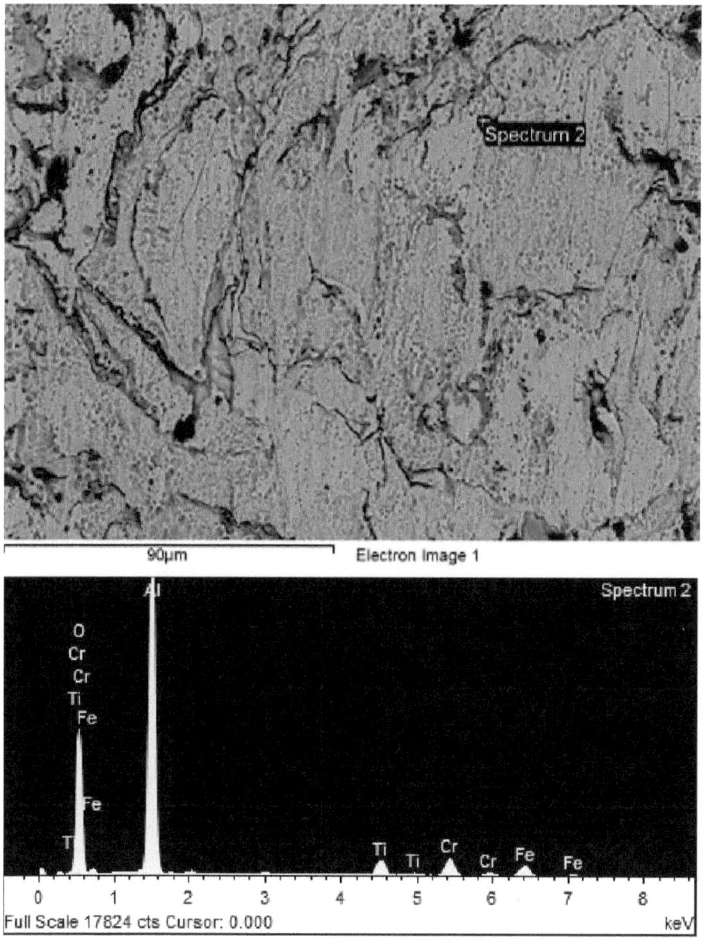

(c)

Element		O	Al	Ti	Cr	Fe	Totals
Content	wt.%	49.58	35.65	3.70	5.98	5.09	100
	at. %	65.89	28.09	1.64	2.45	1.94	100

Fig. 4.29 Results electron microscopy scanning of the coating Fe–Cr–Al–Ti after the heat resistance test: **a** microstructure; **b** spectrum, **c** spectrum interpretation

Table 4.16 Adhesion strength of the coating from core wires [39–41]

CW grade	Developed	PMet 731	EuTronic Arc 509	TAFALOY 26CA
Adhesion strength (MPa)	44	43	46	54

Table 4.17 The total share of pores of long-accumulate mode, %, CW coatings according to the Table 4.14, c

D, micron	0–2	2–4	4–6	6–8	8–10	10–12	12–14	14–16	16–18	18–20	20–22
n[a]	9.7	45.2	67.5	78.7	84.1	90.6	92.8	94.8	95.5	97.0	97.8
A[b]	0.3	4.6	12.5	20.2	26.2	37.3	42.7	49.3	52.4	60.4	64.9
V[c]	0	0.7	2.7	5.4	8.0	14.2	17.7	22.8	25.3	33.1	37.8

[a]Here: n is the numerical share of the pores, A is share area, V is share by volume. The volume of pores was calculated by the formula: $V = \pi D^3/6$, respectively $D = \sqrt{4A/\pi}$ is the diameter of a circle equivalent to the pore area (A)

High adhesive strength and low porosity of chromium aluminum AAS coatings are due to the using combustion products of propane-air mixture as carrier gas in the case of the AAS, thereby providing the initial increase in temperature and the gas flow rate 4–6 times in comparison to AS [31]. Differences in the kinetic and temperature characteristics of the gas flow in the case of the AAS predetermine increasing of velocities and temperatures of particles compared with AS at the end of spraying distance. Because of increasing the velocity of the particles the degree of deformation at the impact with the substrate increases and increasing their temperature improves the physical and chemical interaction "particle-substrate" in the moment of impact, which leads to increased adhesion and reduced porosity of the coating [42, 43].

Chrome aluminum coatings are characterized by the flow of aluminothermic reactions at spraying with the release of large amounts of heat, which increases the temperature of the molten metal drops and provides better interaction of the sprayed coating to the substrate [44, 45].

CW and application technology of heat-resistant AAS coatings were developed on the basis of these solutions to increase service life of power equipment parts working in conditions of gas corrosion [46].

4.2.3 *Electrochemical Corrosion*

The mechanism of electrochemical corrosion is associated with the formation and interaction of micro galvanic elements on the metal surface in moist environment. It is effected by anodic (corrosion oxidation of metal) and the cathode (oxidant recovery situated in a moist environment) processes. With this type of corrosion, the oxidation of the metal and the reduction of the oxidized component of the corrosive environment are localized in certain areas of the surface of the corrosive metal and do not occur in a single act.

The velocity and nature of the electrochemical corrosion process are determined by characteristics of metal, external influences, and design features.

The characteristics of a metal include the following: thermodynamic stability, the position in the periodic system, structure and type of alloy, the presence of impurities, internal stress, etc.

The thermodynamic stability of metal can be approximately estimated by the standard electrode potentials of metals $U_{h,0}$, measured with respect to a normal hydrogen electrode. According to the degree of thermodynamic stability, all metals are divided into five groups [47], according to their place in the electrochemical series.

(1) Metals of increased thermodynamic instability: Li, Rb, Cs, Ba, Sr, Ca, Na, Mg, Al, Ti, Zr, Mn, Cr, Zn, Fe. These metals have less value of the standard electrode potential than the potential of a hydrogen electrode at pH = 7 (−413 mV). Besides corrosion in acidic media—under action of acids and other oxidants such as oxygen, these metals may corrode even in neutral media, i.e., can be oxidized with water when created the necessary conditions. Formally, the following process takes place

$$Me - ne^- = Me^{n+} \tag{4.20}$$

In this case, recovery processes of mentioned oxidants are possible:

$$2H_2O + 2e^- = 2OH^- + H_2, \ \varphi_{H_2|2H^+} = -0.413\,B \quad (pH = 7) \tag{4.21}$$

$$2H^+ + 2e^- = H_2 \quad \varphi_{H_2|2H^+} = 0\,B \quad (pH = 0) \tag{4.22}$$

$$O_2 + 2H_2O + 4e^- = 4OH^- \quad \varphi_{4OH^-|O_2+2H_2O} = 0.816\,B \quad (pH = 7) \tag{4.23}$$

$$O_2 + 4H^+ + 4e^- = 2H_2O \quad \varphi_{2H_2O|O_2+4H^+} = 1.229\,B \quad (pH = 0) \tag{4.24}$$

(2) Thermodynamically unstable metals: Cd, In, Ti, Co, Ni, Mo, Pb, W. Significance of standard electrode potentials is greater than that of the former group metals, but less than zero. Therefore, they cannot be oxidized with water (pH = 7), but will be unstable in acidic media and any media in the presence of oxygen.
(3) Metals of intermediate thermodynamic stability: Bi, Sb, Re, Tc, Cu, Ag, Rh. Have positive values of standard electrode potentials, not exceeding the value of the electrode potential associated with the oxidizing action of oxygen in a neutral medium, Eq. (4.23). Therefore, these metals will be stable in all acidic and neutral media in the absence of oxygen.
(4) Metals of high stability: Hg, Pd, Ir, Pt. Standard electrode potentials of these metals are in the range between the values of the two electrode potentials characterizing the oxidizing action of the oxygen in neutral and acidic environments.
(5) Metals of complete stability: Au. This metal can be oxidized by oxidizing agents listed. Its electrode potential is maximized of the metals considered.

Table 4.18 Thermoelectric series of some metals

Metal	$U_{h,0}$ (mV)		Metal	$U_{h,0}$ (mV)	
	1	2		1	2
Au	+306	+243	Al–Mg–Si	−124	−785
Brass JIC 63	+145	+13	Al 99.5 %	−169	−667
Cu	+140	+12	Gray pig iron	−350	−307
Ni 99.6 %	+118	+46	Cd	−574	−519
08X18H8	−84	−45	Zn 99,975 %	−807	−794[a]
Sn	−175	−184			

1 in the flowing water, saturated with air and not containing chlorine (pH = 6.0, 9.5 °C, 0.1 MPa); *2* in the flowing synthetic sea water saturated with air (pH = 7.5, 25 °C, 0.1 MPa) [48]
[a]Electrolytic zinc on steel, 50 μm

Place in the raw may vary depending on the environment, which acts as the electrolyte (Table 4.18). Potential shift toward positive values can be regarded as an increase in the thermodynamic stability of the metal. During contact between two metals, in the electrolyte environment, the disruption (corrosion) of metal of lower electrode potential occurs. The further metals are arranged in the electrochemical raw, the more basic metal can corrode.

Impurities found in the environment can be adsorbed on the metal surface catalytically accelerating or retarding corrosion. For instance, most iron alloys corrode in sea water much faster than in fresh water at the same concentration of oxygen not containing chlorides. This is due to the fact that chloride ions are adsorbed on the surface of iron to prevent the formation of its protective layers.

Under certain conditions, some metals pass into a passive condition—on their surface layers or films are formed composed of adsorbed oxygen, the oxide of this metal or its salt. The presence of such layers and their structure highly influence on the rate of metal corrosion; in some cases, these layers have a protective effect, as a result the metal corrodes negligibly slowly. In terms of atmospheric air, the passivating films are formed on chromium, nickel, aluminum, zinc, cadmium, and tin. Metals forming such films on their surface are usually used as coatings for protection of other metals from corrosion.

Corrosion protection ways are classified due to electrode potential ratio:

Anodic, when the base metal is coated with the less noble metal with a lower electrode potential. In this case, the base metal (cathode) does not corrode until the destruction of protective coating layer (anode). Furthermore, the oxide film, which is formed on the metal protective layer, may reduce the corrosion rate.

Cathodic, when the metal surface to be protected acting as an anode is coated with a nobler metal having a higher potential electrode (cathode). In this case, the protection will be effective only in that case if the coating does not leak electrolyte. Otherwise, it will cause corrosion of the base metal, while the coating itself is not subjected to corrosion

For both methods of protecting the corrosion resistance depends on the density of the coating, its thickness and adhesive strength, but the mechanism of these factors is different.

At anodic protection, the coating acts as a barrier between the corrosive environment and the substrate. It is destroyed itself, thereby protecting the product. That means to increase the corrosion resistance of the product an increase in coating weight is required. This is especially true concerning the work in active environments where additional protective factors, such as an oxide film, weakly inhibit the interaction of "coating—medium". Technological measures may be to control such mass characteristics of the coating as its density and thickness. In this case, more dense coating is more economical due to the reduction in costs of materials and labor of its application. Furthermore, increasing the adhesion strength leads to more efficient passage of an electric current in the galvanic couple "substrate—coating", which increases the corrosion resistance of the base metal.

In the case of cathodic protection, when electrochemical potential of the coating is higher than that of the substrate (e.g., stainless steel coating on the low carbon steel) to increase corrosion resistance of the product requires reduction of the gas permeability of the coating. This reduces the possibility of access to the interface of the "basic metal—substrate" of the electrolyte which prevents the development of electrochemical processes leading to the destruction of the base metal. Cathodic protection is effective, if the coating completely seals the surface which is exposed to aggressive environments.

Full cathodic protection products will take place in the absence of open porosity, i.e., a continuous path from the outer surface of the coating to the substrate. Otherwise, there is corrosion on the surface of the substrate. As a result, in the center of corrosion the coating swells and peels.

To reduce the gas permeability of the coating, an increase the density, thickness, and adhesion strength of the coating are technologically possible. It is important to keep in mind that the more dense coating is more economic, open porosity here takes place at smaller thickness.

When cathode protection, it is necessary to consider the possibility of side electrochemical phenomena. For instance, zinc plated storage tank for cold water will have a long life. However, when placed copper parts into this container, their presence will cause an electrochemical reaction, accelerates the corrosion of zinc thereby reducing the life of the coating. Similar effect will be caused by the presence of alloying elements and impurities in the metal protective layer. For instance, a layer of aluminum alloy AlMg6Mn unlike pure aluminum will not protect steel surfaces against corrosion in seawater due to the high magnesium content.

Metal structure has different effects on the corrosion rate. Thus, coarsening of the grain does not increase general corrosion but promotes intergranular one.

Reducing the surface roughness improves the corrosion resistance. This is due to the formation of more perfect, dense passivating oxide films on the smooth surface.

In conditions of atmospheric corrosion greater resistance of polished metal surfaces is largely determined by the fact that these surfaces are less exposed to moisture condensation in comparison with the rough surfaces.

External factors of electrochemical corrosion of metals include: the composition of the corrosive environment, its acidity, temperature, fluid velocity, corrosion inhibitors, and stimulants, etc.

The rate of corrosion of metal products in corrosive environments is drastically increased in the presence of mechanical impact. By its origin stresses may be internal, resulting from deformation or heat treatment of metal, or external, caused by externally applied loads. Depending on the nature of the mechanical effects there are following types of corrosion:

- **Stress corrosion cracking**—local metal destruction while exposed to tensile stresses and corrosion attack at the same time. Constant cracks stresses (external or internal) increase the rate of general corrosion of the metal roughly proportional to their value. Development of corrosion cracks occurs as a result of three factors: electrochemical—because of heterogeneity of the metal surface structure (defects of protective films are active anodes); mechanical—the uneven stress leads to an electrochemical heterogeneity; adsorption—the formation of cracks in the metal surface layer is facilitated by the wedging exposure of surfactants adsorbed from a solution. Susceptible to stress corrosion cracking are metal boilers (alkaline fragility), some corrosion-resistant chromium–nickel steels.
- **corrosion fatigue**—the destruction by a simultaneous action on the metal of the corrosive environment and alternate stresses. If alternate stresses exceed the limiting level of metal fatigue, after a number of cycles of alternating loads fatigue cracks develop and the detail is destroyed. When corrosion fatigue the fatigue limit decrease can be seen compared to the fatigue limit in the absence of the corrosive environment (Table 4.19). The mechanism of metal destruction in terms of occurrence of corrosion fatigue is basically the same as that of a stress corrosion cracking but of course must relate to the period of tensile stresses actions.
- **corrosion by friction** or erosion corrosion is the destruction of metal by mechanical abrasive effect on the metal of another metal in the presence of a corrosive environment or its direct action. Corrosion by friction can be presented as two joint processes: electrochemical or chemical interaction of the

Table 4.19 Strength of metals in various environments [23]

Steel	Tensile strength (MPa)	Fatigue limit under alternating loads (5×10^7 cycles) (MPa)		
		In air	In fresh water	In seawater
A570-36	440	250	140	60
NiCr 3.7	630	340	150	110
CrNi 18 25	810	360	310	240

metal with aggressive media and mechanical wear as of surface protective films and of the metal itself.

Such constructive features as closed zones, the concentration of the mechanical and thermal stresses and the contact of dissimilar metals, contribute to the electrochemical corrosion. The corrosion attack intensifies near the interphase surface. So, for example, the process of metal fracture on the ships develops in the area located slightly above the waterline particularly intense. This is facilitated by easier access of oxygen, flushing of protective films from the surface, its periodic wetting with electrolyte followed by desiccation, temperature difference.

Under the terms of the flow, the electrochemical corrosion includes:

- atmospheric corrosion in wet gas atmosphere;
- corrosion in electrolytes;
- soil or underground corrosion of metal buildings;
- galvanic corrosion under the influence of an external power source;
- contact corrosion of metals which means the destruction of the metal in the electrolyte solution due to contact with another metal having a positive electrode potential with respect to the base metal.

The most common is atmospheric corrosion. On the metal surface, the thin layer of moisture containing various gases, especially oxygen, is condensed. The main factors determining the rate of corrosion in the atmosphere are the moisture degree of corrosive metals surface, presence of contaminations, and temperature. These factors vary within a wide range (Table 4.20), for instance, in marine atmosphere there are many salts, particularly sodium chloride; in industrial areas atmosphere there are a lot of sulfur, carbon, nitrogen, and others oxides.

Technical facilities for storage of petroleum products are exposed to electrochemical corrosion although these products are not electrolytes. This is determined by the following factors [50]:

(1) The presence of water vapor inside the tanks due to humidity and temperature changes (Fig. 4.30).
(2) The presence of the dissolved oxygen which content increases with decreasing density of the oil products, increase of the vapor pressure of oxygen (Fig. 4.31), decrease of oil temperature.

Table 4.20 Characteristics of atmospheric types [49]

Type of atmosphere	Industrial		Sea		Country
Relative humidity (%)	79		84		77
Precipitation (mm/year)	670		970		920
pH	4–6		5–7		5–7
Aggressive additives	SO_2	0.1 mg/m^3	Chlorides	2 mg/dm^2	Not limited
Dusts pollution	64 g/m^3		Not limited		

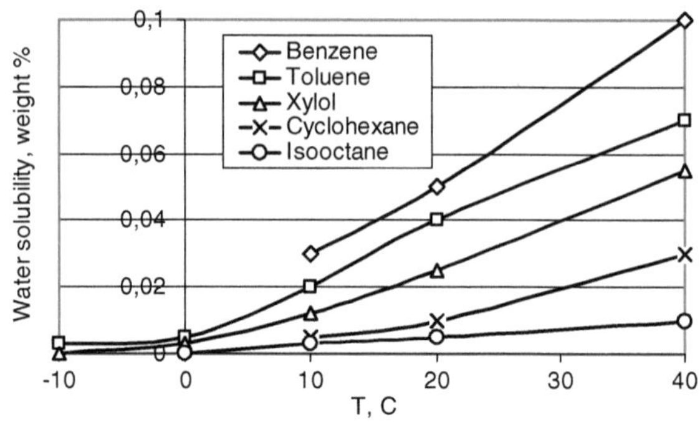

Fig. 4.30 Temperature influence on the water content in hydrocarbons

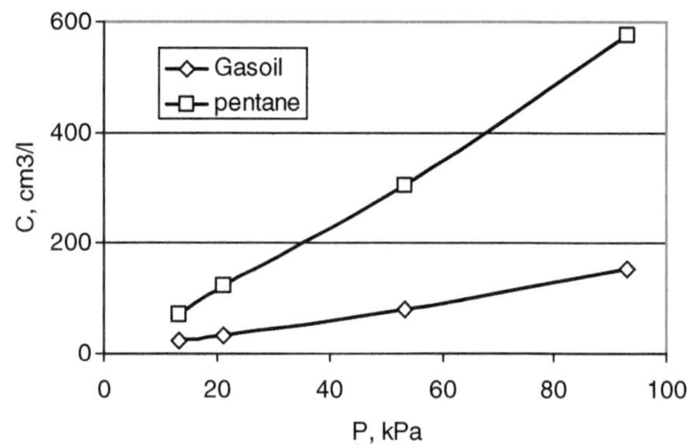

Fig. 4.31 Effect of the vapor pressure of oxygen (P, kPa) and the density of oil products to oxygen solubility in oil products (C, cm^3/l)

(3) The presence of sulfuric and sulfonic acids which are the oxidation products of sulfur impurities such as mercaptans, sulfides, di- and polysulfides, thiopheness.

To reduce the rate of corrosion in these conditions, it is necessary to reduce or exclude the supply of oxygen to the cathode sites of metal surface.

The internal surfaces of tanks, reservoirs, and containers in height and in some areas are operated under different conditions. Therefore, the rate of metal corrosion at the bottom, the lower belt, the sidewall, in the vapor space will vary.

Corrosion is evaluated quantitatively by shrinkage of the metal mass (Δq) per time unit (τ), related to surface area (S) or to metal thickness:

$K = \Delta q/S \cdot \tau$, kg/(m^2/h)—corrosion rate;
$\prod = \mathrm{K} \cdot 8760 \cdot 1000/\gamma$, mm/year—the rate of reduction in metal thickness for the year, where 8760 is number of hours per year, γ is metal density, kg/m^3.

The ten-point scale is also used for corrosion evaluation (Table 4.21) which corresponds to the rate of reducing the metal thickness per year (P) during the uniform process.

Table 4.21 Ten-point scale of corrosion resistance of metals

Resistance group	Corrosion rate *P* (mm/year)	Resistance point
Absolutely resistant	<0.001	1
Rather resistant	0.001–0.005	2
	0.005–0.01	3
Resistant	0.01–0.05	4
	0.05–0.1	5
With lower resistance	0.1–0.5	6
	0.5–1	7
Labile	1–5	8
	5–10	9
Not resistant	>10	10

4.2.4 Influence of Technological Conditions on the Quality of Anticorrosive Coatings

Anticorrosion sprayed coatings are used to protect metals (usually steel) in aggressive environments such as water of varying hardness, steam, industrial atmosphere, fuel environment.

The ratio of electrode potentials has the main influence on resistance when galvanic corrosion. It may change depending on the environment which plays the role of the electrolyte (Table 4.18).

Furthermore, the combination of different materials in the coating changes its corrosion resistance. The protective coatings anticorrosive test when the action of aqueous salt aerosol showed that the resistance of the zinc coating is lower than the coating of the wire type Zn/Al 85/15 [51].

Besides this, the quality of anticorrosive coatings depends on its porosity, density, adhesion. These parameters can be influenced by technological measures:

- Surface preparation prior to coating;
- Thickness of the coating;
- Spraying modes.

Below are the results of experiments to assess this effect. Zn grade Ц1, GOST 13073-77 and stainless wire Св04Х19Н9, GOST 2246-70 (like steel 2ASTM 304) were used as feedstocks. Strength tests were carried out due to the pin breakaway methods and delamination by scratching according to GOST 9.304-87. The surface roughness was evaluated by the roughness heights (*6*), which is measured according to GOST 2789-73 on the profilograph—profilometer type 201. According to GOST 9.304-87, porosity was determined by hydrostatic weighing. Gas permeability was evaluated by the method of Maslovski [27]. The corrosion resistance was evaluated through a rapid test in hygrometer (water vapor atmosphere).

4.2.4.1 The Influence of Surface Preparation

Surface pretreatment determines the strength of parts adhesion with coatings. Its task is to clean the surface of the part from various contaminants, to remove oxide films, to create necessary roughness on the detail surface for mechanical engagement of the spray particles on the roughness of substrate surface and to create conditions for the chemical interaction in the contact zone.

Of the various methods of surface preparation, the required bonding strength is provided only by abrasive jet blasting (AJB). It requires additional technological preparation (compressed air, supply of abrasive, ventilation), increases the complexity of the coating process, but other methods do not provide consistent quality.

We have performed a comparative evaluation of the bond strength of the base metal with coatings obtained using different methods of surface preparation by the method of squares delamination. Scratching grid to the base metal composed of squares with dimensions of 5 × 5 mm were cut up to substrate through the zinc coating with a thickness of 250 μm. The test results showed that only the use of the AJB provides stable adhesion strength (Fig. 4.32).

Quantitative test using "breakaway" method showed that the increase in the interval between the AJB and spraying reduces the adhesion strength. Zinc coating with thickness of 200 μm has decreased adhesive strength from 52 MPa (immediately after AJB) to 43 MPa (24 h after AJB).

Furthermore, the low adhesion strength causes insufficient heat removal to substrate which may cause coating overheating. Adhesion strength is also influenced by the ambient temperature. When it falls below +5 °C, adhesive strength decreases due to moisture condensation on the surface of the part.

The value of surface roughness has greater influence on the properties of anti-corrosion coatings in comparison with wear resistant. Anti-corrosion coatings usually have a small thickness, 0.2–0.5 mm, commensurate with the height of surface roughness. Insufficient surface roughness even with a small coating thickness may cause delamination and a rough surface with a thin coating can be the

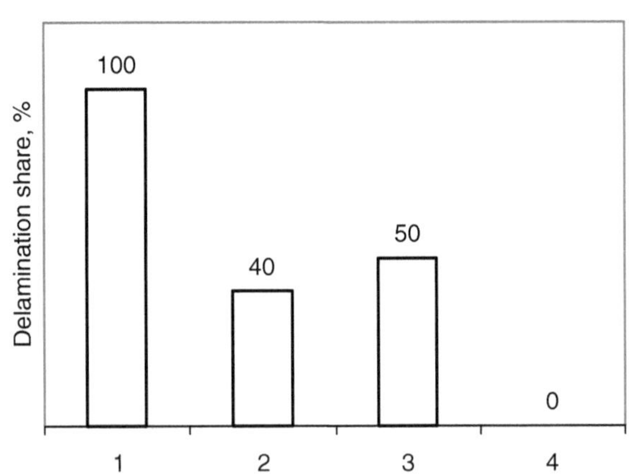

Fig. 4.32 Effect of surface preparation methods on the strength of adhesion (a comparison of the share of squares of cut grid); *1* degreasing; *2, 3* cleaning by abrasive and wire circles; *4* AJB

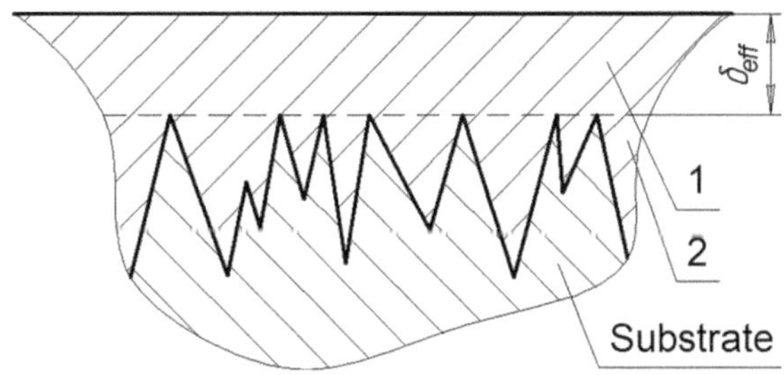

Fig. 4.33 Evaluation of the effective thickness of the coating

reason of through porosity. Increased roughness leads to increased deposition efficiency (DE), since this increases the adhesion strength and a large part of the sprayed particles remains on the surface. On the other hand, the coating material filling the space between heights does not affect the value of the effective thickness of the coating (δ_{eff}) (zone 1), which is defined as the distance from the coating surface to the tops of the roughness heights, Fig. 4.33. Significant increase in roughness and small thickness of the coating increase the share of ballast material (zone 2) in a total coating weight.

Furthermore, at a large size of abrasive particles the resulting vector of the droplet to the normal surface deviates strongly from $\alpha = 0°$. This leads to drops slip along the face roughness reducing DE.

The effect of these features of wear resistant coatings is smoothed since their thickness is 1–2 orders of magnitude greater than the height of the irregularities.

To quantitate the relationship of the roughness with the bonding strength and permeability, a series of experiments were carried out. As material specimens low carbon steel (like A570-36) was used, the feedback material was stainless steel wire (like steel 2ASTM 304), ∅1.6 mm. Spraying was carried out by gun AAS-10 (with the combustion chamber). Jet blasting before, coating was made with ejection gun at an operating pressure of compressed air of 0.6 MPa and grit consumption of 0.05–1.2 kg/s. Fractions of steel grit varied from 0.3 to 2.0 mm. For roughness, the total dependence is such that an increase in the average size of the grit increases surface roughness. However, this dependence is complicated by a variety of factors:

- diameter of delivered abrasive has minimal deviation ±0.2 mm but it may be increased to ±1 mm;
- size of the reusable grit is decreased due to its destruction;
- the height of the roughness is affected by AJB distance, air pressure, and angle of the jet axis to the treated surface. So, A.I. Stroganov et al. showed that for grits ДЧК 0.8 (chipped cast iron grit, fraction 0.8 mm) roughness varies from 0.18 to 0.25 mm, depending on the distance of the AJB, with a maximum at a distance of 100–150 mm [52].

In our experiments, the dependence of the roughness on the size of the abrasive was obtained at a distance of AJB of 50–150 mm, the angle of the jet to the surface 90°, for comparison unblasted surface was used, Table 4.22.

Table 4.22 Dependence of the surface roughness from the grit size and distance processing

Grit size (mm)	0.1–0.3	0.4–0.8	0.5–1.2	0.8–1.4	0.8–2
AJB distance, 50 mm					
Roughness (μm)	12	16	21	28	37
AJB distance, 100 mm					
Roughness (μm)	16	20	25	35	45
AJB distance 150 mm					
Roughness (μm)	14	18	23	34	44

Chipped steel grit

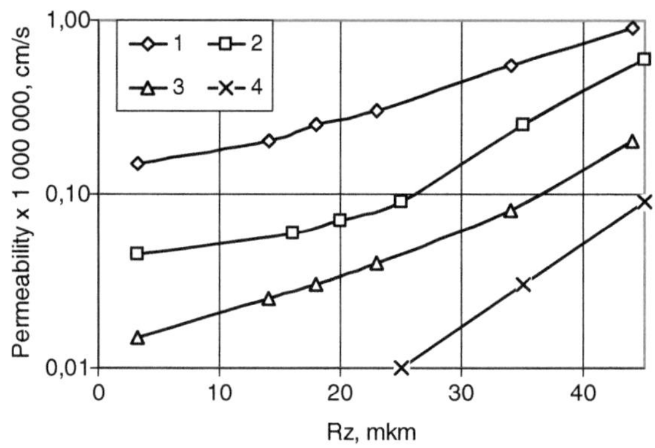

Fig. 4.34 The effect of surface roughness of the substrate on the permeability of coatings. Feedback steel 304 wire. The thickness of layer: *1, 2*—0.15 mm, *3, 4*—0.20 mm; Spraying distance: *1, 3*—150 mm, *2, 4*—100 mm

Analysis of the results of studies on the effect of surface roughness parameters of the coating shows the following.

Corrosion resistance of the coating with increase the surface roughness is reduced by increasing its permeability, Fig. 4.34. Increasing the height of surface roughness of the substrate up to R_z40 leads to increased adhesion strength, a further increase in roughness does not affect noticeable on this parameter, Fig. 4.35. View of curve and value of adhesion strength meet the data of Kharlamov and Borisov for plasma spraying [53], that is, indicating similar quality of PS and AAS coatings due to this parameter. DE increases with increasing roughness to R_z40. Further it decreases slightly, Fig. 4.36.

4.2.4.2 Effect of Coatings Modes

The porosity of the coatings increases sharply with an increase in spraying distance over 200 mm, Fig. 4.37.

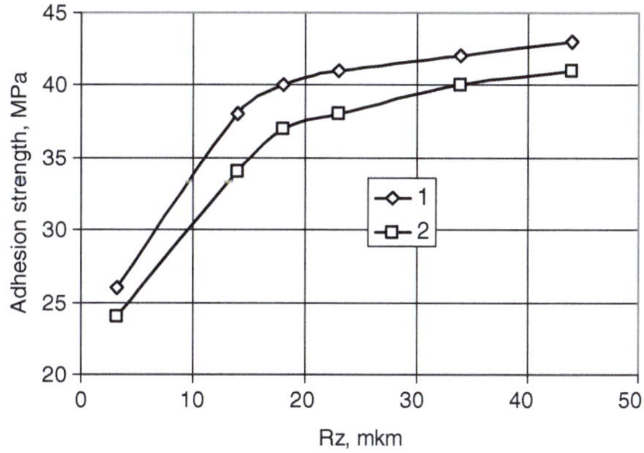

Fig. 4.35 The effect of base surface roughness on the adhesion strength of coatings. Feedback steel 304 wire. Layer thickness 0.15 mm; spraying distance: *1*—150 mm, *2*—100 mm

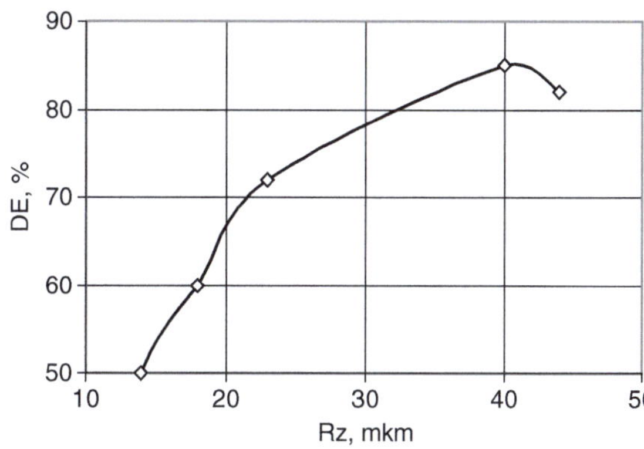

Fig. 4.36 The effect of surface roughness of the Substrate on deposition efficiency, feedback steel 304 wire. Layer thickness—0.15 mm; spraying distance—150 mm

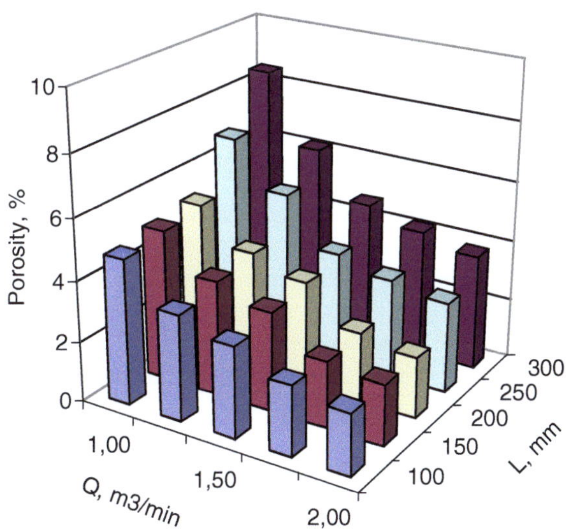

Fig. 4.37 The dependence of the porosity of the coatings from the air flow rate (*Q*) and spraying distance (*L*). Feedback steel 304 wire

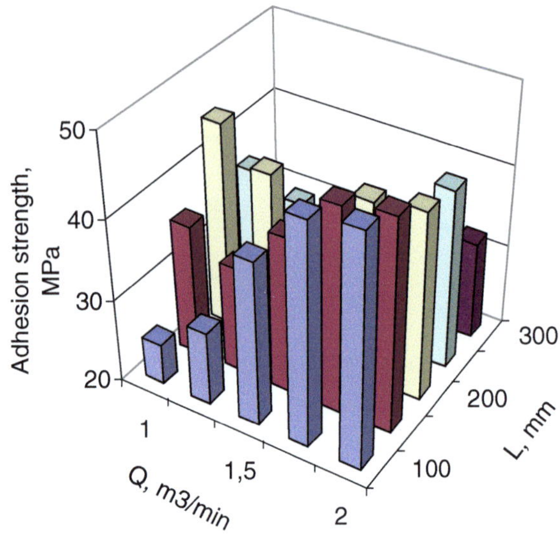

Fig. 4.38 Dependence of the adhesive strength from the air flow rate (Q) and spraying distance (L). Feedback steel 304 wire

This is especially evident at the lowest allowable consumption of compressed air of 1–1.5 m^3/min. With an increase of consumption to the maximum, the porosity decreases monotonically. When increasing the spraying distance of more than 200 mm the velocity and enthalpy of droplets are falling. Therefore, upon impact, the conditions of spreading of the molten metal droplets on the surface become worse. This increases the porosity of the coating. It is technologically complicated to provide small blasting distances, L = 100–150 mm at corners and projections of the surface. To obtain coatings with a minimum porosity, it is rationally to choose L = 150–200 mm, Q = 1.75–2 m^3/min.

The maximum adhesion strength is observed at high air flow rates, Fig. 4.38.

At a distance of 200 mm, this parameter is stable at all air flow values. With the decrease of the flow rate at distance, the adhesion strength decreases due to the increase of thermal stress in the coating. At greater distance, adhesion strength decreases due to the decrease of the kinetic energy of the impacted droplets.

4.2.4.3 Coating Thickness Effect

At the layers, thickness commensurate with an average microroughness height of coatings surface there is a strong change in the open porosity. This, in turn, is accompanied by a change in permeability. With regard to the application of anti-corrosion coatings, we performed a comparative evaluation of these changes for AAS and for typical AS.

At coating thickness of 0.05 mm for AAS open porosity is almost an order lower than in the case of AS, Fig. 4.39. Obviously, this is due to the decrease in the surface roughness as a result of the reduction of coating porosity. Profilograms

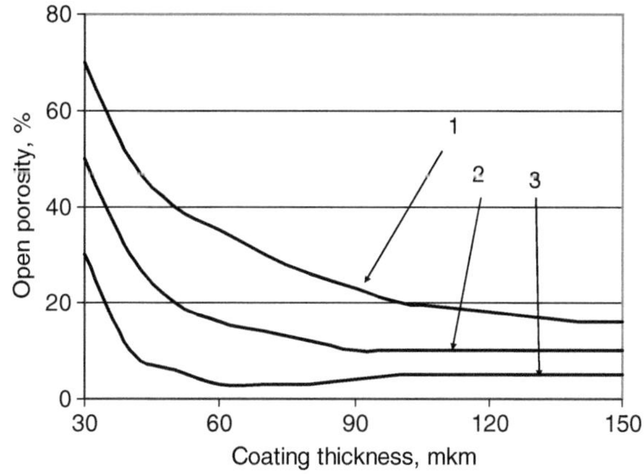

Fig. 4.39 Effect of coating thickness on the open porosity. *1*—AS; *2, 3*—AAS; *1, 3*—*L* = 100 mm; *2*—*L* = 50 mm

analysis shows that the average height of microprojections of AAS-sprayed surface from a distance of 100 mm with a thickness of 0.1 mm at coating thickness is 0.005–0.01 mm, whereas for the AS coatings under similar conditions, this value is 3–10 times higher: 0.03–0.05 mm. Changing the surface roughness is due to the size of droplets impinging on the surface. In comparison to AS [54, 55], the droplet size decreases from 120–260 to 40–80 μm in the case of AAS.

This difference in the open porosity leads to an even greater difference in the gas permeability. For AAS coating, this quantity is 1–2 orders of magnitude smaller than in the case of the AS, when coating thickness is greater than 0.1 mm, Fig. 4.40. With a layer thickness of 0.05–0.1 mm the difference is 4–5 orders of magnitude.

Increasing the thickness of the coating reduces its adhesive strength by increasing the level of residual stress in the coating, Fig. 4.41.

The corrosion resistance of the coating is determined by its porosity and thickness. Porosity for AAS coating is of 1.0–0.1.5 %, whereas for AS coating it is 2–2.5 times higher. Express corrosion resistance testing in the atmosphere of water

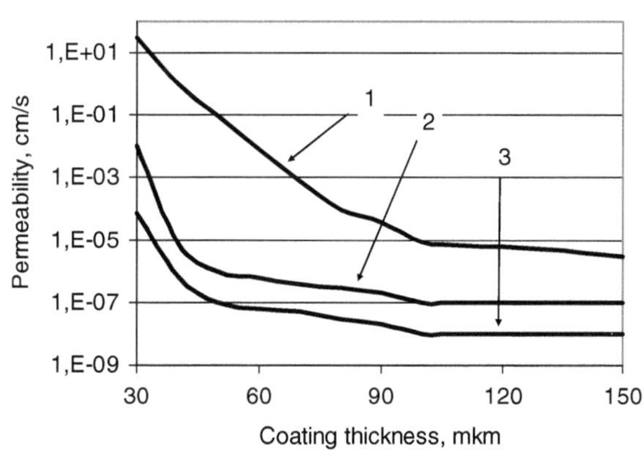

Fig. 4.40 Effect of coating thickness on gas permeability. *1*—AS; *2, 3*—AAS; 1, 3—*L* = 100 mm; *2*—*L* = 50 mm

Fig. 4.41 The dependence of adhesion strength on zinc coating thickness [56]

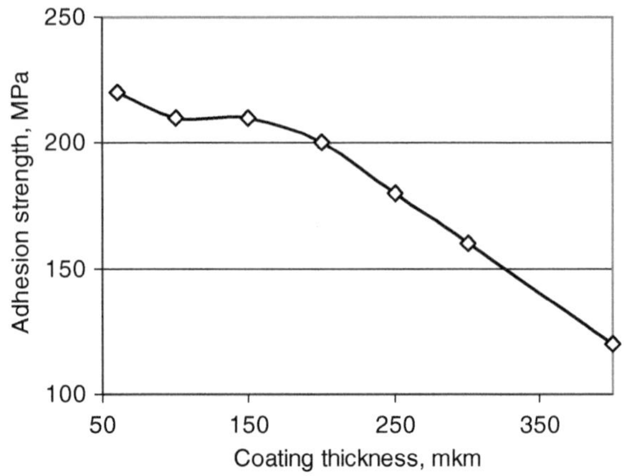

vapor showed that the resistance of AAS coatings with thickness of 150 μm matches the value of AS coating thickness of 200 μm.

Long-term field tests of zinc AS coating with thickness of 60–400 μm were conducted by Lykov [50].

On railway tanks and reservoirs the coatings with thickness of 60 μm were permeable to the medium, due to a lack of continuity. After 2 years of operation with oil the base metal inclusions have been found in such coating. Coatings with thickness of 250, 300, and 400 μm had swellings with diameter of 35–300 μm and local delamination after 1.5 years. The part of swollen zinc coating cracked and metal underneath corroded. In tanks with zinc coating of 100–0.200 μm, the delamination was observed after 4 years of operation, the coating thickness increased slightly due to the oxidation of zinc. It was established that AS zinc coating thickness of 120–150 μm on the inner surface of the vertical and horizontal tanks, is in a satisfactory state after 20 years of operation. Swelling and delamination of zinc layer is more observed in railway tanks than in reservoirs. This is obviously due to the presence of alternating loads. It was noted that the steam processing of such coatings increases their continuity.

References

1. Filippov MA, Filippenkov AA, Plotnikov GN (2009) Wear-resistant steels for castings, monograph. Ural State Technical University, Ekaterinburg
2. Korshunov LG (1991) Wear resistance testing of metals by friction. In: Bernstein ML, Rachshtadt AG (eds) Metallurgy and heat treatment of steel. Metallurgiya, Moscow, pp 387–413
3. Kragelsky IV (1968) Friction and wear. Mechanical Engineering, Moscow
4. Garkunov GN (1991) Triboengineering. Mechanical Engineering, Moscow
5. Khrushchev MM, Babichev MA (1970) Abrasive wear. Nauka, Moscow

6. Dorozhkin NN, Gimelfarb VN, Ragunovich SP, Korobov, YS, Sklyarov GD (1988) Method of wear testing and device for its implementation. USSR Certificate of authorship 1415149
7. Sukharev IP (1977) Hinge joints strength of machines. Mechanical Engineering, Moscow
8. Gimelfarb VN, Pryadko AS, Baranowsky VE, Korobov YS (1988) Durability of thermal coatings under impact-abrasive wear. In: Application of plasma processes and powder coatings in the industry. Russian scientific conference, Sverdlovsk
9. Kulik AJ, Borisov YV, Mnukhin AS, Nikitin MD (1985) Thermal spraying of composite powders. Mechanical Engineering, Moscow
10. Belotserkovsky MA (1999) Antifriction coatings produced by flame spraying of wire materials. In: Protective coatings, Part 3. Minsk
11. Pokhmursky VI, Student MM, Dovgynyk VM et al (2003) Structure and tribological properties of coatings obtained by arc spraying using cored wire. Autom Weld 8:13–17
12. Kudinov VV, Ivanov VM (1981) Plasma sprayed refractory coatings. Mechanical Engineering, Moscow
13. Korobov YS, Nevezhin SV, Filippov MA, Ilyushin VV, Potekhin BA, Gogolev LV (2012) Comparison of the tribological characteristics of Babbitt coatings obtained by activated arc spraying and by alternative methods. Weld Diagn 1:36–39
14. Korobov YS, Gogolev LB, Potechin BA, Ilushin VV, Nevezhin SV (2011) Comparative characteristics of coatings produced from Sn-Sb-Cu Babbitt by arc spraying, gas surfacing and pouring. In: Proceedings of international thermal spray conference and exposition ITSC 2011, Hamburg
15. Bouche NA, Kopytko VV (1981) Compatibility of rubbing surfaces. Nauka, Moscow
16. Gulyaev AP (1977) Material science. Metallurgiya, Moscow
17. Borisova AL, Mits IV, Kaida TV et al (1995) Structure and properties of coatings, based on ferrochrome and ferrochromealuminum, produced by arc spraying from flux-cored wires. Autom Weld 6:3–6
18. Shamko VK, Kovrigo LA (1992) About the fitting calculation for thermal coatings. Avtomobilnaya Promyshlennost 12:23–24
19. Petrov SV, Saakov AG (2000) Plasma of combustion products in surface engineering. Topas, Kiev
20. Easterly F, Verstak A, Baranovski V, Belashchenko V, Schmyreva T (1997) New coatings for protection against corrosion in fossil fuel combustion environments. In: Corrosion'97. NACE International
21. Verstak A, Wang B, Baranovski V, Beliaev A (1998) Composite coatings for elevated temperature erosion-corrosion protection in fossil-fueled boilers. In: Corrosion'98. NACE International
22. Levy AV, Man YF (1986) Erosion-corrosion of chromium steel. In: Corrosion, erosion, wear of materials at elevated temperatures. NACE International
23. Vinogradov SN (2000) Corrosion and corrosion protection. Penza
24. Popel SI, Sotnikov AI, Boronenkov VN (1986) Theory of metallurgical processes. Metallurgiya, Moscow
25. Ryzhonkov DI (1989) Theory of metallurgical processes. Metallurgiya, Moscow
26. Boiler Maintenance Trends Demonstrate Boiler Tube Upgrades Vital for Efficiency (1992) Boiler Maintenance Review 1
27. Maslowski NM (1988) Study of sprayed coatings properties. In: Application of plasma processes and powder coating industry. Russian scientific conference, Sverdlovsk
28. Wielage B, Pokhmurska H, Student M, Gvozdeckii V, Stupnyckyj T, Pokhmurskii V (2013) Iron-based coatings arc-sprayed with cored wires for applications at elevated temperatures. Surf Coat Technol 220:27–35
29. Lai GY (2007) High-temperature corrosion and materials applications. ASM International, Materials Park
30. Zhukov LL (1985) Alloys for heaters. Metallurgiya, Moscow
31. Korobov YS, Boronenkov VN (2012) Arc spraying fundamentals. Physical and chemical laws. Ural University, Yekaterinburg

32. Korobov Y, Fillipov M, Shalimov M, Nevezhin S (2011) Modeling a formation of Fe-Cr-Al coating produced by arc spraying. In: Proceedings of international thermal spray conference and exposition ITSC 2011, Hamburg
33. Kornilov II (1945) Iron alloys: In: Alloys of iron–chromium–aluminum. Academy of Science publishing, Moscow, Leningrad
34. Korobov YS, Shumyakov VI, Pryadko AS (2013) Rational approach to the restoration of parts using thermal spraying. Remont, Vosstanovleniye, Modernisatsiya 3:17–21
35. Korobov YS (2005) Efficiency of using activated arc metallization for the deposition of protective coatings. Weld int 19(7):580–582
36. Tushinskiy LI, Plokhov AV, Tokarev SA, Sindeyev VI (2004) Material's study methods. Mir, Moscow
37. Nikitin VI (1976) Calculation of metals heat resistance. Metallurgiya, Moscow
38. David JY (2008) High temperature oxidation and corrosion of metals. Elsevier, Netherlands
39. P-Met 731/Polymet Data Sheet, Tiba-007, March 2007
40. Tafa Arc Spray Iron-Chrome-Aluminum Wire—Alcro: file 1.9.1.2-Alcro. Technical Bulletin Praxair TAFA. Issue Q10507
41. EuTronic Arc 509/Castolin Eutectic (2003)
42. Abedini A, Pourmousa A, Chandra S, Mostaghimi J (2004) Splat formation in wire arc spraying. In: Thermal spray solutions—Advances in technology and application. Proceedings of ITSC 2004, Osaka, Japan
43. Gedzevicius I, Valiulis AV (2003) Influence of the particles velocity on the arc spraying coating adhesion. Mater Sci 4:334–337
44. Pokhmursky VI, Student MM, Sidorak II, Ryabcev IA, Kuskov YM (2002) Phase composition of the coatings system Fe-C-Cr-Al produced by arc spraying. Autom Weld 10:20–23
45. Pokhmursky VI, Piech BC, Student MM (1986) Fundamentals of protective and restorative coatings by arc spraying from core wires. Phys Chem Mech Mater 6:11–16
46. Korobov Y, Filipov M, Nevezhin S, Karabanalov M (2012) High-temperature oxidation behavior of Fe-25Cr-5Al Arc spraying coatings. In: Air, land water and the human body. Proceedings from ITSC 2012, Houston, Texas, USA
47. Glinka NL (2001) General chemistry. Study course for higher education institutes. Integral-Press, Moscow
48. Krechmar E (1966) Spraying of metals, ceramics and plastics. Mechanical Engineering, Moscow
49. Melnikov PS (1991) Types of corrosion: Guide to plating in mechanical engineering. Mechanical Engineering, Moscow
50. Lykov MV (1978) Corrosion protection of tanks, cisterns, containers and pipes for oil by petrol resistant coatings. Khimiya, Moscow
51. Gulec A et al (2002) Comparison of performance of TWEA sprayed zinc and zinc/aluminium 85/15 coatings for ductile iron pipe protection against corrosion. In: Thermal spray crossing borders. Proceedings of ITSC 2008, Maastricht, The Netherlands
52. Stroganov AI, Drobyshevskiy AS, Gotz AB (1982) Influence of the steel substrate roughness on the adhesion strength to the plasma coating. Poroshkovaya Metallurgiya 10:91–95
53. Kharlamov YA, Borisov YS (2001) Influence of surface microrelief on the bond strength of thermal coatings. Autom Weld 6:19–26
54. Pohmursky VI, Student MM, Piekh VS (1986) Fundamentals for protection and restoration coatings formation by arc spraying from cored wires. Phys Chem Mekh Mater 6:11–16
55. Gubenko BG, Borisova AL, Kostin VA et al (1990) Structural heterogeneity of electric arc coating from steel ASTM 1566. Autom Weld 6:19–23
56. Bernikov OV, Korobov YS (2007) Zinc coating by activated arc spraying. In: Welding: traditions and trends. Ural State Technical University, Yekaterinburg

MIX
Papier aus verantwortungsvollen Quellen
Paper from responsible sources
FSC® C105338

If you have any concerns about our products,
you can contact us on
ProductSafety@springernature.com

In case Publisher is established outside the EU,
the EU authorized representative is:
Springer Nature Customer Service Center GmbH
Europaplatz 3, 69115 Heidelberg, Germany

Printed by Libri Plureos GmbH
in Hamburg, Germany